# *Stone Tools*

## Theoretical Insights into Human Prehistory

*INTERDISCIPLINARY CONTRIBUTIONS TO ARCHAEOLOGY*



---

*Current Volumes in This Series:*

THE ARCHAEOLOGY OF GENDER
Separating the Spheres in Urban America
Diana diZerega Wall

CHESAPEAKE PREHISTORY
Old Traditions, New Directions
Richard J. Dent, Jr.

DIVERSITY AND COMPLEXITY IN PREHISTORIC MARITIME SOCIETIES
A Gulf of Maine Perspective
Bruce J. Bourque

EARLY HUNTER–GATHERERS OF THE CALIFORNIA COAST
Jon M. Erlandson

HOUSES AND HOUSEHOLDS
A Comparative Study
Richard E. Blanton

ORIGINS OF ANATOMICALLY MODERN HUMANS
Edited by Matthew H. Nitecki and Doris V. Nitecki

PREHISTORIC CULTURAL ECOLOGY AND EVOLUTION
Insights from Southern Jordan
Donald O. Henry

PREHISTORIC EXCHANGE SYSTEMS IN NORTH AMERICA
Edited by Timothy G. Baugh and Jonathon E. Ericson

REGIONAL APPROACHES TO MORTUARY ANALYSIS
Edited by Lane Anderson Beck

STONE TOOLS
Theoretical Insights into Human Prehistory
Edited by George H. Odell

STYLE, SOCIETY, AND PERSON
Archaeological and Ethnological Perspectives
Edited by Christopher Carr and Jill E. Neitzel

*A Chronological Listing of Volumes in this series appears at the back of this volume.*

---

# *Stone Tools*

## Theoretical Insights into Human Prehistory

Edited by

GEORGE H. ODELL

*University of Tulsa*
*Tulsa, Oklahoma*

*PLENUM PRESS • NEW YORK AND LONDON*

Library of Congress Cataloging-in-Publication Data

Stone tools : theoretical insights into human prehistory / edited by George H. Odell.
p. cm. -- (Interdisciplinary contributions to archaeology)
Based on papers presented at the Second Tulsa Conference on Lithic Analysis held in the summer of 1993 at the University of Tulsa.
Includes bibliographical references and index.
ISBN 0-306-45198-0
1. Tools, Prehistoric--Analysis--Congresses. 2. Stone implements--Analysis--Congresses. 3. Projectile points--Analysis--Congresses. I. Odell, George H. II. Tulsa Conference in Lithic Analysis (2nd : 1993 : University of Tulsa) III. Series.
GN799.T6S83 1996
930.1'028'5--dc20 95-43087
CIP

ISBN 0-306-45198-0


A Division of Plenum Publishing Corporation
233 Spring Street, New York, N. Y. 10013

10 9 8 7 6 5 4 3 2



Printed in the United States of America

# Contributors


*Nora Franco* • Programa de Estudio Prehistoricos (CONICET), Bartolome Mitre 1970. 5 to 'A' Capital (C.P. 1039), Argentina

*Brian Hayden* • Department of Archaeology, Simon Fraser University, Burnaby, British Columbia, Canada V5A 1S6

*Jay K. Johnson* • Department of Sociology and Anthropology, University of Mississippi, University, Mississippi 38677

*Marvin Kay* • Department of Anthropology, University of Arkansas, Fayetteville, Arkansas 72701

*Toby M. Morrow* • Office of the State Archaeologist, University of Iowa, Iowa City, Iowa 52242

*Stephen E. Nash* • University of Arizona, Tucson, Arizona 85721

*Michael S. Nassaney* • Department of Anthropology, Western Michigan University, Kalamazoo, Michigan 49008–5032

*George H. Odell* • Department of Anthropology, University of Tulsa, Tulsa, Oklahoma 74104-3189

*John W. Rick* • Department of Anthropology, Stanford University, Stanford, California 94305

*Michael F. Rondeau* • 10 Alvares Court, Sacramento California 95833

*Steven A. Rosen* • Archaeology Department, Ben-Gurion University of the Negev, Beer Sheva 84 105, Israel

*Michael J. Shott* • Department of Sociology and Anthropology, University of Northern Iowa, Cedar Falls, Iowa 50614–0513

*Jim Spafford* • Department of Archaeology, Simon Fraser University, Burnaby, British Columbia, Canada V5A 1S6

*Paul T. Thacker* • Department of Anthropology, Southern Methodist University, Dallas, Texas 75275

# Preface

Studies of prehistoric stone tools have progressed rapidly in the past few years and now possess enough analytical and methodological specificity to warrant recognition as a legitimate subfield of archaeology. This development can be documented by the appearance of several books and articles on the subject, and outstanding contributions to lithic analysis are now recognized by a specific award through the Society for American Archaeology. Clearly, the study of stone tools has emerged as a powerful instrument in the interpretation of human behavior in instances in which the written record is spotty or nonexistent.

As interest in stone tool analysis has increased, so has the number of symposia dedicated to the subject, organized to compare data and discuss recent trends. Two such symposia have been conducted at the University of Tulsa. The first, which occurred in summer 1987, emphasized the analysis of large lithic data sets and resulted in a volume of papers edited by Don Henry and me, entitled *Alternative Approaches to Lithic Analysis* (Archaeological Papers of the American Anthropological Association, Vol. 1, 1989). You are now reading the results of the Second Tulsa Conference on Lithic Analysis, held during summer 1993. This conference followed a three-week Lithic Institute to which students from Argentina and various parts of the United States came to study lithic analysis. Since the conference and institute were temporally contiguous and most of the students attended both, there was considerable structural overlap. I am indebted to several individuals and organizations for assuring the success of both of these endeavors.

Institutional assistance was provided in abundance. Funding for the conference and production of this volume was kindly granted by the Social Science Interest Group at the University of Tulsa. TU's Research Office and Arts and Sciences Dean's Office, particularly Deans Kermit Hall (Arts and Sciences), Janet Haggerty (Graduate School), and Al Soltow (Research), supported our efforts from their inception and deserve much of the credit for their success. The Philbrook

Museum cosponsored the banquet and accommodated our every wish on that occasion. The Geosciences Department at TU lent us several microscopes for the purpose of introducing use-wear analysis to students. And the TU Housing Department, particularly Walt Mauer and Trish Kerkstra, cheerfully provided food, lodging, and advice for participants in both events.

On an individual level, the Lithic Institute could not have occurred without the superb teaching and replicative expertise of Errett Callahan (Lynchburg, Virginia) and Jacques Pelegrin (Paris, France). Don Wyckoff entertained us thoughtfully at the banquet and organized an unforgettable trip to procure Edwards chert upon which the students could vent their frustrations. Emotional release was facilitated by Charles Rippy of the Tulsa Zoo, who procured antler of several species from Alaska; and by Philip Wilke who, on a seasonal foraging mission from California to Nebraska, embedded Tulsa in this journey and dropped off a load of obsidian. Tulsa Archaeological Society member Ken McCauley fashioned Rippy's antlers into billets for the students to use. And geologist Van Odell and Mick Sullivan of Ponca City, Oklahoma, led a delightfully informative field trip to the Kay County flint quarries.

Closer to home, Graduate Assistants Patricia Thomas and Russell Townsend were always on hand when needed, and their presence assured that events ran smoothly. Members of the Tulsa Archaeological Society provided airport shuttling services for conference participants, as well as a picnic lunch and transportation for our field trip to the Woolaroc Museum. Our secretary, Dale Phelps, assisted our efforts throughout this experience and produced the layout for the volume. Brian Hayden and Mike Shott freely offered their editorial assistance, thereby improving the final product immeasurably. And finally, my wife Frieda cheerfully put up with this mess and helped with many aspects of it while teaching French at the university. Without the efforts of all these individuals and institutions, these events could never have occurred. I apologize to anybody who, in the melee, may have been inadvertantly overlooked.

GEORGE H. ODELL

# Contents

## PART II. CURATION

## PART III. STONE TOOLS AND COMPLEX SOCIETIES

## PART V. TECHNIQUE AND METHODOLOGY

## PART VI. CONCLUSION

# *Stone Tools*

## Theoretical Insights into Human Prehistory

# Introduction

GEORGE H. ODELL

## RECENT TRENDS IN LITHIC ANALYSIS

Stone tools dominate archaeological assemblages from all but the most recent periods in most areas of the world. It should come as no surprise, then, that trends in lithic analysis have followed trajectories common to archaeology in general. If American archaeology in the first half of this century, for example, can be called "Classificatory-Descriptive" or "Classificatory-Historical" (Willey and Sabloff 1974), then it is a sure bet that these terms also characterize the way that stone tools were analyzed during this period. That is, analytical efforts were expended in providing taxonomic systems and historical reconstructions for the lithic remains that were being unearthed.

The middle of the twentieth century witnessed a burgeoning of analytical methods unparallelled in the history of the discipline. New technologies affected all facets of archaeology but particularly the methodological realm, as dating techniques refined chronologies, spectrographic probes characterized the elemental composition of materials, and new statistical algorithms facilitated innovative research designs and data analyses. These technologies affected research into stone tools as profoundly as research into other material classes, providing a formidable boost to lithic sourcing evaluations, activity analyses, and other kinds of studies.

In addition, a bevy of new techniques and emphases uniquely applicable to the lithic data base appeared. Piece refitting, always a popular sport among ceramic specialists, acquired especial urgency among lithic researchers as

GEORGE H.ODELL • Department of Anthropology, University of Tulsa, Tulsa, Oklahoma 74104-3189.

*Stone Tools: Theoretical Insights into Human Prehistory*, edited by George H. Odell. Plenum Press, New York, 1996.

refinements in spatial analysis allowed the resolution of ever more intricate questions involving occupational boundaries and prehistoric behaviors. Analyses of prehistoric residues on tool edges promised to reveal the specific materials on which tools had made contact, and implements discovered in special circumstances began to be regularly tested for blood, rodent hair, plant phytoliths, and other remnants of prehistoric tool contact. And to achieve more representative functional assessments of entire assemblages, use-wear techniques were developed.

The research necessary to establish these techniques has consumed the efforts of a large body of lithic analysts. Such efforts usually require extensive experimentation and testing apart from their application to the prehistoric record, so the time involved is substantial. In addition, the techniques developed have tended to be particularistic, such that only small parts of an artifact or small samples from a much larger population have regularly been analyzed. Such characteristics have restricted the scope of the conclusions that could be generalized from the information, a problem that affects all of the new techniques mentioned above.

The gestation and development of these techniques have not been without conflict, as alternative possibilities have been introduced and evaluated. Controversy has surrounded blood residue analysis, for example, concerning, first, its existence; then, the most suitable chemical techniques to employ. And the establishment of use-wear credibility, which involved experimentation, blind tests, and the evaluation of post-depositional surface modification, reached almost epic proportions. Greater attention has been given to use-wear analyst credibility, in fact, than to the credibility of the entire cadre of pollen analysts, geomorphologists, paleobotanists, and other collaborators from the natural sciences whose results we are happy to accept at face value. Assessment of credibility has, on the whole, been worthwhile, but it has consumed energy that would otherwise have been expended in more substantive pursuits.

## THEORETICIANS ENTER ORBIT

The particularism inherent in this process caused a predictable backlash among scholars whose interests emphasized the depiction of prehistoric behavior more than the establishment of techniques for future (and less certain) employment. A leader in this movement is Robin Torrence, who stated the position succinctly:

> most recent achievements in lithics studies involve the creation of ever more sophisticated methods for studying stone tools (e.g. use-wear and residue analysis, fracture mechanics, spatial patterning), but the results of analyses using this battery of techniques rarely contribute to our understanding of human behavior because the work is not specifically guided by questions significant to the field as a whole. (Torrence 1989b:1–2)

The issue, then, is theory. So much effort was being spent on the methodological quest that commensurate attention was not being paid to what archaeologists would be doing with the information. And what *was* being done did not appear to be contributing substantially to our knowledge of prehistoric behavior. So concerned was Torrence with this issue that she stated, "Perhaps now stone tools might be worth snatching from the grasp of the specialists who control their study and restored to their rightful place in the center of archaeological studies of past human behavior" (Torrence 1989a:vii).

As so often occurs with reactions, this one is a little bit wrong and a little bit right. It is wrong in implying that the specialized lithic studies conducted during the past 15 years are theoretically bereft and behaviorally bankrupt. Scores of lithic use-wear examples alone could be marshalled to support their theoretical and behavioral relevance: e.g., studies of the technological transition from foraging to agriculture in the Near East (articles in Cauvin 1983; Unger-Hamilton 1988); of ritual behavior (Sievert 1990), craft specialization (Aldenderfer 1990), and intrasite use of space (Lewenstein 1987) among the Maya; of Neanderthal tool-use behavior (Beyries 1987; Shea 1991); and the list goes on. Of course, more theoretically oriented work could have been accomplished if the specialists had not been compelled to expend so much of their energy establishing these techniques in the first place.

On the other hand, Torrence's point is valid in the sense that, at the time she held her symposium on the subject (1982), lithic analysts were not engaged to any great extent in articulating their studies with larger theoretical concerns. That is, although the subject of a study may have been behaviorally relevant, lithic analysts were picking at tiny pieces of an issue, rather than attacking the causes behind the observed phenomena. Consequently, research that was undoubtedly important locally or even regionally was often not easily generalizable on a broad scale. It is difficult to criticize scholars too ardently for theoretical lapses, however, because theory builds on the particularistic work of many people. No field can support a surfeit of theorists whose discourse remains uninformed by real data and fundamental observation. Torrence probably recognized this point but continued to bewail the fact that *few self-proclaimed lithic analysts* appeared to be concerned with the broader implications of the cultural groups being studied.

At the same time major theoretical advances *were* being developed in archaeology from bases in population ecology and ethnography. Optimal foraging theory (Winterhalder and Smith 1981) and the forager-collector model for the organization of hunter-gatherer mobility (Binford 1979, 1980; Kelly 1983) came to be perceived as appropriate frameworks for explaining archaeological site distributions, activities, etc., and it was not long before these models were applied to lithic data (Thomas 1983; Bamforth 1986, 1991; Shott 1986, 1989; Shackley 1990). Other models were subsequently added, notably Bleed's (1986) assessment of tool technological suitability derived from design theory. In fact, lithic data have been employed as frequently as any other kind in the testing of these models, though for the most part not by people whom Torrence would call lithic "specialists." Despite her decrying "the poverty of theory in archaeology relevant to the organization of technology" (Torrence 1989a:viii), this issue is currently being addressed.

## A CALL TO MERGE THEORY AND PRACTICE

Good archaeology requires a constant consideration of theory and practice, of the general and the particular. Although advances in lithic studies have been made in both the theoretical and methodological realms, yawning gaps exist in their articulation. For instance, my own subdiscipline of use-wear analysis, which possesses vast potential for resolving theoretical issues, has been notoriously underemployed for this endeavor. The need to redirect lithic studies toward broader research concerns, particularly within certain areas of the field, is acute.

What better way to address these issues than to gather a cadre of lithic experts in one room and talk about them? I accordingly put out a call for papers to be presented at a conference in Tulsa, Oklahoma, U.S.A., on June 12–15, 1993. The theme of the conference was "Theoretical Concerns in Lithic Analysis."

"Theory" can be defined as a set of systematically connected and verifiable propositions concerning the underlying principles that govern relations within a set of phenomena. As the concept pertains to stone tools, it could mean (1) physical principles pertaining to the tools themselves, as related to their source, manufacture or use; or (2) behavioral principles determining the ways that tools functioned in prehistoric societies. Since my intent was not to stifle debate but to stimulate it, I did not restrict the meaning of "theory" to one definition or another.

The conference itself featured prolonged discussion rather than prepared paper presentations, because papers were written 3–4 weeks before the conference began, sent to Tulsa, and disseminated from there. At the symposium a presenter whose topic was to be reviewed was asked to summarize the ideas in the paper in a relatively brief statement, after which discussion would ensue for 1–1.5 hours. The format of the conference is important for this volume, because the discussion and reactions to various ideas constitute a measure of the contents, directions, and boundaries of the field as of June, 1993. A synopsis of the most important points made in those discussions is presented in the last chapter.

The conference attracted a wide range of practitioners and ideas. Geographically, half of the contributions employ data from the continental United States, whereas the other half include data from Canada, Central America, South America, Europe, and the Near East (2). Although the papers are divided into five parts for purposes of organization, overlaps exist among many of the studies and other divisions might be just as appropriate. As expected, applications to foraging groups are plentiful, but three of the contributions discuss the role of stone tools in complex societies even, in some cases, while being replaced by metal. Projectile points once again prove to be a popular data base, but the authors who analyze them take wildly different approaches, from stylistic to technological to historico-economic.

All contributions are not equally "theoretical," as defined previously, but I would not consider this a drawback. More important is that they share a focus on the behavioral elements of the human groups that produced the tools, rather than on the tools themselves. In so doing the authors have produced a set of innovative studies that we hope will contribute positively toward broadening the field of lithic analysis. In addition, the papers are well grounded in archaeological data, a feature that renders them both realistic and helpful.

The discussions summarized in the final chapter represent opinions expressed by a small subset of lithic analysts at a particular moment in time. Although their scope is limited to the subjects to which they were addressed, their utility lies in what they reveal about the boundaries of those subjects and the connections among them. Although much was stated and a considerable amount of time (relative to other conferences) was devoted to these issues, most discussions had to be terminated while still in progress. So even at this colloquium, much more remained to be aired than was actually stated. The conference and this volume therefore constitute part of a continuing dialog for exploring innovative and productive directions in the field of lithic analysis. If its appearance assists like-minded scholars in this quest and perhaps resolves a few issues along the way, our efforts will have been rewarded.

## REFERENCES


Aldenderfer, M. S. 1990. Defining Lithics-Using Craft Specialties in Lowland Maya Society through Microwear Analysis: Conceptual Problems and Issues. In *The Interpretive Possibilities of Microwear Studies*, edited by B. Graslund et al., pp. 53–70. Societas Archaeologica Upsaliensis, AUN 14, Uppsala.

Bamforth, D. B. 1986. Technological Efficiency and Tool Curation. *American Antiquity* 51:38–50.

Bamforth, D. B. 1991. Technological Organization and Hunter-Gatherer Land Use: A California Example. *American Antiquity* 56:216–234.

Beyries, S. 1987. *Variabilité de l'industrie lithique au mousterién: approche fonctionelle sur quelques gisements français*. BAR International Series 328, Oxford.

Binford, L. R. 1979. Organization and Formation Processes: Looking at Curated Technologies. *Journal of Anthropological Research* 35:255–273.

Binford, L. R. 1980. Willow Smoke and Dogs' Tails: Hunter-Gatherer Settlement Systems and Archaeological Site Formation. *American Antiquity* 45:4–20.

Bleed, P. 1986. The Optimal Design of Hunting Weapons: Maintainability or Reliability. *American Antiquity* 51:737–747.

Cauvin, M.-C. (editor). 1983. *Traces d'utilisation sur les outils néolithiques du Proche Orient*. Maison de l'Orient Meditérranéan, no. 5, Lyon.

Kelly, R. L. 1983. Hunter-Gatherer Mobility Strategies. *Journal of Anthropological Research* 39:277–306.

Lewenstein, S. M. 1987. *Stone Tool Use at Cerros: The Ethnoarchaeological and Use-Wear Evidence.* University of Texas Press, Austin.

Shackley, M. S. 1990. *Early Hunter-Gatherer Procurement Ranges in the Southwest: Evidence from Obsidian Geochemistry and Lithic Technology.* PhD dissertation, Department of Anthropology, Arizona State University.

Shea, J. J. 1991. *The Behavioral Significance of Levantine Mousterian Industrial Variability.* PhD dissertation, Department of Anthropology, Harvard University.

Sievert, A. K. 1990. Postclassic Maya Ritual Behavior: Microwear Analysis of Stone Tools from Ceremonial Contexts. In *The Interpretive Possibilities of Microwear Studies,* edited by B. Graslund et al., pp. 147–157. Societas Archaeologica Upsaliensis, AUN 14, Uppsala.

Shott, M. J. 1986. Technological Organization and Settlement Mobility: An Ethnographic Examination. *Journal of Anthropological Research* 42:15–51.

Shott, M. J. 1989. Technological Organization in Great Lakes Paleoindian Assemblages. In *Eastern Paleoindian Lithic Resource Use,* edited by C. Ellis and J. Lothrop, pp. 221–237. Westview Press, Boulder.

Thomas, D. H. 1983. *The Archaeology of Monitor Valley. 1. Epistemology.* Anthropological Papers of the American Museum of Natural History, vol. 58, part 1, New York.

Torrence, R. 1989a. Preface. In *Time, Energy and Stone Tools,* edited by R. Torrence, pp. vii-viii. Cambridge University Press, Cambridge.

Torrence, R. 1989b. Tools as Optimal Solutions. In *Time, Energy and Stone Tools,* edited by R. Torrence, pp. 1–6. Cambridge University Press, Cambridge.

Unger-Hamilton, R. 1988. *Method in Microwear Analysis: Prehistoric Sickles and Other Stone Tools from Arjoune, Syria.* BAR International Series 435, Oxford.

Willey, G. R., and J. A. Sabloff. 1974. *A History of American Archaeology.* W. H. Freeman and Co., San Francisco.

Winterhalder, B., and E. A. Smith. 1981. *Hunter-Gatherer Foraging Strategies.* University of Chicago Press, Chicago.

# Part I

# Research Design

All archaeology begins with curiosity; and scientific archaeology begins when that curiosity is harnessed to address a problem in a methodical way (see Tuggle et al. 1972). The resolution of problems is also the juncture at which theory begins. As an interconnected system of principles, theory provides both foundation and direction for problem resolution. The problem that the overall plan is designed to resolve provides the operational context in which theory articulates with the archaeological record.

The plan devised for solving a problem is known as "research design." Every analysis has a research design of some kind, no matter how mundane, for this is the principal concept underlying the analysis. Barring a spectacular discovery that dictates its excavation methods, most archaeological work is only as good as the design that guides it.

Research designs can be narrow and specific or broad and enduring, governing years of fieldwork and analysis (Goodyear et al. 1978:161). Strong arguments have been made for conducting projects that are regional in scope (Binford 1964). Such projects consider fieldwork and analysis as codependent, pursued in logical order through a sequence of stages (Redman 1973).

The most provocative archaeological studies are those that have been informed by theoretical considerations of human behavioral issues. An excellent example is David Thomas's (1983) research design for archaeology in the Monitor Valley in the American Great Basin. His analysis combined theories of mobility organization with specific site types based on ethnographic data to provide a behavioral account of the human presence in that valley and its articulation with the environment.

The first chapter of this book presents a research design for lithic analysis. To appreciate it, it is important to understand two things. First, the setting for this study—the interior of British Columbia—has been blessed with recent and relatively comprehensive ethnographies of its aboriginal peoples. These accounts form the basis of the interpretive categories in which Hayden, Franco, and Spafford classify their stone tools. And second, the nature and parameters of the type of site from which the tools derive—a winter pit house occupied in a season of aggregation—is well known both ethnographically and archaeologically.

The feature that distinguishes this analysis, and the reason it is included in this book, is the application of Design Theory to a lithic assemblage. Design Theory postulates specific behavioral responses to problems encountered in a natural or cultural context. Responses to similar problems can be thought of as strategies. Applied to stone tools, this means that constraints (problems) encountered in, say, their production or procurement result in the formulation of strategies designed to obviate those constraints. These strategies include tool design criteria that help resolve difficulties inherent in the tasks in which the implements are to be engaged. Understanding the design criteria enables a researcher to understand why, for example, a tool was made of a particular raw material or blank form, and what role it played in the overall assemblage and in the lifeway of the people who made it. A principal result of Hayden et al.'s analysis is a configuration of activities practiced by the denizens of the Keatley Creek site. This task complex, as they note, is hypothetical and remains to be tested through independent measures such as use-wear or residue analysis.

## REFERENCES

Binford, L. R. 1964. A Consideration of Archaeological Research Design. *American Antiquity* 29:425–441.

Goodyear, A. C., L. M. Raab and T. C. Klinger. 1978. The Status of Archaeological Research Design in Cultural Resource Management. *American Antiquity* 43:159–173.

Redman, C. L. 1973. Multistage Fieldwork and Analytical Techniques. *American Antiquity* 38:61–79.

Thomas, D. H. 1983. *The Archaeology of Monitor Valley. 1. Epistemology.* Anthropological Papers of the American Museum of Natural History, vol. 58, part 1, New York.

Tuggle, H., A. Townsend, and T. Riley. 1972. Laws, Systems, and Research Designs: A Discussion of Explanation in Archaeology. *American Antiquity* 37:3–12.

## Chapter 1

# Evaluating Lithic Strategies and Design Criteria

BRIAN HAYDEN, NORA FRANCO, AND JIM SPAFFORD

## ABSTRACT

A wide range of factors has recently been proposed to explain lithic assemblage organization and tool morphology. These factors include: reliability, maintainability, risk, mobility, versatility, and flexibility. Discussion of all these factors has tended to remain on an abstract level with anecdotal analyses or non-lithic ethnographic observations used for support. The present chapter analyzes a complete assemblage from the Interior Plateau of British Coumbia with the aim of trying to explain assemblage organization and tool morphology. Design theory provides a powerful analytical framework for dealing with these problems. Results demonstrate that basic considerations such as requirements of task performance, raw material availability, and processing volumes play the most important roles in determining assemblage organization and morphology. In trying to apply more recently proposed factors to the explanation of tool morphology, we found many of them to be highly ambiguous and perhaps non-operational. In addition, theoretically expected outcomes of these models sometimes did not match archaeological lithic patterns. In other cases, their usefulness seems akin to considerations of "prestige display" in lithics, i.e., most useful as special case factors and most relevant in carefully defined situations (e.g.,

**BRIAN HAYDEN** • Department of Archaeology, Simon Fraser University, Burnaby, British Columbia, Canada V5A 1S6. **NORA FRANCO** • Programa de Estudio Prehistoricos (CONICET), Bartolome Mitre 1970. 5 to 'A' Capital (C.P. 1039), Argentina. **JIM SPAFFORD** • Department of Archaeology, Simon Fraser University, Burnaby, British Columbia, Canada V5A 1S6.

*Stone Tools: Theoretical Insights into Human Prehistory*, edited by George H. Odell. Plenum Press, New York, 1996.

hunting gear). Nevertheless, all these concepts can be accommodated in a broad design analysis framework, emphasizing constraints, design considerations, and reduction/resharpening strategies.

## INTRODUCTION

What factors are important for understanding the procurement of lithic raw materials, their mode of reduction, the selection and shaping of tools, and the resharpening techniques used? These and related aspects can be broadly referred to as "Tool Formation Processes" (Hayden 1990:89), and can be usefully examined from a "design theory" perspective. Design theory emphasizes various constraints in solving given problems by technological means (Pye 1964, 1968; Horsfall 1987; Budnick et al. 1988). The classic constraints that have been used include: adequate task performance; materials available and their relative costs; technologies available; and economics of various production and use alternatives, including relative use-lives and repair costs. A more complete list for lithic technologies is provided in Figure 1, which also illustrates our conception of the relation between constraints, design considerations, and strategies. We refer to tools that exhibit similar procurement, reduction, and use-life characteristics as belonging to distinctive lithic production and resharpening "strategies." In our case study, we identify six major strategies, including: 1) the expedient block core strategy; 2) the bifacial strategy; 3) the portable long-use strategy; 4) the quarried bipolar strategy; 5) the scavenged bipolar strategy; and 6) the groundstone strategy. Examples showing how various constraints create each of these strategies constitute the body of this chapter.

Recently, there have been a number of additional suggestions in the lithic literature that considerations beyond the traditional criteria of efficiency and task performance also affect tool production strategies and designs. These involve:

- portability for mobile hunter/gatherers (Torrence 1983; Hayden 1987; Parry and Kelly 1987; Shott 1986; Nelson 1991);
- time constraints (Torrence 1983);
- exceptional need for reliability vs. easy maintainability due to high vs. low risk situations (Bleed 1986; Torrence 1989);
- flexibility (changes in tool form for different uses—Shott 1986:19; Nelson 1991:70);
- versatility (the number of uses a tool is designed for—Shott 1986:19; Nelson 1991:70);
- precision (an axe vs. a graver—Aldenderfer 1991a:207);
- longevity (generally supplanting the concept of "curation"—Aldenderfer 1991a:207; see also Hayden 1987).

While several factors, such as the amount of material that can be transported, the absolute amount of time available to carry out tasks, and the precision required

CONSTRAINTS

Task Constraints
(Acceptable Performance)
Task Mechanics
- Precision
- Force
- Nature of Action
Efficiency
Quantity
Time Available
Failure Consequences (Risk)

Material Constraints

Available Materials and Costs
Relative Performance
Relative Wear/Failure Rates

Technological Constraints

Available Technology
Production Costs
Repair/Resharpening/Replacement Costs
Skill Required

Socioeconomic Constraints

Mobility
Transport Capacity
Available Labor
Storage

Prestige and Ideological Constraints

→

DESIGN CONSIDERATIONS

- Size and Weight
- Edge Angle and Form
- Prehension and Hafting
- Length of Use (Use-life)
- Specialization
- Reliability (robustness and "overdesign")
- Ease of repair
- Multifunctionality (versatility)

→

PRODUCTION/ REDUCTION & RESHARPENING STRATEGIES

e.g.
- Expedient Block Core
- Biface
- Long-Lived Flake/Blade Tools
- Bipolar
- Scavenging/Recycling
- Groundstone
- Resharpening:
  - hard hammer (notching, continuous retouching, burinating, etc.)
  - billet
  - pressure
  - grinding

**Figure 1.** A schematic representation of the various kinds of principal constraints on tool design and their relationship to other design considerations and production/resharpening strategies.

to perform a task satisfactorily, can be viewed as basic constraints or aspects of acceptable task performance, the other aspects mentioned in the above list form a class of considerations distinct from basic constraints or the production strategies previously mentioned. Following Nelson (1991:66), we shall refer to them as "design considerations." They include the purposeful consideration of reliability, maintainability, versatility, flexibility, and longevity. Considerable claims have been implied for some of these concepts as constituting fundamental keys to understanding lithic technological organization (Nelson 1991; Torrence 1989). We therefore will pay particular attention to these concepts in the case study to follow. For those unfamiliar with them, we provide a brief review in the following section. Portability will be considered to be subsumed under size and weight design considerations (Fig. 1), and to be most affected by task mechanics, mobility, transport capacity, and available materials.

## DESIGN CONSIDERATIONS

Reliability is perhaps the most central concept in the discussions that follow, since it has been related to high risk conditions and seems to have material design implications. Of the seven criteria Bleed (1986:739) used to characterize reliable tool systems, only a few seem to be directly inferable from archaeological lithic assemblages, at least among those in our case study. These include: overdesigned (stronger than necessary) parts; carefully fitted parts and good craftsmanship; specialist manufacturing and maintenance; and possibly maintenance outside the context of use. Intuitively, it seems that repair of such specialized gear should also require specialized tool kits; however, Bleed argues the opposite for reasons that are not entirely clear. Subsequent discussions in the literature have not fundamentally modified Bleed's characterization. Reliability qualifies as a design consideration because, although a basic reduction and resharpening strategy may account for the general nature of the flake and the tool, the craftsman additionally and intentionally emphasizes aspects such as thickness, care in manufacture, and sturdiness.

Maintainability is a much more difficult concept to deal with, partly because virtually all chipped stone tools except utilized flakes involve some maintenance and eventual replacement, and partly because of subsequent attempts to elaborate on this concept. Again, Bleed's (1986:739–741) original eight criteria contain only a few operational ones for archaeological assemblages. These include: lightness and portability (although virtually all flake tools fit this description); modular or serial design (difficult to demonstrate); simple design; easy maintenance by people with low levels of lithic skills; use in a range of functions; and possibly the occurrence of repair or maintenance during use. Perhaps due to the perception that anything which is not manifestly reliable ought to be maintainable, the maintainable label seems at times like a catch-all category or a default value. Since

Bleed's initial presentation, several authors have suggested that reliable and maintainable systems may actually measure two different aspects of tools (i.e., tools can be neither or both—Torrence 1989; Nelson 1991). Nelson has attempted to isolate a few of the components of maintainable designs which she calls flexibility and versatility (discussed next). According to the rationale for categorizing tools as maintainable, maintainable designs are supposed to be useful under task conditions that are more or less continuous but somewhat unpredictable, or where size and weight constraints are important. They are therefore thought to characterize foragers rather than collectors.

Versatility is considered by Nelson (1991) as an aspect of maintainability, because it is supposed to measure the degree to which a tool can be applied to a variety of needs, a characteristic Bleed (1986:741) proposes for maintainable tools. The term "versatility," however, was initially proposed by Shott (1986:19) to refer to the number of different tasks to which tool classes could be applied. From Shott's (1986:35,38) operational definition of this concept as the number of employable units on a tool, it would seem that he is lumping together in his considerations different types of use, as well as the same use but on different edges of a single tool. He does not explain how either of these cases differs from a single repeatedly resharpened edge that was used for different tasks or the same task repeated many times. We feel that in order to separate tools used for single types of tasks from those used for a diverse array of tasks, it is essential to use more precise measures than the number of employable units per tool. These could include different types of retouch on the same tool (although obviously not foolproof, this is probably a better measure than the number of employable units), or different types of use-wear.

We also feel that "versatility" is a poor descriptive term, especially when used in conjunction with "flexibility." As used by Nelson and as implied by Shott, it is the same as a much more established and descriptive term: "multifunctionality." We prefer to use the more established, and we feel clearer, term "multifunctional." Multifunctionality is included in design considerations, because it certainly can be a deliberate feature of tool making, as spurred endscrapers or other recurring formal tool combinations demonstrate. On the other hand, the occurrence of several different uses on the same flake can also simply be the opportunistic use of available or scavenged flakes for immediate needs, not involving any real design considerations (a totally different strategy). Distinguishing between these conditions is beyond the limited goals of this paper. It tangentially raises the very thorny issue of how tools are to be categorized, described, and interpreted in terms of production strategies. In our analyses, we have tried to describe the primary use of tools as their last use (since our primary interest was identifying activity areas) or on the basis of the least common retouch type, noting other types of retouch as "secondary uses." Shott and Bleed both concur that multiple functions of tools should characterize tools in which portability (due to mobility) is an important constraint.

The term "flexibility" was also introduced by Shott (1986:19,35) and was originally intended to measure multifunctionality, but involving "a wider range of applications." Nelson (1991:70) modified this definition considerably by suggesting that flexible tools undergo changes in their form in order to achieve multifunctional demands. In terms of lithic tools, it is not clear to us how the fine distinctions Shott proposes between versatility and flexibility can be operationalized or standardized. Nor is it obvious how Nelson's definition can be distinguished from recycling or scavenging. Given this situation, we propose that "flexibility" in Nelson's sense is thus more appropriately dealt with in terms of a procurement and resharpening strategy (e.g., our fifth production strategy). Because of these considerations and the opaque descriptive nature of the term, we also favor dropping it from the vocabulary of lithic analysis. It would be especially difficult to demonstrate that this was a design consideration. On the other hand, recycling and scavenging behavior have well established and important roles in lithic analysis, and we will certainly deal with this aspect of stone tools as a procurement and production strategy.

Longevity, or use life, or "curation," is also an established and important concept for understanding lithic assemblages. Because of the multiple and vague implications of the term "curation," conference participants decided to abandon that term in favor of more precise descriptive components of the "curation" concept. In terms of design considerations, the important aspect to be examined here is the intentional choice of flake sizes and shapes, as well as materials and resharpening techniques, which lend themselves to repeated resharpenings and extended use lives.

## THE KEATLEY CREEK CASE

In order to examine the role of the constraints and design considerations discussed above in lithic tool morphology, and to determine whether lithic strategies can be identified, we decided to undertake an exploratory analysis of the lithic assemblage from the Keatley Creek site in the Interior Plateau of British Columbia. The overall analysis of this assemblage has been going on since 1987 and has many goals including an analysis of spatial patterning within and between housepits (Spafford 1991); the analysis of debitage (Prentiss 1993); the understanding of site formation processes; the analysis of prestige items and activities (e.g., Hayden 1990); and the understanding of tool formation processes, including tool design, residue, and use-wear analysis. Some of these studies are still in progress, and some of the studies, like the present one, are heuristic, exploratory studies meant to be preliminary assessments of how well certain approaches seem to work in furthering our understanding of lithic tools.

The Keatley Creek site is located on a high terrace above the Fraser River, about 26 km upstream from the town of Lillooet. It is an unusually large winter housepit village with 115 house-size depressions, some of which exceed 20 meters in diameter. The first evidence of house construction occurs in the Shuswap horizon (4800–2400 bp), and the site appears to have been abandoned abruptly about 1100 bp, at the beginning of the Kamloops horizon. There is strong evidence for basic cultural continuity in the region (at the synthetic culture descriptive level of Interior Salish groups) from Shuswap to Historic times, with continuity in many tool types persisting into the ethnographic present (e.g., Teit 1900, 1906). Thus the uses of a number of important archaeological tool types can be inferred with confidence. The use of some other types can be inferred with relative confidence on the basis of their morphological characteristics.

On the basis of ethnographic descriptions of activities in pithouses and seasonal rounds, it is possible to construct a list of basic problems that would have required technological solutions in winter pithouse villages. It is also possible to identify many of the constraints affecting the solutions to those problems. Thus, most of the necessary elements are in place for a design theory analysis of the Keatley Creek assemblage: we have the problems to be solved (the activities), the constraints involved, the archaeological tools (the solutions), and a good estimation of how the tools were used. Presenting all of the details of these design elements is too large an undertaking for the present volume (the full analysis is in preparation), so we will choose single examples from each of the major strategies to illustrate our approach and to identify strengths and problems with some of the proposed design considerations.

There are no significant time constraints that can be identified for the use of any of the tools in the assemblage; thus this constraint will not be considered. The only case in which a good argument can be made for significant time constraints involves the butchering of salmon in riverside camps during the seasonal peak of their runs; however, no tools from winter villages have been identified as having been used for salmon butchering. Similarly, while a number of tools obviously served multiple functions and/or were recycled, we have no way to determine whether this was an intended aspect of the initial tool design (per Nelson 1991) or simply an opportunistic use or re-use of an available edge. In Shott's original (1986) discussion, the intentional design nature of multifunctional tools is immaterial and these measures are simply meant to describe entire assemblages or tool classes in order to demonstrate how multifunctionality varies with mobility. Although we cannot determine if any of our tools were intentionally designed to be multifunctional, we present our data on multifunctionality for their comparative value. Finally, in terms of labor constraints, we assume that each family is more or less self-sufficient, although some wealthy and powerful families may have had access to greater amounts of labor, and some specialization probably existed. Having established some of the basic contextual parameters of the Keatley Creek assemblage, we now turn to the examination of the principal lithic strategies represented in that assemblage.

## EXPEDIENT BLOCK CORE STRATEGY

In this strategy, cores (Figure 2) are kept at the habitation site. Flakes are removed and modified according to immediate needs, and usually discarded after the immediate task is completed, unless large, still-usable flakes are involved. Core material is obtained from the most easily available sources, and there are generally no needs for especially durable materials.

Types included in this strategy are: expedient knives, scrapers, utilized flakes, notches, denticulates, borers, piercers, and perforators (Figure 3). We will consider expedient knives as an example. Expedient knives are flake tools with continuous low-angled retouch along one or more edges, made by light or moderate pressure flaking and typically extending 1–5 mm from the edge.

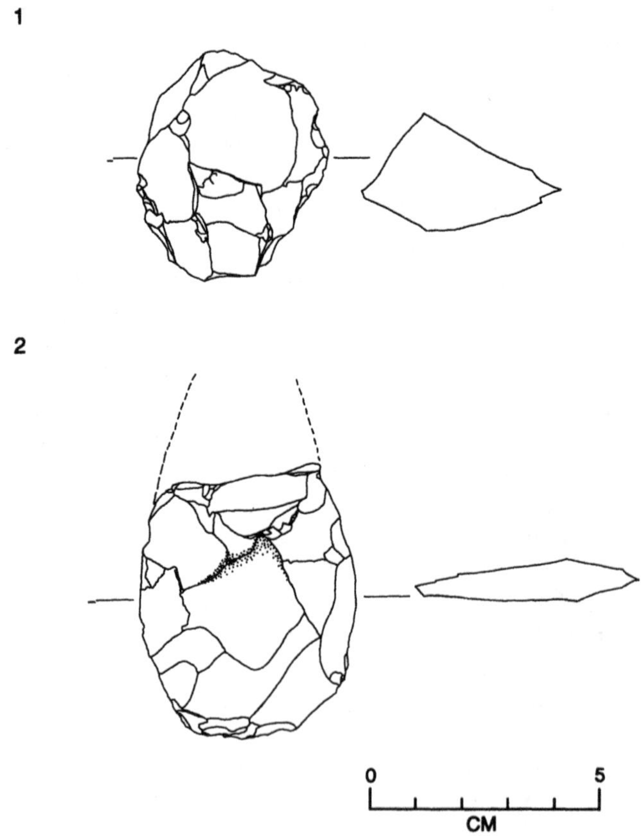


**Figure 2.** No. 1: A typically heavily reduced block core. No. 2: The base of a typical biface from Keatley Creek. Stippling indicates polish on ridges.

## Constraints

*Expedient Knife Task Constraints:* Expedient knives were probably used in some part of the occasional butchering activities thought to be represented at the site (cutting meat, hide, tendons, or filleting) or in cutting rawhide thongs or buckskin for making clothes. Because of the complementary distributions of utilized flakes lacking invasive retouch on housepit floors (Spafford 1991), utilized flakes appear to have been primarily used in other activities, perhaps in basket making. Given the concentration of expedient knives close to the walls, these tools were probably not used in heavy butchering, but instead in light tasks such as cutting up hides for thongs, clothes or other purposes. Given the expedient nature of most of these tools, the minimal amount of retouch along their edges, and their high frequencies, it seems unlikely that their position near the walls was for storage. Their discard location makes more sense as the result of a commonly occurring activity in these areas, perhaps associated with the heavy concentrations of debitage and microdebitage also found near the walls and clearly not related to storage. Expedient knives should be very frequent at the site, due to their expedient nature, the limited number of resharpenings involved (because many successive resharpenings would increase edge angles more than desired), their short use-lives with consequently high discard rates, and the use of these tools in infrequently or sporadically occurring activities. At best, probably one deer or hide per month per housepit would have been processed, representing a low processing volume.

*Material Constraints:* The only requirements should be the use of fine-grained raw material.

*Technological Constraints:* Minor or insignificant manufacturing time characterize expedient knives, although it would be necessary to procure or manufacture a pressure flaking tool in order to keep edge angles acute by removing low-angled resharpening flakes. There are no special needs concerning flake types. Any kind of flake with acute angles and a straight edge would be adequate: hard hammer flakes, billet flakes, blades, and bipolar flakes. However, producing blades can be a technique wasteful of effort and raw material. Systematic blade production requires the preparation of cores and the removal of many preparation flakes.

Moreover, considerable skill, training, and time are necessary to systematically produce blades (see Parry and Kelly 1987; Clark 1987; Nelson 1991:68). The risk of ruining blade cores, and therefore wasting a large amount of high quality raw material, plague the flintknapper at every step in the reduction process. And finally, blade cores require much more specific sizes and shapes of raw materials, as well as high quality materials, thus increasing procurement costs considerably wherever the optimal size and shape of raw material is difficult to find, which was certainly the case in the Lillooet and neighboring regions. For all these reasons, systematic production of blades for knives at winter villages would not be a good design solution. In fact, as Parry and Kelly (1987) and Morrow (1987) have

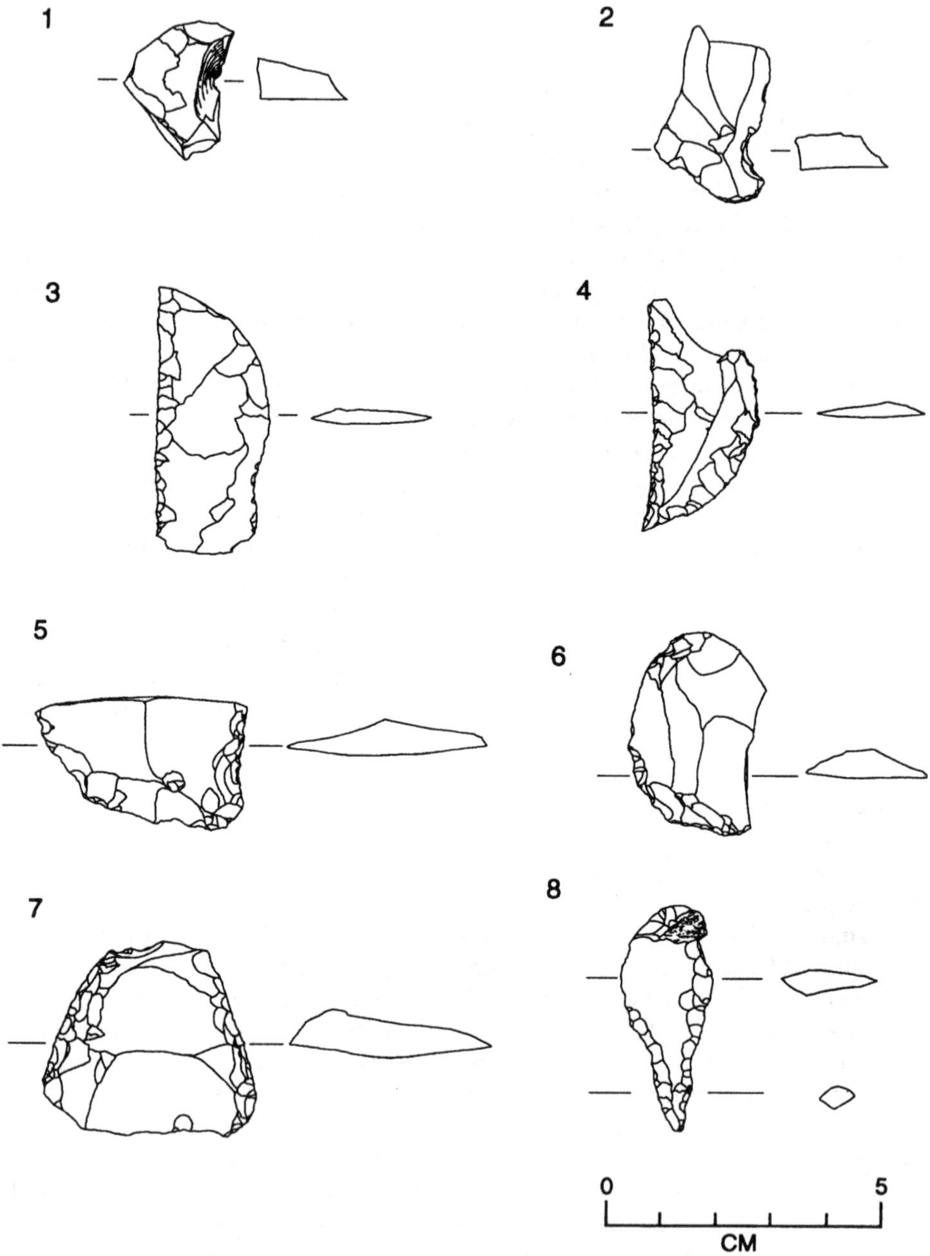


**Figure 3.** Flake tools made from expedient block cores at Keatley Creek. Numbers 1–2 are notches; numbers 3–4 are expedient knives that have been pressure retouched numerous times and exhibit much more extensive retouch than most examples (number 4 also has a delicate piercer fashioned on its distal end); numbers 5–7 are heavily retouched scrapers (note the difference in edge angle and retouch compared to the expedient knives). Number 8 is a boring tool.

argued, the high investment and risks associated with blade production may only make sense under high mobility circumstances when at least one part of the seasonal round intersects abundant sources of raw material of suitable size and shape, since blades clearly do provide more cutting edge per weight of successfully processed stone material. Another reduction strategy, bipolar reduction, produces a great deal of shatter and small flakes and would be wasteful of larger core material in the production of knives.

Billet flakes conserve raw material and they produce acute angles more consistently than hard hammers. Some researchers argue that thin biface reduction flakes are even better than bifaces for cutting hide in skinning (Frison 1989). The ratio of utilized billet flakes to nonutilized ones can provide a general indication of the degree to which biface flakes were used for tools. If a high percentage of billet flakes was utilized, it can be tentatively inferred that billet flakes were often saved or even produced for use as cutting tools. If billet flakes do not show traces of utilization very often, they were probably simply a by-product of the resharpening or manufacturing of bifaces. However, use-wear analysis is required to fully evaluate this hypothesis. Examining the proportion of billet flakes with use retouch can provide an initial indication (Table 1). It is clear that with the exception of size 2 flakes, high proportions of billet flakes were being selected for use. Size 2 is too small in general for this work.

What can be said about billet flakes utilized for producing retouched tools? For expedient knives, many of the flakes (45% of those identifiable) were originally billet produced. The same is true of utilized flakes with acute edges like expedient knives (36% made on billet flakes) and bifacially retouched expedient knives (45%).

Given the much more abundant occurrence of block core flakes in the assemblage, this seems to indicate a preference for the use of billet flakes over hard hammer flakes for expedient knives. This preference is probably related to the more acute original edge angles of most billet flakes and to the desirability of acute edge angles for some kinds of activities (e.g., cutting hide, as Frison mentions). When dull, and given continued use in this activity, some of the billet flakes could be resharpened by pressure flaking into expedient knives. Cutting hide results in a very high wear rate (Frison 1989), with edges lasting only a few minutes. Cutting meat, however, results in very low wear rates.

Although the production of billet flakes may not have been the main reason for bringing bifaces to the site, it is clear that the convenient availability of flakes from biface manufacture and resharpening played an important role in the strategy

**Table 1. Utilized Billet Flakes**

| | Size 2 (1-2 cm) | Size 3 (2-5 cm) | Size 4 (> 5 cm) |
|---|---|---|---|
| Total billet flakes | 1563 | 658 | 7 |
| Billet flakes with utilization retouch | 21 (1.3%) | 230 (35.0%) | 6 (85.7%) |

of tool production and butchering or hide cutting activities. Given the efficiency of block core strategies for producing useful flakes for these tasks, it seems most likely that billet flakes were being used as byproducts rather than being specifically made for use in these tasks.

*Socioeconomic Constraints (Transport):* There are no constraints involving the tools themselves, since these tools do not need to be transported away from the site. Constraints do occur in the transport of raw material to the site. In this case, the transport constraints would be significant, since people would also be carrying gear and as much food as they could carry for winter storage from the mountain areas (where the lithic resources are also located) to the Keatley Creek village. Because of this, people could probably carry only minimal quantities of stone.

## Strategies and Design

*Raw Material Strategies:* Given the need for fine-grained materials, people could be expected to have used the closest available source (trachydacite and chert: 15–20 km distant). Given the low constraints on flake types, any shape and most sizes of raw material could be used. In addition, we suggest that the more wear-resistant cherts and chalcedonies would be saved for tools involving greater requirements for durability and long use-lives. Consequently, the main material expected to be used for butchering at winter villages is the less wear-resistant trachydacite. In fact, the percentages of all types of expedient knives made of this material vary between 91–96%.

*Acquisition/Procurement Strategies:* The trachydacite utilized was not available through the winter occupation. However, it could easily have been acquired during fall hunts and spring plant gathering in Hat Creek Valley about 15 km through the mountains. Caching raw material before winter time in the housepit village could therefore be expected.

*Reduction Strategies:* Given low constraints on tool form and low availability of raw material during the winter, the best strategy would involve the reduction of block cores from which a large range of flakes could be obtained. This would involve the fewest constraints on the available size and shape of raw materials, thereby making greatest use of them. In addition to the production of flakes from block cores, the use of billet flakes obtained as resharpening byproducts or specifically for butchering tasks from bifaces would have been used whenever possible due to the superior quality (thinness, acuteness, straightness) of these flakes.

*Tool Form and Resharpening:* There are only minor constraints concerning form: adequate low angles and straight edges, plus the need to be held comfortably (tools needed to be more than 2 cm long). Thus, tool design simply involved selection of straight, acute edged flakes with edges longer than 1–2 cm, as well as the use of the most appropriate resharpening technique. The minimal size of the utilized flakes resembling expedient knives is less than that of other types

**Table 2. Minimum and Mean Dimensions of Expedient Knives (Includes Only Whole and Chipped Tools)**

| Type | N | Minimum | Mean (cm) | S. dev. |
|---|---|---|---|---|
| Normally retouched expedient knives | 340 | 1.1 | 3.31 | 0.97 |
| Bifacially retouched expedient knives | 89 | 1.3 | 3.62 | 1.34 |
| Utilized flakes resembling expedient knives | 144 | 1.1 | 2.98 | 0.78 |
| Inversely retouched expedient knives | 139 | 0.8 | 3.29 | 0.09 |

(Table 2). This could indicate, perhaps, a range of sizes below which flakes were not retouched and were simply used in small tasks if the edge angle was appropriate.

The sizes for all these tools, as for the assemblage in general, were relatively small, with a mean of 3.3 cm and a standard deviation of 0.9. There is a general correspondence between these measures and the small size of the block (multidirectional) cores found at the site.

Resharpening strategies involved pressure retouch to maintain an acute edge angle and minimize raw material waste. Larger flakes were probably reutilized and resharpened to conserve raw material.

*Longevity:* The extent of resharpening varies widely, from a fraction of a millimeter to bifacial retouch that covers much of both faces. This indicates that many tools were used only for brief periods before being discarded or abandoned. Although the larger, more extensively retouched pieces may have been stored between tasks, many of the less retouched pieces seem to have been expedient tools in the most characteristic sense. This may have been caused by the small size and thinness of most of these tools, but it may also reflect the difficulty in maintaining very sharp edges over many resharpenings. Given generally short use-lives, there is probably no significant benefit in hafting such tools.

*Variables of Design:* Maintainability was probably emphasized, because all of these tools can be easily replaced. They can also be easily and quickly resharpened by people with the most basic lithic skills.

*Multifunctionality:* Larger flakes, in particular, could be used on more than one edge (either in the same or different tasks). Although there is no evidence for creating multifunctional tool designs, a surprising 48% of expedient knives exhibited additional types of retouch, which seem to indicate alternative functions. This unusually high percentage is also reflected in other tool types produced with this strategy such as notches (44%) and piercers (46%), although utilized flakes (29–35%) and scrapers (26%) exhibited less frequent alternative uses. For a strategy used in a manifestly non-mobile context, these extremely high rates of multifunctionality accord poorly with Shott's (1986) postulated relationship between mobility and multifunctionality. We suspect raw material availability and transport constraints play a more basic role in this case.

*Frequency:* The frequency of these kinds of tools, accounting for almost 20% of all retouched tools, is relatively high when compared with the rest of the assemblage. Their high frequency is undoubtedly related to the moderately frequent need for butchering and hide cutting tools, plus their short use lives and expedient nature, as well as the infrequent performance of other activities requiring tools in winter villages.

*Specialization:* Because of the lack of time constraints, the episodic and relatively moderate volumes of material being processed, and the relatively simple nature of the task, there is no need to develop any specialized or extra-efficient tool for butchering and hide cutting. Therefore, the simplest, lowest-cost, effective design solution was employed.

In this case, the nature and frequency of the task, as well as limits on raw materials, made it desirable to use billet flakes from bifaces whenever possible, and to sharpen them using pressure retouch. The result was small, easily maintainable and replaceable expedient tools used and retouched to varying extents under conditions in which few constraints of time or risk existed.

## BIFACE STRATEGY

The bifacial strategy makes most sense in the context of high mobility (as tools used in traveling to seasonal camps) and high constraints on the amount of stone material that can be transported on such trips. The advantages of bifaces include their presumed multifunctionality, their economy of raw material use, and the potential utility of resharpening flakes. It is important to note that some authors, like Shott, refer to all bifacially retouched pieces (including flakes, projectile points, and handaxes) as "bifaces." In contrast to this excessively general use of the term, we use "biface" only to refer to relatively large, bifacially reduced tools which are clearly not projectiles, drills, or other specialized flake tools (Figure 2).

### Constraints

*Biface Task Constraints:* Bifaces are usually considered multifunctional or versatile tools, i.e., they can be employed in a wide variety of activities (cf. Winters 1969; Ahler and McMillan 1976; Johnson 1987; Bamforth 1991:230; Nelson 1991). Because of this, they are often viewed as useful tools when there are strong constraints in the quantity that can be transported. They thus make most sense in high mobility situations (see Bamforth 1991:226-9; Sassaman 1992:256–7) such as at seasonal hunting camps. They are especially effective where collector strategies are pursued, involving the transport of large amounts of food for storage, thus reducing one's ability to carry tools or raw materials. "Disk or bifacial cores maximize tool material; they provide a variety of flake forms for use as tools, yet these can be thin while having extensive, usable edge length (high edge-to-weight ratio). . . . In addition, the biface can be changed to a variety of forms and resharpened with minimal reduction of the stone; therefore few need to be carried" (Nelson 1991:74).

The use of bifaces could have had other advantages, such as for sources of raw material (Kelly 1988; Ingbar 1990; Nelson 1991). At Keatley Creek, for instance, billet flakes could be obtained from bifaces and were used as expedient knives for butchering or hide cutting (Figure 3). Nearly exhausted bifaces could also be sharpened into more specialized bifacial knives (Morrow 1987:141). In addition, broken bifaces could be recycled as cores to obtain a few more flakes or as wedges (pièces esquillées).

Bifaces could also have been useful in some activities carried on at seasonally sedentary sites, like the Keatley Creek village. Activities may have included woodworking (e.g., arrows, leisters, net hooks) and butchering. The occurrence of broken bifaces in the center areas of housepits (Spafford 1991) indicates that they were probably used in activities requiring considerable space or producing copious debris. Jones (1980) considers bifaces to be effective tools for butchering animals. He has efficiently used handaxes (which we consider functionally equivalent to bifaces) made on quartzite, phonolite, and basalt or trachydacite for skin-cutting, skin removal and meat-cutting. Except for the initial cutting of the hide on medium-sized animals, he believes that these tools are more effective, longer-lasting and more comfortable to hold than simple acute-edged flakes.

We do not expect high amounts of meat processing or work on wooden tools to have been undertaken at winter pithouse sites. Moreover, other kinds of tools could have been used in these tasks. On the other hand, it is clear that bifaces were used inside pithouses, because of the many broken fragments that occur in the centers of the floors. Interestingly, bifaces seem to have been resharpened in the sleeping areas between the hearths and the pithouse walls (where billet flakes concentrate), whereas the bifaces seem to have been used in the center floor area, where broken biface fragments concentrate (Spafford 1991; Prentiss 1993). However, biface resharpening flakes could have been used as raw material sources for making other kinds of tools, such as expedient knives.

*Material Constraints:* Fine-grained raw material would be easiest to manufacture into a biface and would provide better cutting edges for butchering. Trachydacite and chert (and chalcedony) would consequently be the best materials to use. Table 3 displays the different manufacturing stages and the quantities and percentages of raw materials employed. The data show that the vast majority of bifaces were made of trachydacite (especially the fine-grained variety), chert and chalcedony.

Table 3. Raw Materials Used in Biface Manufacturing

| | Stage 4 (Type 131) | Stage 3 (Type 134) | Stage 2 (Type 193) | Stage 1 (Type 192) |
|---|---|---|---|---|
| Fine-grain trachydacite | 83 (82.2%) | 27 (75.0%) | 31 (77.5%) | 11 (91.7%) |
| Coarse-grain trachydacite | 10 (9.9%) | 3 (8.3%) | 3 (7.5%) | 1 (8.3%) |
| Chert | 3 (3.0%) | 3 (8.3%) | 1 (2.5%) | 0 (0.0%) |
| Chalcedony | 4 (4.0%) | 3 (8.3%) | 4 (10.0%) | 0 (0.0%) |
| Quartzite | 1 (1.0%) | 0 (0.0%) | 1 (2.5%) | 0 (0.0%) |
| | 101 | 36 | 40 | 12 |

*Technological Constraints:* The manufacturing time, tools, effort, and skill required for bifaces are probably the highest of any chipped stone artifact type in the assemblage. No estimates are available for wear rates of bifaces.

Flake sizes need to be large and thin enough to be able to be reduced afterwards. No bipolar flakes, blades, or billet flakes would be suitable. Large hard hammer flakes or direct reduction flakes from cores are the most suitable forms.

*Socioeconomic Constraints (Transport and Mobility):* In general, complex hunter-gatherers and collectors using logistical settlement patterns have high constraints on tool transport due to the need to transport food in bulk and to carry increased amounts of technological gear. There are, consequently, constraints on the weight and bulk of individual tools carried to sites, especially if food necessary for survival during the winter was also being transported. Bifaces might be especially important foraging tools in early spring before lithic resources could be replenished. Theoretically, the constraints associated with high mobility and/or high transport loads (e.g., on logistical hunting forays) should lead to an emphasis on multifunctional tools (Shott 1986). At present, we cannot empirically determine how specialized or multifunctional Keatley Creek bifaces were, although we suspect they were multifunctional.

## Strategies

*Raw Material Strategies:* Fine-grained materials are best for controlled flaking and sharp, acute edge angles. The closest sources of adequate-sized trachydacite, chert or chalcedony should have been preferred. Chert and chalcedony may have been the most sought after, because they are more wear-resistant. However, the size of the nodules available may have been critical. Large chert nodules are probably rarer than large trachydacite nodules, in this region.

*Acquisition/Procurement Strategies:* Due to possibilities of breakage during the manufacturing process (Johnson 1989) and/or the existence of internal flaws, it would make most sense to perform the initial stages of biface reduction (stages 1 and 2) at or near the quarry. Roughed out bifaces were probably taken to Keatley

Creek at the beginning of the winter occupation or cached there previously. Because of the effort and skill required in the manufacture of thin bifaces, because of the high risk of breakage (especially during resharpening), and because of their suitability for use in different tasks, they probably constitute personal gear carried by individuals. In fact, this constitutes the best case that can be made for a personally owned tool in the entire chipped stone assemblage, and it is unlikely that, as a general rule, thin bifaces would have been lent to other people, given their costs and risk of breakage. Sassaman (1992:257) explicitly relates the use of bifaces to hunting activities and therefore views them as men's tools. Most of the bifaces (78% of all bifaces in the sample) recovered at the site represent later manufacturing stages (Callahan's 1979 stages 3 and 4). This supports the idea of a very long-lived tool adapted to conditions of transport constraints on stone materials.

*Reduction Strategies:* These would consist of the removal of bifacial billet flakes to avoid rapid consumption of raw material, to minimize the weight of raw material in transport, and to maintain adequately low edge angles on tools.

## Design

In design terms, the resharpening mode using billets largely determines the overall shape of the tool, except for the proximal and distal ends. Hayden (1987) argued that thin biface morphology and billet flaking make sense primarily in terms of tactics to conserve low edge angles on tools while maximizing the number of resharpenings and use of raw material (in comparison to hard hammer resharpening). This design and resharpening strategy, although costly in terms of manufacturing time, effort, skill, and tools, provides important benefits where there are significant constraints on the transport of tools or on raw material availability, together with moderate or high processing requirements. The size of the tool is potentially important in understanding its role in butchering. Bifaces would be better for primary butchering than flake tools due to their larger size and greater weight (Jones 1980).

The distal tips theoretically could be shaped in any fashion, although most examples from Keatley Creek are pointed. Pointed tips would have been useful for tasks requiring gouging tools, such as the hollowing out of indentations for the placement of fire drills, thus adding to the versatility of bifaces.

The proximal ends were probably shaped to facilitate holding or hafting, although this has not been studied in detail. Hafting would have extended their use-life even further by enabling relatively small stubs to be used. It would also have increased the weight and ease of manipulating the tool in butchering tasks-attributes emphasized as important by Jones.

*Longevity:* Bifaces were clearly designed for prolonged use and many resharpenings. As noted previously, this makes sense given high mobility constraints with low material availability (seasonal or geographic) and high transport loads.

*Other Variables of Design:* It is probable that one of the main design characteristics emphasized was multifunctionality (versatility), although there are no morphological features per se that would lead one to postulate this. The inference of multifunctionality derives primarily from comparative use-wear (L. Keeley, personal communication 1992) and contextual or theoretical considerations. Thin bifaces are certainly not reliable tools, given their high rate of breakage and their fragility, although the gearing up investment and manufacturing efforts seem more typical of reliable tools, as well as their assumed context of use (time constrained hunting). Moreover, maintenance and use seem to have occurred in different locations at Keatley Creek (Spafford 1991), a characteristic of reliable tools. Although Bamforth (1991:230) has argued that bifaces are maintainable tools, it is questionable whether they should be considered maintainable, since it is not clear what comparable alternatives would have been employed if a biface broke. In addition, there is no reason to believe that extra bifaces were carried by individual hunters while hunting or that such spares could have been quickly or easily inserted into hafts (contra Keeley 1982). On the other hand, bifaces are clearly made for multiple resharpenings and are portable. Thus, whether bifaces should be considered unusually reliable or maintainable tools, or whether these distinctions are meaningful in this case, is open to debate. Bifaces have considerable potential for recycling; however, this is primarily a function of their size. It is doubtful that recycling (flexibility) considerations played much role in the actual tool design, and there is certainly no operational way to demonstrate such an assumption at this point. Recycling of bifaces could easily have been an opportunistic afterthought.

*Frequency:* There are only 205 bifaces or fragments in the sample (3.4% of identifiable tools). Frequencies are expected to be low, given long use-lives and use away from village sites. On the other hand, recycling and breakage into small fragments probably artificially elevated these frequencies.

*Specialization:* Jones (1980) thought bifaces made excellent butchering tools due to their weight and prehension characteristics. They may thus be considered specialized tools in this regard, although their potential for multifunctional roles is great. It is difficult to assess their status as specialized tools at this time. Certainly, the degree of investment of time, energy, and skill in their production is characteristic of specialized tools, but this relationship may be more complex than is often assumed.

## PORTABLE LONG-USE STRATEGY

The goal of this strategy is to carry specialized flake tools in high mobility contexts that will last as long as possible and thus avoid the need to carry excess stone weight. Thus, the most durable materials are often reserved for these tools, they usually have high resharpening potentials, and because of their specialized nature and resharpening requirements, we suspect they were probably most often produced near quarry sites, since a large proportion of the core reduction debris would not be suitable for making them. In the Keatley Creek assemblage, tools manufactured according to this strategy include endscrapers, specialized "key shaped" scrapers, and drills (Figure 4); projectile points can be considered a special case. We examine drills as an example of this strategy.

## Constraints

*Drill Task Constraints:* In order to be able to bore small, deep holes in moderate-to-hard materials with reasonable efficiency, tools must be narrow, have a tip that will cut or abrade, and be capable of relatively fast rotation. Although these tools were probably infrequently used, the high rates of rotation associated with them and the fact that they were probably intensively used for varying periods of time, possibly up to several hours per event, undoubtedly involved frequent resharpenings.

*Material Constraints:* Optimal materials would be those that were tough, durable, fine grained, and easily flaked. Trachydacite or chert/chalcedony would

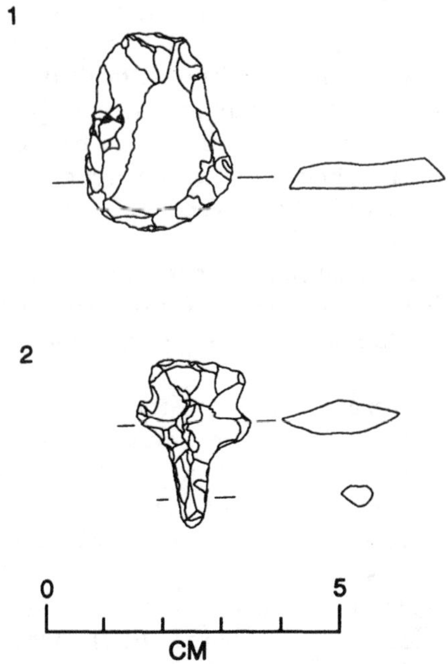


Figure 4. Number 1: An endscraper; number 2: a drill from Keatley Creek.

be the best choices, with chert and chalcedony having more advantages in terms of toughness, non-brittleness, and durability.

*Technological Constraints:* Relatively large, long, thin, straight flakes or triangular sectioned blades would be necessary for making drills. Because of the elongated, narrow bit with sub-circular cross-section and hafted nature of drills, they must have been some of the most time-consuming and difficult chipped stone tools to manufacture in this region. Wear rates were probably unusually high, due to the small working edges, strong pressures, and highly auto-abrasive environments. Frequent resharpenings must have been required.

*Socioeconomic Constraints (Transport):* Unfortunately, we do not know if drills were used exclusively at winter village sites, or whether they might also have been carried about on seasonal rounds and used at hunting or fishing sites. It would certainly have been a minor transport cost to carry the drill bit during seasonal moves, but whether there would have been a need for drills (e.g., in repairing fish net frames, snowshoes, or other items) is unknown. Nevertheless, transport constraints do not appear to have had much influence on tool design. Drills are highly portable and may have been transported as part of personal gear.

## Strategies and Design

*Raw Material and Procurement Strategies:* There would have been a clear selection for chert/chalcedony and perhaps extra efforts to procure this material, either by traveling farther, by searching for suitable raw material longer, or via exchange. In fact, 24% (N=33) of the drills were made of chert or chalcedony.

*Reduction Strategies:* Bifacial thinning generally provides thinner flakes; however, these tend to have greater curvature than hard hammer flakes. Nevertheless, if relatively large, straight billet flakes did occur, they probably would have been selected as blanks for drills (or projectile points). Relatively thin, straight hard hammer flakes could probably have been produced more easily and more frequently. Most drills have been so extensively modified that it is impossible to determine the type of flake from which they were made. The infrequent use and manufacture of drills would not warrant the development of a specialized blade technology.

*Tool Design and Resharpening:* Task constraints impose narrow limits on the morphology of the bit. In order to facilitate rapid rotation, it would be advantageous, if not necessary, to haft drill bits. Moreover, either due to the need to drill deep holes and/or the desire to prolong the use-lives of these tools, drill bits were relatively long. Resharpening would have been performed by pressure flaking, due to the delicate nature of these tools.

*Other Design Variables:* Drills must be viewed as highly maintainable tools, since they are designed for repeated resharpening and replacement. They are not particularly robust or overdesigned. Indeed, given the task constraints, it is difficult to see how drills could be overdesigned or more specialized. Although risk and time constraints cannot be considered significant, it is difficult to imagine how drill morphology might change even if risk and time became important considerations. Thus it is not clear how these conceptual constructs help in understanding tool morphology in this case. Other factors such as task constraints, amounts of drilling involved, and rate of material consumption seem far more important for understanding drill design. Similarly, versatility and flexibility considerations do not advance understanding of drill morphology.

*Frequency:* Because of the infrequent need for drilling small, deep holes, as well as their long-lived, resharpenable status, drills should be relatively rare in winter village assemblages. In fact, they represent only 0.6% of the tool assemblage.

*Specialization:* It is difficult to imagine a more specialized chipped stone tool. Although our data indicate that 29% of drills had secondary uses, these are all cases of uncertainty in categorization, i.e., drills viewed as possible bifacial perforators or pendants. The high degree of specialization in this case was probably not caused by very high processing volumes, but by narrow task constraints. Perhaps because of the more specialized morphologies and repeated resharpenings of most tools in the portable, long-use strategy, multifunctionality is poorly developed (8% of key-shaped scrapers, but 30% of endscrapers, which may include many cases of hafting retouch). This is somewhat contrary to the expectations developed by Shott (1986), who maintained that the most portable tools should exhibit the greatest degree of multifunctionality.

## QUARRIED BIPOLAR STRATEGY

This is a strategy oriented to the special needs for large, coarse-grained, spall tools, which could be left as site furniture or discarded after use.

### Constraints

*Spall Tool Task Constraints:* Spall tools found at the site are generally similar to hafted ethnographic specimens (Figure 5) recorded by Teit (1900, 1906) and Albright (1984). These ethnographers reported that large spall tools of coarse-grained rock were used to stretch hides in the tanning process. Coarse-grained stone is desirable for the final softening procedure in making buckskin, because it will not cut through hides with the application of the very high pressures required to stretch the skins in breaking down the lignin fibers. These tools are thus good for stretching hides and touching up hides during membrane removal, a time-and-effort consuming procedure. High intensity, but sporadic and low-frequency, use probably characterized these tools, as they involved the same considerations as expedient knives, i.e., scarcity of deer and hides.

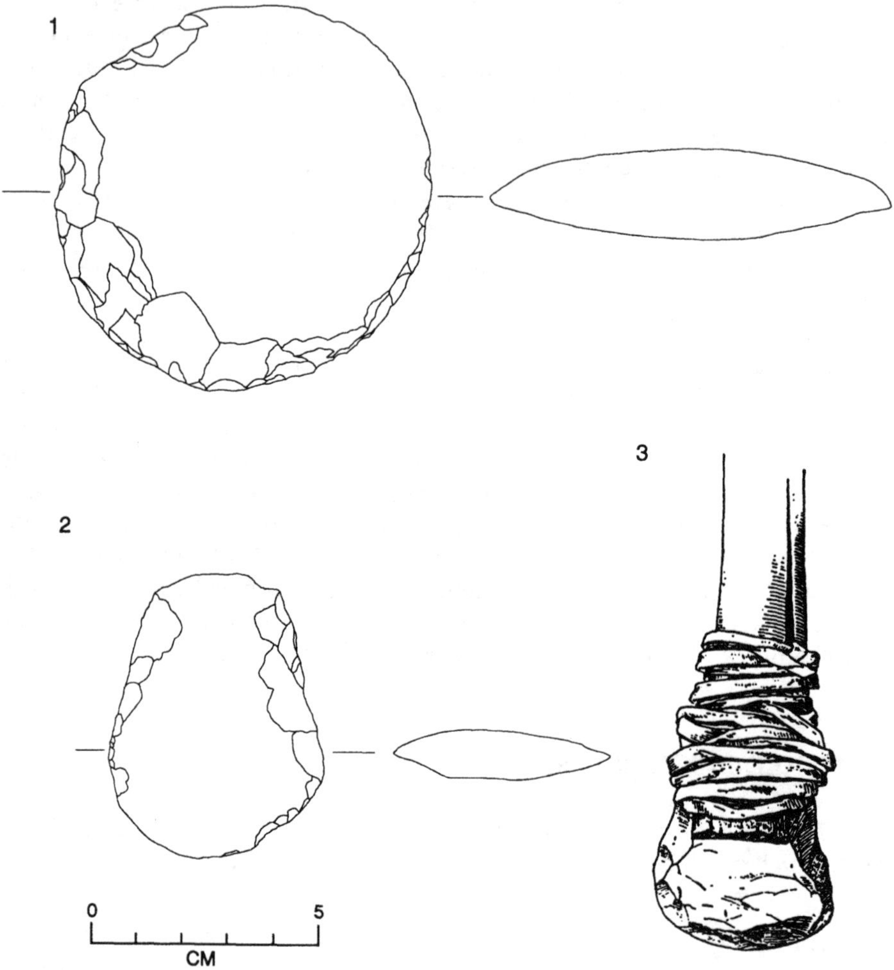


Figure 5. Numbers 1–2: Two quartzite spall scrapers from Keatley Creek, probably used for the finishing stages of hide scraping and stretching, as was the ethnographic example illustrated by Teit (1900) (number 3). The ethnographic specimen is about half the scale of the archaeological specimens, and the haft was probably a meter or more in length.

*Material Constraints:* Coarse-grained stone is desirable for this task, as it grips and removes any remaining wet endoderm on the hide. Quartzite has a coarse grain, does not crumble, and is flakable. Few stone types in the area have similar characteristics, although some other coarse-grained igneous and metamorphic rock types were also used.

*Technological Constraints:* Little time or effort were required for the manufacturing of the stone component, although proper hafting involves substantially more time and effort. Difficulty is involved in shaping hafts to fit specific flakes of these large sizes and in binding flakes to withstand high stresses. Retouch would have been needed only if the original edge was too sharp or jagged, or for hafting modification. According to ethnographic information, the use-life of this kind of tool was very long, extending over generations (Albright 1984), thus minimizing average yearly manufacturing time.

*Flake Type:* Large flakes are required to maximize the effect of stretching on skins. The main source of coarse-grained materials large enough to produce these flakes is rounded quartzite river cobbles and boulders. Bipolar splitting is the most effective and perhaps the only means of producing large flakes from them, although occasional flakes produced by direct hard hammer percussion might also be suitable.

*Wear Rate:* Wear formation is very slow for these tools and is an almost insignificant consideration.

*Socioeconomic Constraints (Transport):* There are a number of indications that spall tools were highly conserved (Albright 1984) and were treated as site furniture, rather than transported on seasonal moves. The unusual weight and size of the spalls, not to mention their long, stout hafts, would have been a heavy burden to carry. Since these tools were used only in winter villages and perhaps at fall mountain hunting camps, and since they had very long use-lives, it would have made far more sense for people to have cached spall tools at the sites of their use. The high percentage of whole retouched spall tools (N=37, with 76% whole or minor damage) supports this suggestion, since there are no obvious reasons to have abandoned whole and still usable tools at the site. Alternative solutions such as pointed sticks (Teit 1900) could even have been used for stretching hides in mountain locations. Excavations at Keatley Creek revealed a number of clearly cached spall tools in Housepit 7. Thus, as site furniture similar to anvils, there would have been little active transport of spall tools. Given the strong transport constraints, their size and weight favored leaving them at winter housepits.

## Strategies and Design

*Raw Material Strategies:* The basic strategy for hide stretching tools was to use coarse-grained materials that would not cut into hides. Quartzite should have been most favored. In fact, 41.5% of the sample is quartzite, 19.5% is fine- and coarse-grained trachydacite, 9.8% is coarse-grained andesite, 4.9% is olivine, 2.2% is shale, and 22.9% is indeterminate material.

*Acquisition/Procurement Strategies:* Quartzite raw material could have been obtained in the form of river cobbles near the Fraser River. However, there would have been some restrictions on the availability of the quartzite, especially during the coldest part of the winter season from frozen ground and snow cover.

Therefore, people would have needed to procure quartzite materials in advance or to have kept tools from previous seasons. Most tools were probably kept from year to year because, in spite of the easy access to this raw material (the distance to the river is only one kilometer), ethnographically the use-life of these tools was very long (Albright 1984). One may wonder why this type of tool was kept, if the procurement cost was low. We suspect that their longevity is probably related to the difficulty of obtaining raw material in the winter. It may also be related to the difficulty involved in shaping hafts to fit specific large flakes, and possibly the effort associated with splitting cobbles to obtain a tool with an edge that worked precisely right for a given individual. Moreover, as with anvil stones and abrading stones, there was simply no reason to discard these tools from one year to another.

Trachydacite was not locally available near Keatley Creek. Spall tools could have been brought into the site as finished tools, or they could have been manufactured from cobbles that were brought to the site as cores. Some of the trachydacite utilized was fine-grained (12.2% of the sample). As this does not seem to have been the preferred material for this type of tool, it may have been utilized at some times because there was no better material immediately available.

*Reduction Strategies:* There was a need to split cobbles in order to obtain the largest possible flakes. Because of this, and because of the round shape of river cobbles and the toughness of quartzite, bipolar reduction was the best strategy, although direct percussion may also have been used in some cases. No other reduction strategies are capable of producing suitable flakes.

*Tool Design and Resharpening:* There is a need for dull edges or edges that grip slippery surfaces. Long-handled hafted tools are much more efficient for "staking" (stretching) skins, because much more pressure can be applied in this fashion. This pressure is critical for stretching skins and making buckskin. In addition, large, broad edges are important for softening large areas at once (Hayden, n.d.). Thus large, broad flakes of coarse-grained material, suitable for lashing to long wooden hafts, constitute the main tool design, although other technological solutions made entirely of wood could also have been used.

*Other Variables of Design:* Reliable designs have strengthened parts, are overdesigned, have a sturdy construction, and possess careful fitting of the parts. In the case of spall tools, there was a high investment in shaping the hafts to adequately fit the tools. Hafting use-wear is frequently evident on spall tools. The tools were specialized and robustly designed, apparently due to the nature of the task and the high pressures exerted. There was virtually no maintenance involved for the stone parts of these tools, and they are probably the most cumbersome chipped stone tool to transport in the entire assemblage. The impression that spall tools were designed with reliability in mind is strengthened by the lack of indications that spall tools were maintained. Many were never retouched or rarely needed resharpening. Nor was breakage frequent. There is no indication that recycling potential (flexibility) or multifunctionality (versatility) played any significant role in tool design or use. Only 9% show any other possible signs of alternative use, probably because of the specialized nature of the material and the tool itself.

While all of these characteristics have been suggested as typifying "reliable" designs, there is little correspondence to the risk factors that Bleed (1986), Torrence (1989), and Nelson (1991) suggest produce reliable designs. Thus reliability in this case was probably incidental to the basic task mechanics. Reliable designs are supposed to be more suitable when there is a premium on resource capture and processing time. This is clearly not the case for spall tools. Any need for emergency tool replacements might have been achieved simply by keeping extra parts in storage or by using alternative solutions such as plain wooden sticks. However, this cannot be determined from tool morphology. On the other hand, Bleed (1986:741) also argued that reliable designs are optimal when there are predictable times of need and downtime, as well as in situations for which bulk and weight are not critical. This corresponds more closely to the use context of spall tools.

*Specialization:* Hafted spall tools are among the most specialized and non-versatile tools in the assemblage. This specialization does not appear to have been caused by risk or time constraints, but to basic task mechanics.

*Frequency:* Because of moderate-to-low processing volumes and extremely long use-lives, spall scrapers are expected to be rare, and they are (N=41; 0.7% of the assemblage).

## SCAVENGED BIPOLAR STRATEGY

Although clearly present in the Keatley Creek assemblage, our original research design did not provide for the quantitative collection of data on this aspect of lithic-related behavior. However, it can be subjectively stated that large tools and flakes, as well as bifaces and residual block cores, often seem to have been "recycled" via simple intentional breakage (in order to use broken edges) or via bipolar reduction to create new flakes. This strategy is related to Nelson's (1991) "flexibility." Unfortunately, we cannot provide a detailed analysis of this strategy, although its importance is expected to have increased as lithic reserves decreased during the seasonal occupation of the pithouses at Keatley Creek.

## GROUND STONE CUTTING STRATEGY

This strategy is used under conditions of high-volume processing involving cutting tools and/or to display control of wealth and power. We will examine nephrite adzes as an example.

## Constraints

*Nephrite Adze Task Constraints:* Ethnographic information for the area mentions the utilization of ground stone adzes for heavy woodworking (Teit 1900). This kind of tool could also have been employed for heavy butchering, and appears to have been used to shape antler prehistorically on the basis of adze marks found on large pieces of antler at Keatley Creek.

Adzes were probably used in the construction of the pithouse wooden roof superstructures, as well as of interior planking for sleeping platforms and other furniture. On the basis of Teit's accounts and drawings, the number of logs needed in the construction of a medium-sized housepit can be estimated at about 312 (24 large logs, 44 medium logs, and 244 smaller poles). High, heavy-duty cutting requirements might also have been involved during the year in building deer fences in the mountains, removing large amounts of bark for cambium, canoes, baskets, roofing, and constructing drying racks, net frames, bows, log ladders and the log or plank sculptures documented by Teit (1906) and others. Thus, very large quantities of wood would have been episodically processed. Ground stone adzes may not always have been more efficient in cutting wood than chipped stone equivalents (Hayden 1987); however, where cutting requirements were extremely high, it would have been more costly in terms of effort, time and scheduling to return to quarry sites at short intervals to replace exhausted tools or materials. In this respect ground stone cutting tools had major advantages over chipped stone tools. The cutting tasks themselves simply required sharp, semi-abrupt edges with considerable mass in order to render penetration effective.

*Material Constraints:* A durable, tough raw material suitable for reduction by abrasion and sustaining sharp cutting edges would have been optimal. At Keatley Creek nephrite was utilized. Other igneous and metamorphic rock types in the area were also used and would have been easier to manufacture, although few would have produced as effective and strong a cutting edge as nephrite. Smaller chipped stone quartzite flake adzes and large bifacial quartzite core adzes may also have been used as an alternative tool by poorer families for some wood working or barking activities. Thus more than simple practicality may have been involved in the use of nephrite for adzes.

*Technological Constraints:* By all measures, it would have been exceedingly time consuming to manufacture ground stone adzes, especially with nephrite. To cut 1 mm of nephrite with traditional techniques requires an hour of work. In addition, the maintenance of cutting tools by edge-grinding involves a considerable amount of labor (cf. ethnographic data in Hayden 1979, 1987). However, other chipped stone alternatives may have involved greater total costs. Nephrite is extremely tough and durable and would be unusually long-lasting, with low wear rates and low resharpening requirements, thereby reducing average yearly costs. The only requirements for shape would be the need for raw material large enough to manufacture into an adze through grinding.

*Prestige and Ideological Constraints:* Hayden (1987) initially related the importance of edge-ground adzes or axes to high wood cutting requirements. In some instances, such as with the high manufacturing time involved in making nephrite adzes, edge grinding may also have been related to the existence of free time or to the control over others' labor in the form of slavery or the ability to commission work. The goal of prestige technologies is to use control over labor to produce desirable items that are too labor-costly for most people to be able to afford, thus displaying individual power and wealth. Given the inordinate amount of labor involved in producing nephrite adzes, they are prime candidates for prestige artifacts. Other types of stone may have been nearly as effective and have involved much lower production costs, but would not have had as much prestige display value. In the Lillooet region, the existence of social hierarchies including slave labor was documented ethnographically (Teit 1906), as was the existence of some occupational specialization (Romanoff 1992). Unusually long nephrite adzes were clearly used as wealth and prestige display items (Smith 1900; Emmons 1923:26–27).

*Socioeconomic Constraints (Transport):* Despite their considerable weight, especially when hafted, nephrite adzes were probably not left at winter village sites (in contrast to spall tools), because of their high procurement and manufacture costs and the need for cutting tools at other seasonal locations. Given their very high value, they were probably part of personal gear. Ground stone adzes are good solutions to transport constraints, because their replacement rate is far lower than that of chipped stone heavy wood working tools. They thus require the transport of only one groundstone tool, rather than several chipped stone versions, including replacements.

## Strategies and Design

*Raw Material and Procurement Strategies:* At the Keatley Creek site, several fragments of adzes made of nephrite were recovered. The source of this raw material was probably the Fraser and Bridge River lag deposits, where occasional nephrite cobbles and boulders are found today. Thus some raw material could have been obtained while other activities—such as fishing—were carried out. However, if more material was required than could be found opportunistically, it would have been a time-consuming endeavor. In addition, the use-life of ground nephrite adzes is very long (probably spanning more than one generation), thereby minimizing the average yearly procurement cost.

*Reduction Strategies:* Grinding using sands (preferably garnet sands), garnet sandstone, water, and wood or cord is the only effective traditional technique for shaping nephrite, since this material is extremely tough and does not flake readily or predictably.

*Tool Design and Resharpening:* Due to the intensive and long duration of processing numerous logs, hafting provides critical advantages in easing the fatigue and trauma to hands involved in hand-held chopping tools. The manufacturing and maintenance costs of the haft are more than offset by savings in fatigue and perhaps an increased efficiency involved in processing large quantities of wood. Resharpening the cutting edge would have required a non-permanent type of haft. The most practical would probably have been a friction fit accompanied with binding. Smooth surfaces provide far superior friction fit hafts than irregular surfaces, such as those typical of chipped stone tools. Thus, in addition to edge grinding to prolong use-life (and reduce consumption of raw material), all surfaces could be expected to have been ground. Grinding is also the only means of shaping nephrite in a controlled fashion. Sizes (width) should be a function of the size of the wood or antler being worked and the mass required to penetrate wood effectively. Initial lengths would probably have been as long as possible without creating loading conditions leading to breakage, but maximizing the resharpening potential and use-life of the tool.

*Other Design Variables:* Nephrite adzes appear to be highly reliable on the basis of morphological criteria. They are made of much tougher materials than is strictly necessary, they were robust, they required elaborate advance manufacturing, and they required specialized repair or resharpening. However, there is a more fundamental question involved in understanding the strategies behind the manufacture of nephrite adzes. This involves whether the apparent "reliable" character of these tools was only incidental to, or a by-product of, a more basic concern with material conservation strategies used in the face of large processing requirements, or even more importantly, of a basic concern with displaying wealth and power. There is no indication that they were used in high-risk contexts, while transport and material costs were clearly high.

*Frequency:* Aside from loss, discard of this kind of tool at the site is probably due to breakage and abandonment of small or flawed pieces that could not be easily reshaped to form smaller adzes. Because nephrite adzes were conserved and were almost certainly part of personal gear, they should be very rare in winter village assemblages.

*Specialization and Multifunctionality:* Although ground stone adzes could be used for cutting many things, they were probably developed for a specialized task: procuring and processing large amounts of timber. Few arguments can be marshaled to support the idea that this multifunctionality played a significant role in tool design or manufacture. Any multifunctionality attributable to adzes makes more sense as the opportunistic use of conveniently available tools developed for other purposes. Nor does their potential for recycling appear to have had any influence on raw material choice or tool design, although broken adzes might have been reshaped into small chisels or even prestige ornaments. Frequencies of these artifacts are too low to evaluate such probabilities. Since very specialized materials were used, and since a great deal more time and effort and specialized hafting designs were involved in the manufacture of these tools (in contrast to more generalized forms of wood chopping tools), adzes should probably be considered highly specialized tool types.

## DISCUSSION

The technological organization of a group responds to different environmental conditions, such as the distribution and predictability of resources and their periodicity, productivity, patchiness and mobility. It involves resource acquisition, manufacture, manipulation, and discard or loss (Nelson 1991). Our focus in this discussion will be on the acquisition, manufacture and manipulation of stone resources and the evaluation of basic strategies in these domains.

The groups that have inhabited the Lillooet region probably acquired most of the raw material they needed for their tasks during their seasonal round. Trachydacite and chert, the raw materials utilized for the manufacture of most of the tools, are not available near the site. We believe that the different core strategies documented at Keatley Creek were emphasized in order to manage the problem of limited raw material availability according to the minimal requirements of various classes of tasks involved.

Blade cores do not occur at the site, except for bladelet cores from prehousepit components. We propose that most prepared cores, such as Levallois cores and especially blade cores, are actually wasteful of raw material (contra Sheets and Muto 1972; Clark 1987; Nelson 1991:68) because of the high risk of failure at all stages, the initial need to shape cores, and the need for specific sizes, shapes, and high quality raw material. These factors also increase search and procurement times and the investment in training and making the tools required for successful blade production. On the other hand, once produced, blades have the advantages of having relatively long, sharp cutting edges per unit weight with little unusable edge, of being relatively thin and straight, and of facilitating multiple resharpenings for tools made on distal or proximal ends such as burins, endscrapers, borers, piercers, drills, and points.

Therefore, we suggest that, among mobile hunter-gatherers, blade technologies should occur: a) where processing large volumes of material involves the distal end tools just mentioned; b) where there is an unusual need for large numbers of straight, thin, long flakes as in high volume butchering and filleting; and c) where high mobility places a premium on edge:weight ratios and where this coincides with seasonal visits to high quality lithic sources with abundant suitably sized nodules for blade making.

In the Keatley Creek case none of these conditions applied in the Late Prehistoric (pithouse) tradition. The most sensible use of the generally small-to-medium sized raw material available within the seasonal round would have been as block cores from which almost all flakes larger than 2 cm could have been used. This reduction strategy relying on block cores would have provided maximum flexibility in terms of the production of different sizes and shapes of blanks for most flake tools, including small expedient knives, drills, notches, and endscrapers. The production of highly varied types of flakes depending on situational needs cannot be achieved with blade cores. It is therefore not surprising to find that block core reduction was the dominant strategy used at the site. The overall small size of tools also supports the interpretation that block core reduction provided the most efficient use of raw material for most needs. Whether it would have been economical to reduce cores at quarries and simply carry away suitable flakes to the winter village, or to carry cores to winter villages to maintain flexibility of blank production, is unclear. Both strategies were probably used, perhaps with an emphasis on quarry site production of those blanks with the greatest size and shape constraints (endscrapers, drills, key-shaped scrapers, bifaces). In fact, while cores occur at the site, refitting attempts so far have failed to produce a single conjoinable pair, indicating substantial off-site flake production (as well as clean-up of debris for outside discard). At this point, however, we have not been able to distinguish expedient tool production at the site from the introduction of flake blanks from quarries.

## Major Strategies

The term "strategy" can be used at a detailed level (e.g., the resharpening or procurement strategy) as well as a broader, more encompassing level. This results in some confusion, and perhaps different terms ought to be applied to the different levels of problem-solving approaches, or at least to the different labels applied to different types of strategies (e.g., reduction strategies).

In addition to the six reduction strategies we have discussed, other strategies occur in assemblages elsewhere in the world (e.g., prepared blade strategies). However, we feel that the identification of these six major types of strategies is a reasonable initial step in the systematic analysis of a specific assemblage for our heuristic, exploratory goals.

Both the tool and debitage analyses (Spafford 1991; Prentiss 1993) indicate that the housepit assemblages at Keatley Creek are dominated by the expedient block core and tool strategy. Some analysts have suggested that the use of expedient cores may be a result of the lack of time stresses when living off stored foods (Torrence 1983); to increased dependence on plants involving less risk than hunted foods (Torrence 1989); or to relatively sedentary occupations involving the stockpiling and constant availability of raw material (Parry and Kelly 1987; Johnson 1987). We suggest that the expedient reduction of block cores is the most efficient use of raw material in terms of procurement, reduction, and the employment of minimal amounts of raw material in any given task. We argue that there would have been considerable constraints on the amount of raw material that could have been brought to the winter villages due to the need to transport large quantities of food to be stored and of gear from the mountains to the winter village, and that stone was therefore used in an extremely economical fashion. This is indicated by numerous factors: 1) the unusually small size of the tool assemblage as a whole and the remarkably small size of many of the expedient tools and cores; 2) the high rate of breakage (frequently intentional) and re-use of edges formed by breaks (Prentiss 1993); 3) the high degree of culling of all usable flakes (Prentiss 1993); 4) the high frequency of multiple-edge use (subjectively observed); 5) the high ratio of tools to debitage (0.12–0.16 in Housepits 3 and 7); 6) the great variability in the size range of tools and the extent of their resharpening; and 7) the frequent recycling of broken bifaces and exhausted cores through bipolar reduction or other means. While the original research design did not anticipate the recording of all these variables, many of them became evident during analysis and we can make relatively confident statements about them on a subjective basis.

The next most common strategy involved the intensive use of bifacial reduction flakes as expedient tools, and this too makes sense under conditions in which raw material is scarce. Although it is clear that bifaces were used in specific activities in the center of the large housepits during the winter occupations, the overall design of biface tools makes most sense in terms of high mobility (see Sassaman's 1992 summary). Bamforth (1991) has shown that in California there was a clear emphasis on bifaces in hunting campsites, with a complementary emphasis on expedient "utilized flakes" (probably including types similar to the expedient knives at Keatley Creek) at the larger, more sedentary logistical sites. This fits our view of how the overall lithic technology was organized in the Keatley Creek case. The use of bifaces in pithouses probably represents the incidental use of a handy and convenient tool for butchering, woodworking, or some related activities, rather than the condition under which bifaces could be expected to be most adaptive (i.e., under high mobility and transport constraints).

## Evaluating Design Considerations

Trying to assess the role of design considerations in tool formation processes is a new and largely unimplemented analytical program. The difficulties are not to be underestimated. However, by focusing on the various components of tool formation processes in the framework of design theory, a much more robust understanding of lithic assemblages is gradually emerging. It is worth iterating that, contra Torrence's (1989) criticism of Hayden's (1987) approach as being non-situational and inherently progressivist, the approach advocated here (and which Hayden has always advocated) is entirely responsive to specific situations and contexts. In previous publications (Hayden 1987), some attention was drawn to very broad changes in technology over long periods of time and an attempt was made to understand how conditions and constraints could have changed over time to result in such broad-scale technological changes. This was never meant to be viewed as inherent progress, or to be divorced from practical adaptive behavior. Nor has Hayden ever adopted the rigid maximization models that Torrence imputes. In fact, he recognized the important role of risk in modeling behavior long before Torrence began to discuss the topic (Hayden 1983).

Perhaps one of the most confusing aspects of the recent theoretical treatment of stone tools has been the discussion of specific elements such as reliability without reference to where they fit into broader frameworks. This has been exacerbated by the implicit or explicit claims that particular new concepts such as risk constitute the keystones to understanding the totality of lithic assemblage organization. One result has been a lack of clarity in distinguishing various levels of analysis such as those identified in Figure 1. Nelson (1991) attempted to clarify the relationships of some of these concepts, although design was emphasized almost to the exclusion of other factors. We have found the framework presented in Figure 1 of our analysis particularly helpful in this respect. However, dealing with design considerations still poses a number of problems which we will highlight for the use of researchers working in this area in the future.

One initial problem area is the unusually vague referent of many of these concepts. Reference is rarely made to specific tools or specific attributes, but rather to tool classes, assemblages, or even more nebulous tool systems. Nelson (1991:58) even makes the remarkable statement that reliable designs cannot be recognized in specific artifact types, which makes us wonder whether the concept can have any operational meaning at all. Modifications and changes to definitions have not helped in dispelling these mists. The role of specialized versus multifunctional tools in the reliable/maintainable scheme is not clear, either. Bleed (1986) mentioned that multifunctional tools should be maintainable, implying, but never stating, that specialized tools should be reliable. As the analysis of drills has shown, a number of clearly maintainable tools (on morphological criteria) are extremely specialized and have virtually no recycling potential (flexibility), in contrast to Nelson's (1991) expectations for maintainable tools. Moreover, there may be other aspects of the "maintainable" category that could be more profitably examined as independent design considerations, just as we feel that multifunctionality is most profitably dealt with on its own rather than being subsumed under the overly global catch-all term of "maintainable." At present there has been no clear exploration of whether maintainable and reliable are mutually exclusive design considerations. In general, we consider the tendency to lump too many discrete concepts under global concepts such as curation and maintainability as one of the major impediments to making progress in understanding lithic assemblages. Other concepts, such as reliability, are more bounded and potentially productive.

This leads to a second problem area, one of the most serious shortcomings of recent work dealing with design, that is, operationalization of definitions. Shott (1986) and Aldenderfer (1991b) are among the few analysts to have made such an attempt, and even here, there are considerable problems. The newly proposed design concepts are highly abstract and their operationalization has so far been highly subjective, relying for identification on the relative emphasis of design considerations in tools or tool classes. Since all flake tools are light and portable and require some maintenance or resharpening (except perhaps utilized flakes), how can maintainability be identified? How can overdesigning be identified? Or parallel subsystems? Or specialist manufacturing? Are spall tools or nephrite adzes overdesigned? How can the opportunistic use or recycling of a tool be distinguished from a "strategy" in which the tool was designed to be multifunctional or recycled? How can intentionality be identified, or circularity of reasoning be avoided? Even using Bleed's own examples (a reliable sealing harpoon point versus a maintainable dart point), it is difficult to determine how these could accurately be identified as reliable or maintainable from an archaeological context. The subtle changes of emphasis that are said to characterize various design considerations only add to the overly subjective assessment of these analytical approaches.

A third problem area involves the underlying relevance of design concepts. Even in cases in which a reasonable measure of the concept may be attainable (e.g., longevity, multifunctionality), no well-tested or verified explanations of the conditions under which these values change have emerged. Instead, there exist numerous abstract models to explain longevity (i.e., curation), reliability, specialization, and other characteristics. These are usually validated by reference to trends in non-lithic materials systems (or by reference to other unverified abstract theoretical constructs), but are rarely examined successfully using lithic archaeological assemblages. As a result, significant exceptions exist to virtually all the models that have been proposed. For instance, Binford argued that long-lived (curated) tools were developed only during and after the Upper Paleolithic, yet Acheulian bifaces were clearly transported over long distances and used over long periods of time. Similarly, Nelson (1991:74) and others argued that, under conditions of high mobility, tools should be light and portable; yet mobility during Acheulian times was probably as high as any hunting and gathering group reliant on foot transport, and their tools (bifaces) are among the heaviest to occur in the archaeological record. In the context of this analysis, Shott (1986) argued that multifunctionality should increase under conditions of high mobility. Yet we found that multifunctionality was highest, indeed exceptionally high, in the block core reduction strategy which was used under conditions of essentially no mobility, i.e., in pithouses during winter stays. In this case, availability of and access to raw material seems to have been the main factor favoring multifunctionality. Similarly, tool-to-debitage ratios in pithouses were unusually high, a condition also argued by several authors to be associated with high mobility, and clearly not the case at Keatley Creek. Grounding and testing the many newly proposed models of assemblage variability is perhaps the most pressing area for research in lithic studies today.

According to the current explanatory models of reliable designs (Bleed 1986; Torrence 1989; Nelson 1991:66–67), the severity of consequences involved in failure to accomplish a task (severity of risk) ought to be related to the emphasis on reliable designs. While this seems logical, problems arise in attempting to apply it to prehistoric contexts. For instance, according to this model one might expect highly reliable tools to be used on hunting forays, especially during seasons when food reserves were low. On hunting forays, however, instead of robust, over-designed, sturdy, strengthened tools, we find thin bifacial tools used in butchering, which are relatively easily broken by accident, use, or resharpening. In yet other examples, such as nephrite adzes, there are morphological indicators that these tools should be considered reliable designs due to excessive robustness, advance preparation, lack of evidence of maintenance during use, effort investment, special materials used, and high degree of specialization. Yet referring to these tools as "reliable" does not correspond to the conditions of risk or time constraints that have been proposed as causing reliable designs. Given these examples, it is not clear that these concepts will be useful or productive in understanding lithic assemblages.

A fourth problem with many current approaches to design has been the exclusive focusing on sometimes chimerical design considerations to the exclusion of other basic factors such as task mechanics and volume of materials being processed. There has been little attempt to deal with such issues as whether the selection of durable raw materials has been caused by high risk or the need for long-lived tools. In the case of bifaces, drills, and keyshaped scrapers in the Keatley Creek assemblage, we feel that the arguments are heavily weighted in favor of longevity considerations, rather than any aspect of risk.

While the concepts of reliability and risk are stimulating and provocative, they have not yet proved their utility in analysis. In our view, other basic constraints and design considerations have been ignored, but provide far greater explanatory power in understanding most tool morphologies. On the other hand, there may be certain key instances in which concepts such as reliability provide critical understanding for specific tool types in a manner similar to prestige and ideological constraints. In the vast majority of cases, prestige considerations such as those presented in the case of nephrite adzes play little role in tool formation processes or assemblage structure. However, in rare cases, prestige becomes an absolutely critical element, and failure to recognize such a factor would seriously flaw any analysis. This situational invocation of special design considerations and constraints does not imply that prestige is the key to understanding the total organization of lithic assemblages. We suggest that reliability and risk may play similar roles with respect to prestige in lithic organization, that is, they may only be important for explaining a few unusual cases and need be invoked only as exceptional or special features where this seems particularly warranted. As Ellis and Lothrop (1993) point out in their critical evaluation of these concepts, the ethnographic observations on which most of these models were built considered only food procurement tools (mostly non-lithic), whereas most archaeological lithic tools are maintenance or manufacturing tools divorced from immediate risk consequences. We have found in the Keatley Creek analysis that, most of the time, other more basic constraints and considerations appear to have overriding influences. The intentional design concept of "flexibility," as defined by Nelson, seems difficult to identify or to use in terms of empirical applications. In our analysis its only meaningful application is in the examination of core types (block versus prepared cores).

Clearly, the holistic tool formation type of analysis which we have attempted here is still in its developmental phase. Yet the results seem promising enough to warrant more detailed experiments, comparisons, analyses, and data gathering that could transform our initial formulations into much more robust conclusions about the organization of lithic technology and the strategies that were used prehistorically to deal with technological problems.

## ACKNOWLEDGMENTS

We would like to thank Margaret Nelson, George Odell, and Michael Shott for critical comments on this chapter. Our gratitude also extends to the Fountain and Pavilion Indian Bands and to the many people involved in the Keatley Creek excavations for their cooperation and help. We would like to acknowledge the financial support of the Social Sciences and Research Council of Canada, the British Columbia Heritage Trust, and the Simon Fraser University Special Research Program.

## REFERENCES

Ahler, S., and R. McMillan. 1976. Material Culture at Rodgers Shelter. In *Prehistoric Man and His Environments*, edited by W. Wood and R. McMillan, pp. 163–199. Academic Press, New York.

Albright, S. 1984. *Tahltan Ethnoarchaeology*. Archaeology Department, Simon Fraser University, Publication No. 15.

Aldenderfer, M. 1991a. The Analytical Machine: Computer Simulation and Archaeological Research. In *Archaeological Method and Theory*, vol. 3, edited by M. Schiffer, pp. 195–247. University of Arizona Press, Tucson.

Aldenderfer, M. 1991b. Functional Evidence for Lapidary and Carpentry Craft Specialties in the Late Classic of the Central Peten Lakes Region. *Ancient Mesoamerica* 2:205–214.

Bamforth, D. 1991. Technological Organization and Hunter-Gatherer Land Use: A California Example. *American Antiquity* 56:216–234.

Bleed, P. 1986. The Optimal Design of Hunting Weapons: Maintainability or Reliability. *American Antiquity* 51:737–747.

Budnick, F., D. McLeavy, and R. Mojena. 1988. *Principles of Operations Research for Management*. Irwin, Homewood, Illinois.

Callahan, E. 1979. The Basics of Biface Knapping in the Eastern Fluted Point Tradition: A Manual for Flintknappers and Lithic Analysts. *Archaeology of Eastern North America* 7:1–180.

Clark, J. 1987. Politics, Prismatic Blades, and Mesoamerican Civilization. In *The Organization of Core Technology*, edited by J. Johnson and C. Morrow, pp. 259–284. Westview Press, Boulder.

Ellis, C., and J. Lothrop. 1993. Technological Change during the Paleoindian and Archaic Periods in Northeastern North America. Paper presented at the annual meetings of the Society for American Archaeology, St. Louis, Missouri.

Emmons, G. 1923. *Jade in British Columbia and Alaska, and Its Use by the Natives.* Museum of the American Indian, Heye Foundation. Indian Notes and Monographs No. 35, New York.

Frison, G. 1989. Experimental Use of Clovis Weaponry and Tools on African Elephants. *American Antiqutity* 54:766–784.

Hayden, B. 1979. *Paleolithic Reflections: Lithic Technology of the Australian Western Desert.* Australian Institute of Aboriginal Studies, Canberra.

Hayden, B. 1983. Review of *Hunter-Gatherer Foraging Strategies,* edited by B. Winterhalder and E. Smith. *Canadian Review of Sociology and Anthropology* 20:101–103.

Hayden, B. 1987. From Chopper to Celt: The Evolution of Resharpening Techniques. *Lithic Technology* 16:33–43.

Hayden, B. 1990. The Right Rub: Hide Working in High Ranking Households. In *The Interpretative Possibilities of Microwear Studies,* edited by B. Graslund et al., pp. 89–102. Societas Archaeologica Upsaliensis, AUN 14, Uppsala.

Hayden, B., n.d. Using Stone Tools in Producing Buckskin. Report on file with the author.

Horsfall, G. 1987. Design Theory and Grinding Stones. In *Lithic Studies Among the Contemporary Highland Maya,* edited by B. Hayden, pp. 332–377. University of Arizona Press, Tucson.

Ingbar, E. 1990. The Hanson Site and Folsom on the Northwest Plains. In *Ice Age Hunters of the Rockies,* edited by D. Stanford and J. Day, pp. 169–192. Denver Museum of Natural History, Denver.

Johnson, J. 1987. Introduction. In *The Organization of Core Technology,* edited by J. Johnson and C. Morrow, pp. 1–12. Westview, Boulder.

Johnson, J. 1989. The Utility of Production Trajectory Modeling as a Framework for Regional Analysis. In *Alternative Approaches to Lithic Analysis,* edited by D. Henry and G. Odell, pp. 119–138. Archaeological Papers of the American Anthropological Association, no. 1.

Jones, P. R. 1980. Experimental Butchery with Modern Stone Tools and Its Relevance for Paeolithic Archaeology. *World Archaeology* 12:153–165.

Keeley, L. 1982. Hafting and Retooling: Effects on the Archaeological Record. *American Antiquity* 47:798–809.

Kelly, R. 1988. The Three Sides of a Biface. *American Antiquity* 53:717–734.

Morrow, C. 1987. Blades and Cobden Chert. In *The Organization of Core Technology,* edited by J. Johnson and C. Morrow, pp. 119–150. Westview Press, Boulder.

Nelson, M. 1991. The Study of Technological Organization. In *Archaeological Method and Theory,* vol. 3, edited by M. Schiffer, pp. 57–100. University of Arizona Press, Tucson.

Parry, W., and R. Kelly. 1987. Expedient Core Technology and Sedentism. In *The Organization of Core Technology,* edited by J. Johnson and C. Morrow, pp. 285–304. Westview Press, Boulder.

Prentiss, W. 1993. *Hunter-Gatherer Economics and the Formation of a Housepit Floor Lithic Assemblage.* Ph.D. dissertation, Department of Archaeology, Simon Fraser University.

Pye, D. 1964. *The Nature of Design.* Studio Vista, London.

Pye, D. 1968. *The Nature of Art of Workmanship.* Cambridge University Press, Cambridge.

Romanoff, S. 1992. The Cultural Ecology of Hunting and Potlatches among the Lillooet Indians. In *A Complex Culture of the British Columbia Plateau,* edited by B. Hayden, pp. 470–505. University of British Columbia Press, Vancouver.

Sassaman, K. E. 1992. Lithic Technology and the Hunter-Gatherer Sexual Division of Labor. *North American Archaeologist* 13:249–262.

Sheets, P., and G. Muto. 1972. Pressure Blades and Total Cutting Edge: An Experiment in Lithic Technology. *Science* 175:632–634.

Shott, M. J. 1986. Technological Organization and Settlement Mobility: An Ethnographic Examination. *Journal of Anthropological Research* 42:15–51.

Smith, H. I. 1900. *Archaeology of the Thompson River Region, British Columbia.* Memoirs of the American Museum of Natural History, volume 1, part 6, New York.

Spafford, J. 1991. *Artifact Distributions on Housepit Floors and Social Organization in Housepits at Keatley Creek.* Unpublished M.A. thesis, Archaeology Department, Simon Fraser University.

Teit, J. A. 1900. *The Thompson Indians of British Columbia.* American Museum of Natural History, Memoirs 2(4), New York.

Teit, J. A. 1906. *The Lillooet Indians.* American Museum of Natural History, Memoirs 2(5), New York.

Torrence, R. 1983. Time Budgeting and Hunter-Gatherer Technology. In *HunterGatherer Economy in Prehistory: A European Perspective,* edited by G. Bailey, pp. 11–22. Cambridge University Press, Cambridge.

Torrence, R. 1989. Retooling: Towards a Behavioral Theory of Stone Tools. In *Time, Energy and Stone Tools,* edited by R. Torrence, pp. 57–66. Cambridge University Press, Cambridge.

Winters, H. 1969. *The Riverton Culture.* Illinois State Museum, Reports of Investigations, no. 131. Springfield, Illinois.

# Part II

# Curation

Hunter–gatherer economic strategies have dominated human existence for more than 99% of its duration, so it is no wonder that archaeological, and particularly lithic, research has concentrated on hunter-gatherer groups. Most of this research has been culture historical in nature, as archaeologists have striven to understand specific portions of the prehistoric record. Driven by a desire to involve concerns of culture process rather than static observation, Lewis Binford's (1977, 1979, 1980) ethnographic studies of the Nunamiut Eskimo offered a provocative theoretical framework for all such systems. Binford redirected the dialog about hunter-gatherers toward considerations of mobility strategy, a behavioral concept.

Adaptive strategy is concerned with problem solving (Nelson 1991:58) and is amenable to study through technologies, such as stone, that were developed for the furtherance of such tactics. The road to ascertaining prehistoric technological strategies involves an initial emphasis on lower-level facts such as the tasks in which tools were involved and the characteristics of their manufacture. These parameters enable the achievement of higher-level understandings of prehistoric tool-using systems.

Ultimately we wish to possess several kinds of information, in hierarchical fashion. For example, to determine accurately that a piece of stone was employed for chopping wood is good first-order data; to determine that it was held in a handle and to understand what that handle looked like is more informative yet; to be able to assess its role in the panoply of the culture's woodworking implements is better still; and to ascertain the nature of woodworking in the culture and its articulation with other tasks constitutes a level of information rarely achieved. Yet there exist still higher levels of integration, in which we begin to understand the true "function" of tools, i.e., how they fit into a culture's broader adaptive strategies. Knowledge at this level is about as close as archaeologists are likely to come to understanding the "essence" of prehistoric people through their stone tools, and it is the promise of such knowledge that has rendered theoretical considerations of residential strategy so seductively alluring.

Prehistoric mobility organization and adaptive strategy meant planning on a multitude of levels about which we can now only hazard an educated guess. The attempt to associate recovered artifacts with specific strategies is fraught with pitfalls, for the artifact frequently represents only part of an implement, and that implement may have been engaged in activities a few steps removed from the overall goal of the strategy. This would have been the case, for example, with the hafted burin that manufactured the bone tip that was placed in the harpoon that was used to hunt seal during a 10-day foray away from a base camp during a period of moderate dispersion. We will never know all the specifics of these prehistoric cultural groups, but framing the question with respect to adaptational strategy may at least provide us with an appropriate model for discussing the issue.

Although stone tools represent a window on residential strategies, that window is by no means perfectly transparent. Mobility organization is highly correlated with specific tool use behaviors that we do not yet fully comprehend. One of these is "curation," which necessitates considerable planning (Kuhn 1992) and is closely associated with logistical mobility (Binford 1979; Bettinger 1987:126–127). Important to this strategy is personal gear, including tools, which should be used for a relatively long time and more conservatively than other types of gear (Kuhn 1991:78).

Curation has been defined in several ways, but all of these definitions emphasize the special use or care of certain tools, reflected in, for example, special preparation, extra maintenance, multiple uses, or transport to another location. It is obvious that the special concern that people showed for specific tools is important to our understanding of the concept, but many different behaviors have been subsumed under the same term. In addition, not all societies that practiced curation employed all of the strategies implied or did so to the same extent. As Nelson (1991:88) put it,

> Curation . . . may be employed more or less extensively, different kinds of items may be curated, and various aspects of curation (transport, reuse) may be emphasized, depending on how economic and social concerns are prioritized.

She also confirmed that curation and expediency are not mutually exclusive systems, but planning options (Nelson 1991:65). Regarded in this way, each option is multifaceted, containing elements that may or may not be employed, depending upon situation and need.

The complexity represented by the concept of curation has caused some confusion in the archaeological literature—so much so that two of the participants at this conference, George Odell and Stephen Nash, chose to analyze the subject. Both authors recommend either a moratorium on the term, or, if used, a statement concerning which definition is being applied in each specific case. These recommendations met with general approval at the conference, but engendered lively critique from scholars who continue to find the concept useful.

The third chapter, by Paul Thacker, places curational practices into a strategic framework for Gravettian and Magdalenian groups inhabiting the same region of Portugal. Thacker shows how these groups differed in their lithic raw material procurement, reduction strategies, and site distributions, and relates these parameters to differences in their overall adaptation to the landscape.

## REFERENCES


Bettinger, R. L. 1987. Archaeological Approaches to Hunter-Gatherers. *Annual Review of Anthropology* 16:121–142.

Binford, L. R. 1977. Forty-seven Trips: A Case Study in the Character of Archaeological Formation Processes. In *Stone Tools as Cultural Markers*, edited by R. Wright, pp. 24–36. Australian Institute for Aboriginal Studies, Canberra.

Binford, L. R. 1979. Organization and Formation Processes: Looking at Curated Technologies. *Journal of Anthropological Research* 35:255–273.

Binford, L. R. 1980. Willow Smoke and Dogs' Tails: Hunter-Gatherer Settlement Systems and Archaeological Site Formation. *American Antiquity* 45:4–20.

Kuhn, S. L. 1991. "Unpacking" Reduction: Lithic Raw Material Economy in the Mousterian of West-Central Italy. *Journal of Anthropological Archaeology* 10:76–106.

Kuhn, S. L. 1992. On Planning and Curated Technologies in the Middle Paleolithic. *Journal of Anthropological Research* 48:185–214.

Nelson, M. C. 1991. The Study of Technological Organization. In *Archaeological Method and Theory*, vol. 3, edited by M. Schiffer, pp. 57–100. University of Arizona Press, Tucson.

Chapter 2

# Economizing Behavior and the Concept of "Curation"

George H. Odell

## ABSTRACT

Stone tool curation is a concept employed to explain certain aspects of prehistoric hunter-gatherer behavior, and its effect on lithic assemblages can be similar to that of responses to lithic raw material scarcity. These two concepts are examined here with respect to their validity and viability, and are applied to lithic data from the Lower Illinois Valley. Because of operational difficulties, only two of the five most commonly employed components of curation could be applied to the Illinois data, and the results of these two have led to different conclusions. Definitional ambiguities suggest that the term "curation" should not be employed without specifying its precise meaning, and should probably be restricted to those definitions emphasizing settlement and mobility organization. In comparing the effects on lithic collections of curation and scarcity-induced economizing behavior, certain variables in the Illinois Valley data were found to be useful in distinguishing either of these from the other.

## INTRODUCTION

One of the principal interpretive problems in archaeology involves making human behavioral sense of stone tool assemblages. All stones are not tools, and all

**GEORGE H. ODELL** • Department of Anthropology, University of Tulsa, Tulsa, Oklahoma 74104–3189.

*Stone Tools: Theoretical Insights into Human Prehistory*, edited by George H. Odell. Plenum Press, New York, 1996.

tools are not equal—a recognition that historically spawned the assignment of certain objects to "type collections" to be considered separately, and that promoted the concept of "tool" as a piece of modified stone. For years archaeologists were content to interpret lithic assemblages as representing sets of activities relatively unconnected to the system from which they derived. However, the development of paradigms involving sedentism and the organization of forager mobility have allowed some analysts to widen their analytical scale beyond the individual site, to the adaptive systems in which prehistoric hunter-gatherers interacted. The lithic variables employed to answer questions deemed important have changed accordingly—from, say, producing trait lists (Cole and Deuel 1937; Webb 1974) to investigating variables such as the frequency of retouch (Bamforth 1986; Hayden 1987; Kuhn 1991) and tool breakage (Roper 1979; Sullivan and Rozen 1985).

## BEHAVIORAL INFLUENCES ON TOOL ASSEMBLAGES

Perhaps the greatest current influence on interpretations of New World lithic assemblages is the now well-known forager-collector model of hunter-gatherer mobility organization. This model, developed through ethnographic fieldwork (Binford 1977, 1979, 1980), has been tested on samples of modern hunter-gatherer cultures (Kelly 1983; Shott 1986) and applied to archaeological material (Torrence 1983, 1989; Lurie 1989; Shott 1989a). In simplified form, the model presents a dichotomy between residential and logistical mobility. Foragers exercising residential mobility moved as a group to resources at appropriate times, exploited those resources, and then moved as a group to a new location. Collectors exercising logistical mobility tended to establish multi-purpose base camps in ecotonal areas from which they sent out task forces to exploit specific resources. Their base camps can be expected to have been relatively sedentary, at least on a seasonal basis (Binford 1979; Lurie 1989).

This model has been criticized for being too simplistic to be able to interpret the multi-faceted decisions that individual groups must make when faced with variable environments and social constraints (Chatters 1987; Nelson 1991:84ff.). These criticisms, which Binford (1980:12) himself anticipated, are well taken and, if the mobility organization of hunter-gatherer peoples is perceived solely with regard to site types, the results will not be responsive to hunter-gatherer decisions and solutions, which often cross-cut our modern tendency to categorize them (Nelson 1991). Nevertheless, these strategies were practiced by some hunter-gatherers and have validity within the groups studied, with some degree of generalization possible. I am perpetuating the distinction in this paper because a thorough discussion of mobility issues necessitates discussing their effects on contrasting strategies, and some debate has already occurred within this paradigm. I will use the terms embodied in the model heuristically, recognizing that most cultures vary to some degree from the examples from which the model was derived.

The principal reason for the popularity of this model is that it seeks to explain the archetypal conundrum, "What made prehistoric hunter-gatherers tick?"—at least in the areas of economy and settlement. To pose the issue is to pose the related question of which factors drive human adaptation and cause change. Important considerations affecting the adoption of a particular mobility strategy are weather patterns, latitude, landforms, exploitable biomass, and the degree of risk involved in exploiting individual elements in the environment. The latter factor has been recently emphasized as a causative influence (Myers 1989; Torrence 1989), because it integrates the degree of accessibility of the other factors with the human forces. As Binford (1980:15) put it, for example, the greater the seasonal variability in temperature, the greater the quantity of critical resources and the greater the probability that incongruities will occur in their distribution. Such incongruities embody a risk of failing to encounter sufficient resources at the right time, to which one anticipated response is the adoption of increasingly reliable strategies, particularly those involving logistical mobility. Since tools constitute a group's articulation with its environment, they provide mechanisms with which to manipulate that environment and obviate risk, and can be expected to have been responsive to changes in mobility strategies.

Tools in prehistoric times were employed to solve specific physical problems, in a complex of strategies. In one way or another these strategies often involved the concept of "curation," which Shott (1989b:24) has defined as "the realized utility of a tool" or, following Binford (1973; 1979:263), "the practice of maximizing the utility of tools by carrying them between successive settlements." The concept has also been perceived in terms of efficiency (Binford 1979; Torrence 1983; Bamforth 1986). Since all types of curation have the net effect of prolonging the amount of time that an implement remains operable within a cultural system, considerations of tool use-life are also relevant to the concept (Ammerman and Feldman 1974; Schiffer 1976:60; Shott 1989b).

The dimensions and ramifications of curation are many, creating considerable confusion in at least three areas: 1) what, exactly, the strategy involves; 2) how it is used in mobility organization; and 3) how it shows up in the archaeological record. This latter point is important, because archaeologists can interpret the situation only with the evidence at hand which, in many cases, consists solely of stone tools. However, the appearance of these tools also responds to factors other than mobility organization—factors such as foraging patterns, the buffering of risk, and constraints in the amount and kinds of material available for tool use (Kuhn 1991:76). Bamforth (1986), in fact, argued that tool raw material availability was closely related to curation practices, and that two components of curation, maintenance and recycling, were often responses to shortages of raw material, rather than to curation.

An important question with regard to our evidence, then, is this: how can we discriminate between the effects of raw material availability and the forces of curation? Since both are specific adaptive responses and a cultural group may choose to employ some forms of curation but not others (Nelson 1991:88), it is obvious that they are best evaluated on a case-by-case basis. Nevertheless, to establish certain adaptational generalities would be helpful for scholars interpreting prehistoric behavior in disparate regions. If progress is to be made on these issues, we must examine both concepts in order to extract meaningful behavioral correlates that can be employed operationally on archaeological assemblages. This paper will attempt to address these two related issues.

## THE CONCEPT OF "CURATION"

The concept of "curation" was, to my knowledge, first inscribed in the archaeological literature in the form in which it is currently employed by Lewis Binford (1973:242–244). In that article Binford was primarily occupied with defending his "functional argument" for the Mousterian against all comers but, in addressing the relationship between tool use and discard, he introduced the notion of a "curated technology." As he perceived it at that time, such a technology involved transportation, preservation of tools for future use, and efficiency. Later he and others offered additional insights to round out the concept, but the net result of these contributions has been a general confusion about what "curation" really means and how the term should be used. As Chatters (1987:341) put it, "Analysts lack a clear conception of what constitutes tool curation." Let us investigate the concept to see what is wrong.

Bamforth (1986:48) concluded that "tool curation is a complex set of behaviors that cannot be explained by any single factor"—a statement with which I heartily agree. I also agree with many of his individual points, but have a slightly different perspective and am concerned with evaluating the degree to which these points can be operationalized. Since he and Binford have already considered most of the relevant issues, it would be useful to couch this discussion initially in terms that these authors have already presented. Bamforth (1986:39) classified curation into five different aspects: "production of implements in advance of use, design of implements for multiple uses, transport of implements from location to location, maintenance, and recycling." I will discuss each of these ideas in turn, evaluating their operational viability and behavioral meaning. Simply put, I will be asking the questions, "how easy is it to measure?" and "how useful is it once it has been measured?"

### Production of Implements in Advance of Use

The production of implements in advance of use was one of the behavioral correlates that Binford originally associated with curation, and he subsequently confirmed its importance with respect to the Nunamiut perception of personal and household gear (Binford 1979:269). Later Torrence (1983:11–13) utilized this aspect of the term exclusively, Kuhn (1992:189) employed it in describing the Mousterian provisioning of places, and Nelson (1991:62–63) proclaimed that it was a critical variable differentiating curated from expedient technologies.

Two kinds of sites or concentrations have provided evidence for the planned manufacture of tools in preparation for their use: the lithic workshop and the cache. The status of a site as a *workshop*, established by an abundance of debitage and paucity of finished tools, is not always obvious, because the locale may also have served in other capacities. A good example, recently excavated in Wyoming, appears to have been a "gearing up" station at which implements were manufactured for eventual employment in activities such as hunting and mass butchery, contemporary evidence of which exists in the region (Francis 1991).

The *lithic cache*, a concentration of stone tools presumably sequestered for use in time of need, demonstrates long-range planning as a risk-abatement strategy, and the role of technology within that strategy (Binford 1979:257; Kornfeld et al. 1990). Although foodstuffs might be buried for eventual future exigencies, tools or other objects would have been just as useful and ultimately more preservable (see Thomas 1983:81–82, for a description of their occurrence in the Great Basin). A particularly clear example of a cache, dislodged by agricultural equipment in east Texas, served to document the most common types of plow damage (Mallouf 1981).

Workshops and caches illustrate the production of tools in advance of use, though they may not be common in any particular region and may be difficult to detect. Workshops would have been appropriate either for highly nomadic people (e.g., the Wyoming bison hunters cited above) or for logistically organized groups such as the Nunamiut gearing up for a special-purpose foray (Binford 1977). Caches were certainly employed by individuals or by small, logistically organized groups while away on forays, but they were also employed by more residentially mobile peoples, as Kornfeld et al. (1990) point out.

At the level of the individual implement, *hafting* may be considered a form of tool production in advance of use. Keeley (1982:798–799) and Shott (1986:39) have considered the practice of hafting indicative of a curated tool, and Binford (1977:35) implied the same thing when stating that "composite tools might be expected to occur more commonly in curated assemblages." Oswalt (1976) demonstrated its important role in preparing reliable tools in logistically organized Eskimo societies; and among prehistoric cultural groups, Lurie (1989:52) considered this practice an indicator of complexity, which should have occurred more frequently in logistically than in residentially organized systems. Although potentially informative, very few actual prehistoric hafts have been documented, since the organic component is so rarely preserved. Certain tools such as projectile points, of course, contain visible portions that have been interpreted as "haft elements" and analyzed accordingly (e.g., Shott 1986). This indicator is useful for those classes of tools that contain a haft element, but are not applicable to the rest of the assemblage.

Since traces of hafting damage have been documented experimentally and archaeologically on stones inserted into organic shafts (Odell and Odell-Vereecken 1980; Odell 1981a; Stordeur 1987; Shea 1988; Owen and Unrath 1989), lithic use-wear analysis has offered the possibility of extending interpretations of hafting to implements that do not possess visible haft elements. It appears that such damage is produced only under conditions favorable for its formation, so not all hafted tools can be expected to reveal interpretable haft damage (Moss and Newcomer 1982:292). Since it is likely that only a subset of tools in any assemblage was hafted, interpreting entire assemblages on this parameter requires large sample sizes. This characteristic has not characterized most recent use-wear studies, but the situation may change as the utility of this endeavor becomes more fully realized.

From the considerations discussed above, I conclude that the manufacture of tools in advance of use constitutes an assessment that is not commonly made within archaeological assemblages. On a site level it can be interpreted from workshops or caches, but these are relatively rare. On the artifact level the recognition of hafting has proven useful on tools containing haft elements, but these usually constitute a small subset of the tool repertoire. Recognition of hafting within entire assemblages is possible through use-wear techniques, but the number of practitioners who have been able to observe samples large enough to be statistically interpretable are very few, and those among this number who have tried to record hafting traces are fewer still.

The ability of these indicators to substantiate the production of tools in advance of use is variable. Workshops and caches can inform on the presence of such behavior, but interpretations of specific mobility strategy are difficult to make on this basis alone, since both site types might logically be produced by the more mobile elements of any adaptive strategy. The hafting of individual tools possesses greater interpretive potential, as the practice should theoretically increase with technological complexity and an increase in logistical mobility strategies.

## Design of Implements for Multiple Uses

Although the concept is plausible, I have been largely unsuccessful in tracking down the sources that Bamforth used in associating the design of tools for multiple uses with curation. The best candidates I have encountered are the discoidal cores carried as personal gear by the Nunamiut for the removal of flakes when needed and for the *exhausted core*, which invariably terminated its use-life in the form of a round scraper (Binford 1979:262). Binford, however, did not specifically state that this practice was characteristic of curation as he perceived the concept at that time.

Kelly later discussed the various ways that *bifaces* could be employed by hunter-gatherer groups: "as cores; as resharpenable, long use-life tools; or as shaped, function-specific tools which are part of a reliable technology" (Kelly

1988:731). Each of these functions bears important implications for a group's mobility strategy and should be considered as part of its overall organizational response. The generalized form of a biface which "allows it to be modified into other tools, such as scrapers" (Kelly 1988:718) and the use of bifaces as cores to produce flakes, which themselves are generalized forms, are quintessential multifunctional tools (see also Winters 1969; Ahler and McMillan 1976), and it is clear that Kelly perceived them as such. He also associated the use of bifaces as cores with tool curation, though he did not specifically define curation in those terms.

Despite my failure to locate a specific source that defines curation with respect to the design of implements for multiple uses, my canvass of the literature was not particularly comprehensive. It has indicated, however, that Aboriginal use of certain tool forms, particularly bifaces, is multifunctional in most respects; and that two of Kelly's proposed biface modes (as cores and as long use-life tools) are associated with "curation" as understood by contemporary archaeologists. These considerations argue for retaining the notion of multiple-use tools for testing later in this paper.

Bamforth's wording of the concept—"design of implements for multiple uses"—reveals a possible definitional ambiguity: whether the emphasis should be on *the act of designing,* which is conscious premanipulation similar to the preparation of tools discussed previously, or on *the fact of multiple use.* To emphasize design assumes that some care was taken to produce a generalized tool form, predisposing the tool to the potential for being utilized in a variety of ways. To emphasize use itself is more objective but may be influenced by factors other than those commonly understood to be curational, to be discussed shortly. In addition, the context in which this concept was developed involved considerations of forager efficiency and planning (Binford 1977, 1979; Bamforth 1986), which emphasized the conscious, deliberate aspect of the definition (the "design") in preference to the more impulsive one (the "use"). For these reasons, applications of the concept employed here are design-oriented (see Hayden et al., this volume), though it will be shown that the actual uses of the samples chosen are also multifunctional.

## Transport of Implements from Location to Location

Taking implements from one locality to another is a special situation that Shott (1986) called the "carrying costs" of mobility, and is a dominant sense in which Binford (1973, 1979) first used the term. Most archaeologists would agree that favoring certain tools over others by transporting them is a significant behavioral practice that constitutes curation (Shott 1989a:288; Nelson 1991:65; Kuhn 1992:189), and what archaeologist has not excavated a site in which a few tools can be demonstrated to have originated in distant locales? In one such study, Henry (1989) employed dimensional data from scar facets and whole tools on north-central Oklahoma sites to demonstrate a tendency for prehistoric peoples to have conducted substantial knapping episodes near raw material sources, and transported tools and highly reduced blanks to their habitations. A similar phenomenon might be expected at quarries, where old, worn-out implements would have been discarded in favor of a fresh supply. A good example is the Mt. Jasper rhyolite quarry in New Hampshire, a curated, 99% exotic, workshop assemblage of primarily dulled and broken rejects (Gramly 1980).

The problem, of course, is not in finding a few such implements, but in finding *all* of them, or at least enough of them to represent their occurrence accurately. In some areas chert sources are relatively discrete, so that distance to raw material source is an easy factor to control (Johnson 1989:132). Other areas, however, do not benefit from point-provenience sourcing. For example, the Land of Burlington Limestone in the center of the United States contains chert deposits that extend over much of a three-state area. Within this formation, chert from different regions usually looks the same (i.e., is inter-regionally homogeneous), whereas chert within each region and even within individual sites is highly variable on several parameters (Rubey 1952; Luedtke and Meyers 1984; Odell 1984). To add to the difficulty, some of the Mississippian limestone formations within this area and on its periphery (e.g., Chouteau in Illinois, Reeds Spring and Keokuk in Arkansas and Oklahoma) yield cherts that are visually quite similar to Burlington. Attempts to discriminate cherts within this area on a sounder technical basis than the typical warm thumb technique currently in use have so far proven only mildly successful (B. Luedtke, pers. commun. 1987). The ease of measuring the transport of tools therefore depends on the ease of determining, within a relatively narrowly proscribed radius, the origin of the raw material from which the tools were made. Some regions are more fortunate than others in this regard.

Relating the transport of raw materials to mobility strategy is a difficult question. Since most hunter-gatherers were mobile to some extent, and since raw material procurement could be embedded in special-purpose forays as well as in residential moves, the existence of distantly procured tool materials does not automatically favor one kind of settlement strategy over another. To be most effective, lithic sourcing information should be combined with other data bases. For example, Holen (1991) combined raw material identifications from two early historic Pawnee sites in Nebraska with source area and ethnohistoric information to assess the procurement strategies of two different Pawnee bands. Holen benefited greatly from historic sources, without which the task would have been much more difficult.

## Tool Recycling

In evaluating tool use among the Nunamiut Eskimos, Binford (1977:33) mentioned their practice of returning certain tools to their village for "recycling into other items." Binford did not detail the prevailing recycling practices among this group of people, but at one point he mentioned that "12 additional items were broken but were returned to the village for repair or recycling" (Binford 1977:34). Somehow, Bamforth (1986) got the impression that unbroken, usable (though not necessarily used) tools indicate recycling. Nowhere have I found substantiation for such a notion, and Binford himself indicates that it was usually the *broken* ones that were taken back to the village and recycled into something else. Bamforth expended considerable energy chasing this variable and, in his California data set, came to a conclusion that would have made more sense had the initial assumptions been correct:

> early sites contain disproportionately large numbers of broken and retouched implements . . . , indicating a decrease in tool maintenance and recycling. This is the precise opposite of Binford's (1977:35) prediction about the relationship between the complexity of the subsistence-settlement organization and the degree of reliance on curation. (Bamforth 1986:46)

I suggest that recycling is a concept that is too difficult to characterize adequately in interpreting the archaeological record. It is clearly not described by the traits discussed above, but the opposite traits are also inadequate. That is, unused tools could hardly have been recycled; and broken, utilized tools may or may not have been recycled, but it is difficult to distinguish on this basis alone. However, if recycling means that an item was transformed into a different functional entity, then one might capture its essence by recording the number of different tasks per tool. The problem with this is that it assumes that any reshaping of the tool was not comprehensive enough to destroy traces of previous use, an assumption that is probably not valid in most cases. Multiple traces on such a tool could have been the result either of recycling or of the preordained design of that tool for multiple uses. Wear on a broken portion of a tool would likewise be ambiguous, since it would not indicate design for multiple use, but would not necessarily constitute "recycling," either.

Alternatively, one could conceive of recycling as remaking an implement into a different kind of tool; such a process might be indicated by the presence or amount of retouch. But how does one distinguish retouch for this purpose from retouch for the purpose of resharpening or maintenance, as will be discussed presently? This issue has continually perplexed lithic analysts and, whereas distinctions in some cases might be made on the basis of the steepness of retouch, the presence or amount of edge modification remains an inaccurate indicator of the concept of recycling. Since I have virtually exhausted the logical ways that recycling can be measured and have failed to find one that works, I conclude that we would be better off acknowledging the concept but working on something else.

## Tool Maintenance

Tool maintenance has been considered curation because it prolongs the use-life of a tool (Bamforth 1986; Shott 1986:40, 1989b:24). However, Bamforth

(1986) has convincingly argued that it could just as logically be considered a response to tool raw material shortages for the same reason. I will refer to the concept here as a form of curation, but its role depends on the specific situation in which it was involved.

To maintain a tool is to sharpen it, reshape parts of it for insertion or manual prehension, or remake it into another kind of tool (recycling). Any of these activities, applied to chipped stone, involve secondary modification of the margins, so *tool retouch* has served as the primary indicator of this phenomenon. The problem with this concept is that there are several reasons for retouching a tool: to shape it for the hand or a haft; to prepare a projection, as with drills, gravers and projectile points; to blunt an edge so that hominid meat is not mixed with the jerky; and to sharpen a dull edge. Judging by the recent literature, however, one would think that the only reason to retouch a tool was to sharpen it. Hayden (1987:34) has propounded this position, stating,

> I view most (but not all) tool morphologies as the result of resharpening rather than any manufacturing process in which an artisan had a preconceived form in mind, sat down to make a biface, or a scraper, and then tucked it away for vague future uses. I have rejected this latter point of view largely as a result of my own and other archaeologists' work with contemporary lithic using groups such as the Australian Aboriginals, the New Guinea Highlanders, and the Maya. . . . Individuals who continue to make stone tools in traditional ways simply do not have normative ideas of morphology when they make most tools.

I detect a tyranny of ethnography going on here. First, most of my midcontinental American lithic assemblages contain a substantial bifacial component, but among the groups mentioned by Hayden (i.e., Australian Aboriginals, New Guinea Highlanders, and Maya), only the Maya possessed a noticeable bifacial technology and this type was not ubiquitous even in their assemblages. Second, many of my assemblages contain a few recognizable chipped stone tool types such as drills and projectile points that exhibit a concern for formal specificity; this feature is even more prevalent in other areas of the world (e.g., the European Upper Paleolithic). The assemblages on which Hayden was working, however, are remarkable for their lack of formal specificity. From these considerations I would submit that most of the ethnographically studied groups from which Hayden draws his information possess lithic assemblages that are dissimilar from many prehistoric assemblages in significant ways.

And third, while a tool user may not have had a specific mental template in mind when commencing his activities, he certainly possessed a "comfort template"—i.e., he knew when an implement did not fit his hand and when it hurt, and it was easy enough to eliminate an offending edge or projection by whacking it off. Alternatively, if the tool was to be hafted, many situations would have called not for substantial shaping, but for an occasional blow to facilitate the fit. These considerations, among others, convince me that sharpening dull edges was not the principal reason for retouching a tool in prehistory. If I am right, then a lot of analyses will have to be re-thought.

Two indices possess considerably more likelihood of having been responsive to the prehistoric sharpening of tools than the presence or extent of retouch. The first is the *ratio of total length:haft length* of bifacial implements possessing a haft element. As Shott (1986:44) explains,

> the shorter a tool is in relation to the length of its haft, the more it has been resharpened and reduced from its original size. Therefore, as resharpening and reduction increase, the resharpening ratio declines; lower values of the ratio are associated with greater resharpening.

He goes on to state that comparisons of assemblages using this index assume that such tools started out their use-lives with similar ratios. There is no guarantee that this assumption is justified if the assemblages being compared are temporally distant from one another, but it should be with closely affiliated groups. Additionally, he assumes (probably justifiably) that, once manufactured, haft elements were not reduced further. Characteristics of a ratio like this are: 1) in referring to extent of reduction on an interval scale, it does not require the analyst to interpret the function served by that modification, or to assume that all such modification was caused by sharpening behavior; and 2) it is only applicable to pieces, mostly bifaces, possessing a haft element, which were presumably all hafted; it cannot be applied to retouched tools lacking such elements, whose status as hafted implements could easily be called into question.

The second believable indicator of tool sharpening is the characteristic *alternate bevelling* of hafted bifaces that occurs in several Midwestern locales in the Early Archaic and occasionally later periods. Its relationship with curation, however defined, is poorly known; in the Illinois Valley it appears to represent a response to lithic resource availability (Wiant and Hassen 1985).

Binford (1977) made a strong case for the importance of tool maintenance within the highly curated technology of the Nunamiut, and discussed the relationships among artifacts that should obtain in "curated and maintained technologies"—in this case, those involving logistical organization. Shott's (1986) application of his resharpening ratio to archaeological material from the American West, however, is at variance with Binford's from at least one perspective. That is, Shott found that the more mobile Plainview assemblage exhibited a lower ratio, suggesting more extensive sharpening and therefore a greater degree of tool maintenance among presumably residentially mobile Paleo-Indians, whereas Binford surmised that more tool maintenance should occur in a logistically organized base camp. At present it is difficult to reconcile these two positions. Both studies pertain to situations of relatively extensive mobility, but neither has compared assemblages from residentially and logistically organized societies with one another using a variable that is likely to show tool maintenance. Such a comparison might prove very informative.

Part of the problem relates to a difference between tool maintenance as an event and as a process. As an event, it occurred when margins became dull, the insert slipped in its haft, or the tool needed to be refitted. These events could happen in residential settlements or in base camps in preparation for a logistical foray. As a process, it refers to the concept of the "maintainable" tool, which "can be made so that if it is broken or not appropriate for the task at hand, it can quickly and easily be brought to a functional state" (Bleed 1986:739). Such tools are redundant and often modular in design and, in Binford's system, relate closely to forager organizational patterns and expedient tool use. Thus we have a terminological inconsistency in which "tool maintenance" is clearly concentrated at logistical base camps "gearing up" for special-purpose expeditions; but "maintainable tools" refers to specific qualities of the equipment more common among residentially organized foragers.

I would suggest that, although tool maintenance occurs wherever people use tools, the quality of "maintainability" means easily refurbishable, without much effort. Since more effort was probably expended in this activity among collectors than among foragers, an accurate measure of tool maintenance should reflect this relation. One must be careful to select an appropriate measure, however, and I have already explained my dissatisfaction with the equation of retouch with resharpening. That is, if one measures tool maintenance by calculating total retouch throughout a chipped stone assemblage, then the results, in my opinion, will not accurately reflect either sharpening or maintenance. But if a specific subset of the assemblage such as projectile points, which were likely to have been hafted and sharpened in the haft, were measured for this quality, then one probably is measuring the desired attribute. We need to be specific in describing which usage of "tool maintenance" is being referred to and measured in each case.

## RAW MATERIAL AVAILABILITY AND ECONOMIZING BEHAVIOR

Behavior that economizes stone tools is a conservational response to either a current dearth of tool raw material or a projected future need. Conservation may be manifested through manufacture or use. That is, economizing tool users tend to make more implements than usual out of the available resources; prolong the lives of the tools they have through reshaping, resharpening or rehafting; or utilize tools with greater intensity. If this behavior is detectable in archaeological lithic assemblages, it can be expected to have been manifested in variables related to either tool manufacture or use.

The notion of cost-benefit relationships or economizing behavior in foraging societies has been criticized for not having fully considered embedded lithic procurement in subsistence-settlement strategies. That is, why would procurement have cost much (or anything) if the group had to be in the same area to subsist, anyway (Binford 1979:259–261)? And if it cost little, one could expect little visible evidence in the prehistoric record. Although perhaps generally true, some lithic procurement sites belie this scenario by exhibiting evidence of too much intensive labor to have been simply a stopover embedded in some greater exploitation scheme (Reher 1991). Embedded procurement was clearly not the only option available to hunter-gatherers. Thus to accurately evaluate a specific prehistoric situation, one must assess the raw material needs and resource availability of the local polity.

Even if lithic procurement was embedded, the amount of stone transported to a base camp would have been strongly technology-dependent. Given the dominant technologies in North America, even postulating water transport and easy accessibility of raw material to river systems, large quantities of resources could not have been carted into any locality without a considerable expenditure of labor. Barring life-threatening circumstances, it is unlikely that prehistoric inhabitants in lithic-poor areas would have overanticipated their seasonal needs by a very great margin, nor is there any evidence that they did so. If this is the case, then tool users would have had to remain somewhat conservative of their material in case their access to sources was impeded or journeying to source areas became inconvenient or difficult. Thus I would expect that people living in most lithic-poor locales would have had to practice some form of economizing behavior. Since evidence of tool conservation has been observed in several lithic data sets (Bamforth 1986; Odell 1989; Kuhn 1991), such behavior is detectable in at least certain situations.

I have argued that lithic procurement may have been either an embedded or an exclusive strategy, that tool conservation was operative in most raw material-poor situations, and that such behavior is potentially detectable in lithic assemblages through anomalies in the manufacture or use of tools. But was it a truly important constraint shaping hunter-gatherer mobility decisions? It certainly seems to have been with regard to certain intensively utilized source areas such as the Spanish Diggings in Wyoming (Reher 1991) and among some ethnographically studied groups such as the Pawnee, who apparently organized their bison hunts at least partly around certain regularly exploited lithic sources (Holen 1991). In addition, Bamforth (1986) considered the curational practices of maintenance and recycling to have been responses to raw material shortages rather than an indication of efficiency or specific mobility strategies. On the other hand, Nelson (1991:77) regarded the location of camps away from raw material sources and the failure to stockpile stone at lithic-poor localities as conscious decisions affected primarily by considerations of mobility. This point is underscored by Koldehoff (1987:175), who viewed Mississippian reduction strategies in the chert-poor American Bottoms region as wasteful, influenced more by decisions of mobility than of resource availability.

At this point we begin to approach a chicken-and-egg conundrum the resolution of which is really not very informative. I maintain that, in some situations, behavior responsive to tool conservation, tool resource availability, and mobility organization are detectable in the lithic data base. More importantly, behavior responsive to any of these can be preserved in different parts of the data set, and is therefore distinquishable in these cases. In the following section I will attempt to demonstrate these differences using information from Illinois. In this data set are examples that conform to curational situations, as defined herein, and to shortages of lithic tool-making material. They will enable us to compare causal relationships and attributes of the data that will allow us to discern these relationships.

## EXAMPLES FROM THE ILLINOIS VALLEY

In the late 1970s and early 1980s, highway construction in west-central Illinois occasioned the survey of large corridors of land, the testing of archaeological sites discovered there, and the intensive investigation of those areas determined to be historically significant (Farnsworth and Walthall 1983). I performed a suite of formal, technological and functional analyses on several of the lithic assemblages from these excavations, and subsequently selected for comparison 10 components from five sites in the Lower Illinois Valley in close proximity to one another, dating from about 6400 BC to AD 1100. The data set, which includes 1 late Early Archaic, 4 Middle Archaic, 1 Late Archaic, 3 Middle Woodland, and 1 Mississippian component, has been fully described and compared in Odell (1995). Since the sites were all in good context and carefully excavated, intensive lithic analyses were justified; and since the range of variables and chronological spread of the components were relatively wide, the data set has proven to be responsive to a wide variety of research questions.

Two kinds of concerns are of interest here: prehistoric curation and responses to differential lithic resource availability. As explained by Bamforth (1986) and confirmed in the current discussion, the concept of curation and the classifying of assemblages into "curated" and "expedient" are not particularly enlightening by themselves. Not all of the various definitions of curation are operationally viable or equally useful, and they need to be related to specific technological behavior. Doing so enables us to discern the effects of specific factors or constraints, and to evaluate their influences.

### Responses to Curation

In discussing the five aspects of curation as depicted by Bamforth (1986), I have suggested that three of them are not worth pursuing with the Illinois Valley data set. First, the transport of implements from location to location is difficult to discern because the dominant lithic raw material type in the region is Burlington chert, which occurs naturally over a very broad area and is not able to be visibly differentiated within that area. With the exception of the Middle Woodland period, when raw materials from non-Burlington sources entered the region in substantial quantities, we cannot establish with any accuracy where a specific piece of Burlington chert originated. Second, recycling of tools from one type to another is, with a few highly inferential exceptions, extremely difficult to measure in any region. My data set cannot be manipulated to depict this parameter accurately. And third, I cannot, in good conscience, measure tool maintenance. The trait most often employed to characterize this behavior is sharpening retouch which, as argued previously, usually cannot be distinguished from other functions of retouch. Where it can be distinguished involves the characteristic bevelling frequently observed on Early Archaic projectile points in this region, but this trait is absent in our sample. Shott's resharpening ratio provides an appropriate measure, but my data set does not include the appropriate variables.

Two testable aspects of curation remain. The first, production of tools in advance of use, can be characterized by the frequency of *tool hafting behavior*. A use-wear analysis of 9 of the 10 Illinois Valley components registered a large enough sample of prehensile traces to enable comparison of the components. Observations were recorded according to whether the traces most likely originated from a haft or a hand, or from indeterminate prehensile origin (Odell 1981a, 1994a, 1995; Odell and Odell-Vereecken 1980).

In another article (Odell 1994a) I have presented the use-wear evidence for prehension in Illinois Valley prehistory. In the current paper I will excerpt specific conclusions in order to make a more general point. Figure 1 illustrates hafting wear as a percentage of all prehensile traces recognized. It has been graphed chronologically using the mean radiocarbon date for each component. The results indicate an overall rise in the regression line though considerable variation is present, resulting in a relatively weak correlation coefficient of .524. It might be expected that the incidence of manual prehension should fall while that of hafting rises, but this need not necessarily be the case because of the presence of a third, generalized, variable (indeed, in the Mississippian component both hafting and grasping traces increase). Fall it does, however (Figure 2), and the resulting percentages show a closer fit with the regression line than is the case with hafting wear (r=.690).

The increase in haft damage and decrease in traces of grasping suggest a gradual change in prehensile behavior through time, toward a more frequent production of tools in advance of use. Since hafting usually has the functional effect of enhancing the reliability of implements, its increase is consistent with the proposed growing need for reliable tools among logistically organized collectors.

The second testable aspect of curation is the design of tools for multiple uses. As previously discussed, one tool type for which this quality obtains is the biface. In the Illinois Valley assemblages studied, bifaces universally exhibit a wide

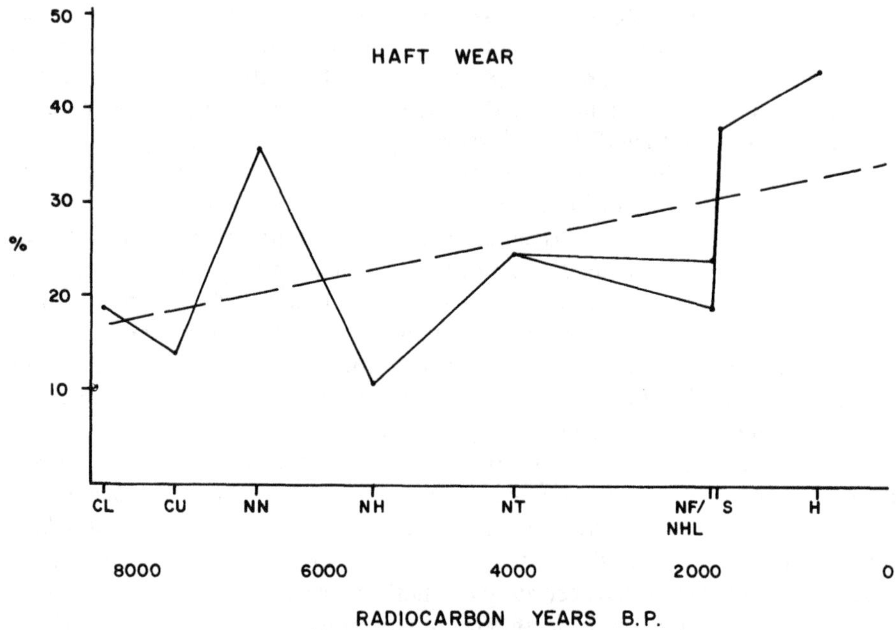


**Figure 1.** Percentage of hafting to total prehensile traces in nine components located with respect to their average radiocarbon date (taken from Odell 1994a). CL, CU = Campbell Hollow Lower and Upper Horizons; NN, NH, NT, NF, NHL = Napoleon Hollow Napoleon, Helton, Titterington, Floodplain Middle Woodland, and Hillslope Middle Woodland components; EZ = Elizabeth sub-Mound 6 component; S = Smiling Dan Middle Woodland component; H = Hill Creek assemblage.

variety of use-wear traces (Odell 1995), indicating substantial variability in tool motion and worked material parameters. Since both design and use of the biface type are generalized, then chronological trends within this type should provide information on changes in curational practices.

Koldehoff (1987:155) asserted that, in most areas of the Eastern Woodlands of North America, biface production decreased through time. This contention is supported by the Illinois Valley data (Figure 3). Koldehoff interpreted this pattern as heralding a decline of curated technologies in favor of expedient ones, apparently correlated with declining residential mobility.

The phenomenon of multifunctionality itself did not necessarily decline, however. Figure 3 also depicts three other tool types—retouched pieces, scrapers and projectile points—that served multipurpose roles in other collections (Ahler 1971; Wylie 1975; Ahler and McMillan 1976; Odell 1981b). Their proportions remained relatively stable through time, suggesting that it was the specific production of bifaces, rather than of multifunctional tools in general, that changed.

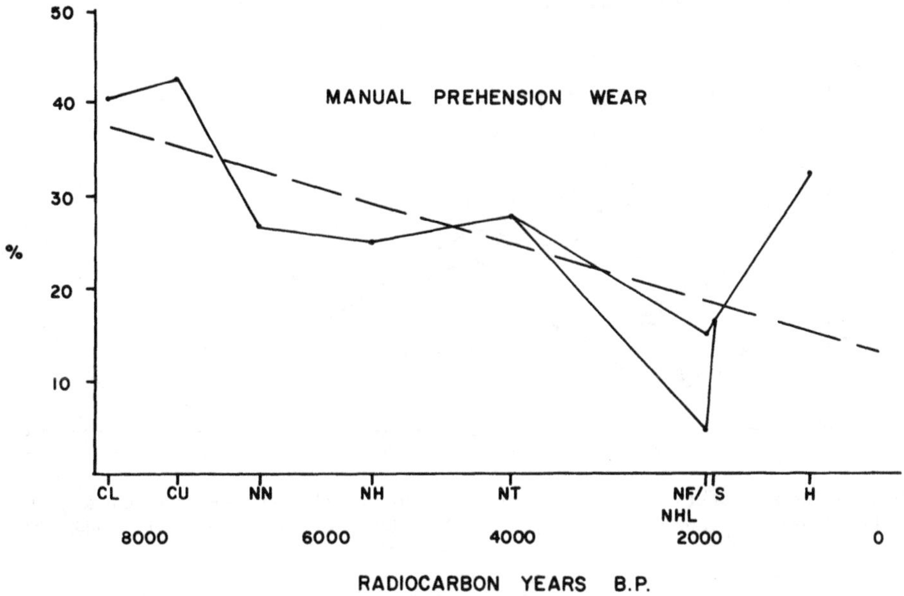


**Figure 2.** Percentage of manual to total prehensile wear in nine components located with respect to their average radiocarbon date (taken from Odell 1994a).

Our investigations have produced concepts of curation that are inconsistent with one another. The preparation of tools in advance of use, measured by the practice of hafting, increased through time; whereas the design of tools for multiple uses, measured by the relative quantity of bifaces, decreased. It would appear that the term "curation" has been employed for separate phenomena that operate very differently in cultural systems. Hafting was potentially practiced for several reasons, but one of the most important was probably to increase the reliability of implements on solitary or small-group forays and single-encounter situations. This strategy, exemplified by the intensive preparation of gear at a base camp in anticipation of a trip to extract some specific resource, is relatively specialized. On the other hand, the design of tools for multiple uses is a generalistic strategy that emphasizes easy maintenance and continuous encounters with potentially different resources. These are two very different strategies, and to refer to each as "curation" is both imprecise and misleading.

## Responses to Raw Material Availability

The Illinois Valley data set employed in this analysis is appropriate for investigating human tool economizing behavior in response to differential avail-

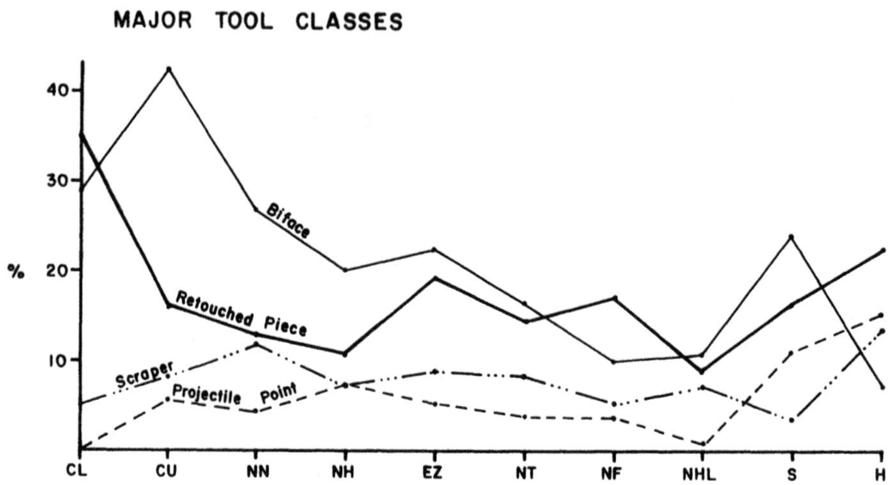


**Figure 3.** The proportions in their respective type collections of several well-represented types in studied Illinois Valley components, arranged chronologically from left to right (taken from Odell 1995).

ability of lithic material, because three of the sites (7 components) occur in areas of easy access to flint sources, whereas two sites (3 components) occur in areas lacking exposed flint-producing bedrock (Figure 4). The latter—Campbell Hollow and Smiling Dan—are located east of the Illinois River where the Mississippian Burlington Formation dips eastward under the floodplain alluvium and is covered by Pennsylvanian and more recent deposits in central Illinois (Odell 1984). If certain variables are responsive to prehistoric tool economizing behavior, their results for the three components east of the river should show strong positive correlations, and should be different from those of sites west of the river. Such results should be distinguishable from responses to chronology, sedentism, or mobility strategies, because the components in the chert-poor area east of the river are temporally spread almost as far as they can be in this data set: Campbell Hollow in the late Early and early Middle Archaic periods, and Smiling Dan in the Middle Woodland. Distinctions based on conservational behavior involving all of the components should therefore exhibit non-chronological correlations.

As argued previously, lithic assemblages in regions characterized by a paucity of suitable raw materials should exhibit some tool conservation in either manufacture or use. If raw material were scarce, a tool maker would be inclined to use up every last available bit of it. One should witness the effects of this behavior in the cores, which should be either less plentiful or smaller than cores in areas of lithic abundance. In the Illinois Valley the truth of this assertion is supported by Figure 5, which illustrates the percentage of cores with respect to the entire type collection, the subset in which cores were placed for this exercise. The fewest cores

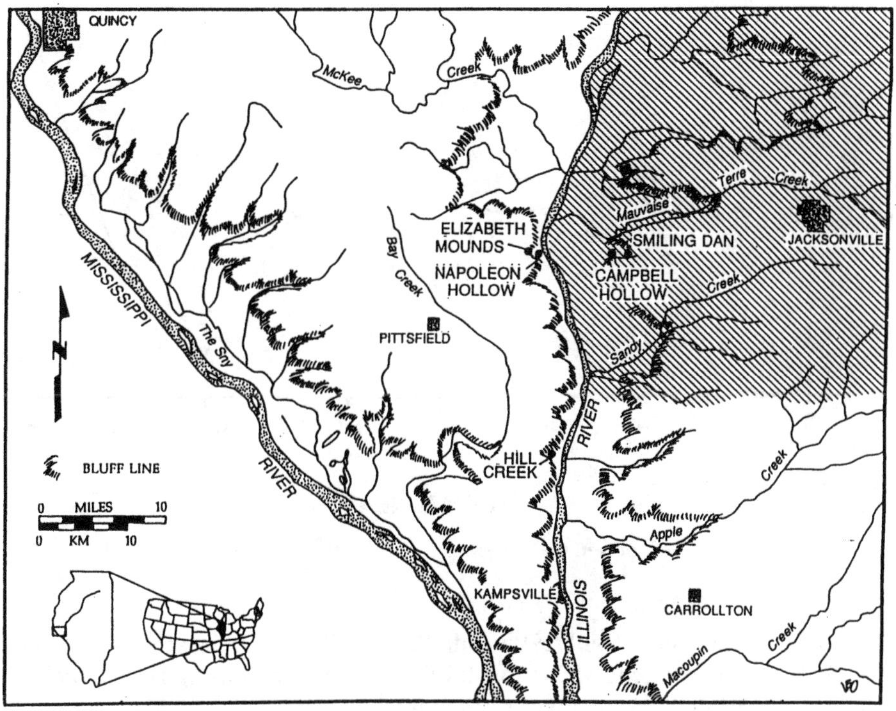


Figure 4. Sites included in the data set presented in this analysis.

were contained in the Campbell Hollow Lower (0) and Upper (2) Horizons located east of the river, the earliest occupations in the sample. Also relatively low on cores is the Smiling Dan site, also east of the river. Evidence there of semi-permanent occupation including substantial structures might at other habitations be associated with large lithic workshops, but the paucity of cores in an otherwise stable and permanent settlement was probably caused by the need to economize raw material.

Two components west of the river also contain relatively few cores, but in each case this anomaly can be explained by site function as depicted through other lines of evidence. For example, the Middle Woodland Hillslope (NHL) assemblage from Napoleon Hollow contains only one core, but this component has been interpreted as a deposit of specialized tools used in mortuary activities (Wiant and McGimsey 1986; Odell 1994b), for which tool production would have occurred elsewhere. The Late Archaic Titterington assemblage at Napoleon Hollow also yields relatively few cores, but this assemblage contains several other differences with base camps occupied before and after it. On this and other sites Titterington peoples were characterized by less residential stability and a greater reliance on

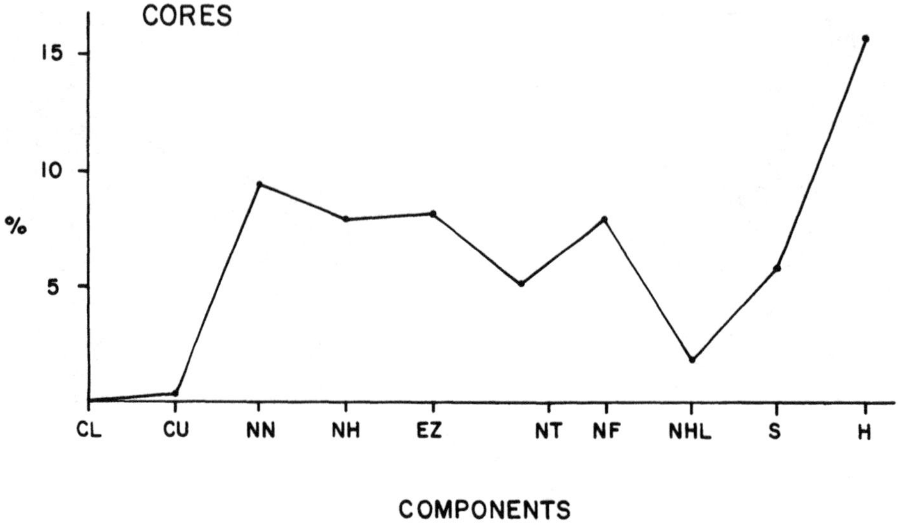


Figure 5. Percentage of cores in the respective type collections (data taken from Odell 1995).

hunting for subsistence (Cook 1986; Odell 1995). This reversion to a more mobile lifestyle brought core production to a level below that of Smiling Dan, which was a more occupationally permanent settlement.

Other forms of conservation can be observed in the types of cores and debitage produced. For example, a core that is too small for comfortable freehand production can be fractured more completely by placing it on a hard surface and bashing away, a process that can provide at least a few usable tools. Known as a bipolar technique (Crabtree 1972:42; Callahan 1987; Jeske and Lurie 1993), it is not always recognizable by distinguishing characteristics of the flakes or cores themselves. Opposing sets of ventral ripple lines, for example, are rare, and distal crushing can be caused by forces other than bipolar flake production.

In observing Illinois Valley debitage I was not concentrating on detecting bipolar activities, but recorded unmistakable specimens as such. My samples include 2 bipolar flakes in the Campbell Hollow Upper Horizon assemblage, 3 in the Lower Horizon, and 7 at Smiling Dan (Odell 1995). The evidence is not overwhelming, but the paucity of bipolar flakes is explained by the afore-mentioned difficulty of recognition and general goals of the analysis. It is significant that all of the components from which bipolar flakes were recognized originated east of the Illinois River, whereas no evidence of bipolar technologies was uncovered in assemblages west of the river.

Since core smashing was apparently a popular economizing measure, it is possible that similar strategies of further reducing existing implements to provide new and different tool forms as required were also practiced. A need for conservation would also have induced people to utilize their implements more extensively. In either case, the net result would have been an increase in tool breakage in areas of scarce raw materials.

I have investigated this phenomenon by comparing the two Archaic sites on either side of the Illinois River—Campbell Hollow and Napoleon Hollow—on two different tool categories: bifaces; and four primarily unifacial types consisting of gravers, unifaces, retouched pieces and scrapers. As a measure of extreme fragmentation among bifaces I selected "biface fragments"—pieces that are essentially too small or battered to enable classification by reduction stage. As illustrated in Figure 6b, the quantity of biface fragments at flint-poor Campbell Hollow is substantially greater than at Napoleon Hollow—a distinction significant at the .001 level on difference-of-proportions tests for every possible combination of components between the two sites (but not within the sites). Fragmentation among unifacial types exhibits the same pattern, Campbell Hollow tools showing considerably more breakage than Napoleon Hollow tools (Figure 6c). Difference-of-proportions tests of every inter-site combination except Lower Horizon/Titterington are significant at the .05 level. Fragmentation of utilized unretouched debris exhibits similar tendencies, though inter-site differences are not significant.

Since the previous investigations concerned only the two Archaic sites on either side of the river, it can be legitimately inquired whether these findings are generalizable to all components. I accordingly tabulated the proportion of complete (unbroken) tools in all chipped stone type collections, and present the results in Figure 7. The pattern is very clear: those components located east of the Illinois River (solid black columns) all exhibit substantially greater fragmentation than the components west of the river (hatched columns).

One might surmise that conservational practices in breakage and manufacture would have had their counterpart in tool use, that people would have tended to utilize their tools more intensely in areas of resource scarcity. I tested this hypothesis in a number of different ways, but could detect few consistent patterns (Odell 1989:166–167). I attribute this failure to the excessive breakage of tools in resource-poor zones, noted above. That is, fragmented tools are smaller than the pieces from which they originated, often restricting the amount of activities in which they can comfortably be engaged. A broken tool is therefore likely to possess fewer detectable used areas than a complete one. For instance, a complete tool utilized on 3 different sectors (or functional units: FUs) that broke in the middle might result in two tools, one with 1 FU, the other with 2 FUs. If the implements had been rendered unusable by this breakage, they would have remained in the discard pile, each possessing fewer FUs than the original tool. And even if the tools were utilized subsequently, their smaller size would probably have precluded them from being used more extensively than the original tool was. Breakage would affect intentional retouch in similar ways, producing results at variance with the original hypothesis.

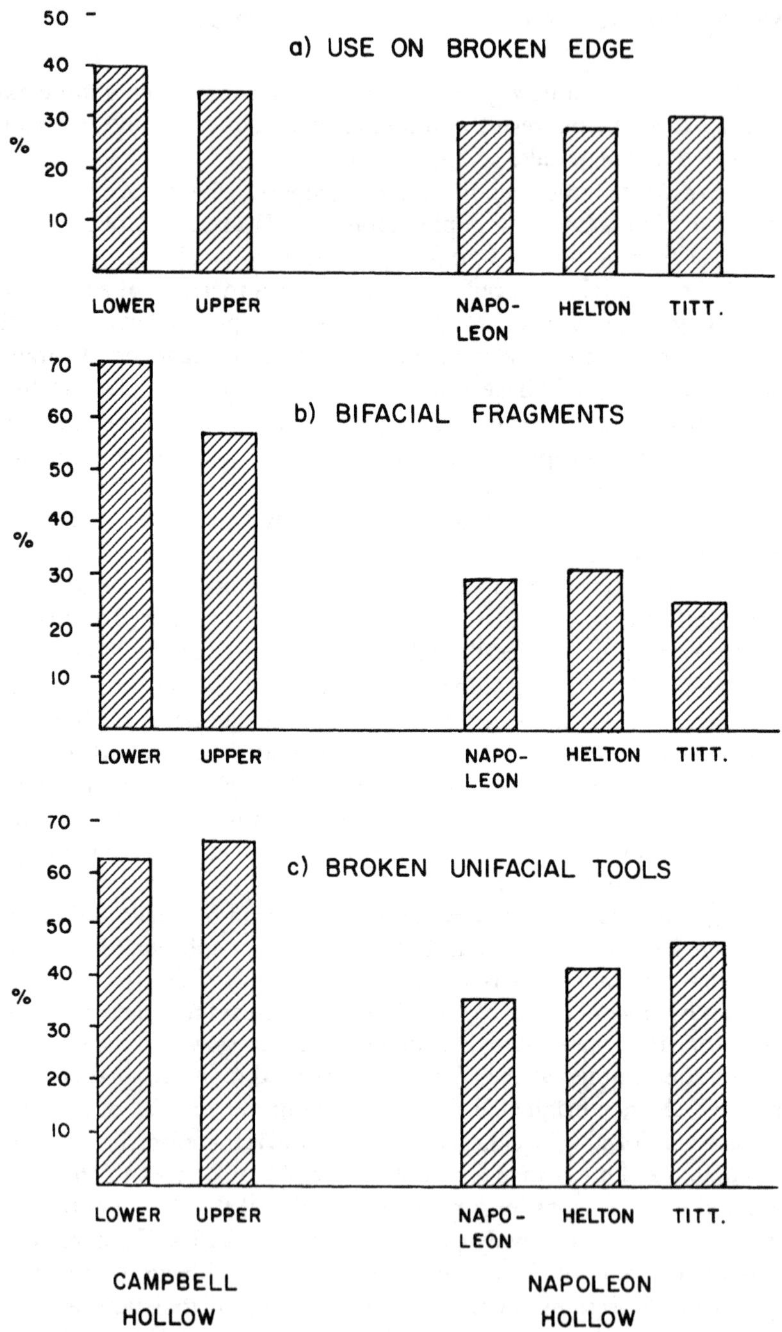


**Figure 6.** Comparisons of components at the Campbell Hollow and Napoleon Hollow sites. Percentages of: a) type collection tools of which a broken edge was utilized; b) stage indeterminate "biface fragments"; and c) broken unifacial tool categories (gravers, scrapers, retouched pieces, unifaces).

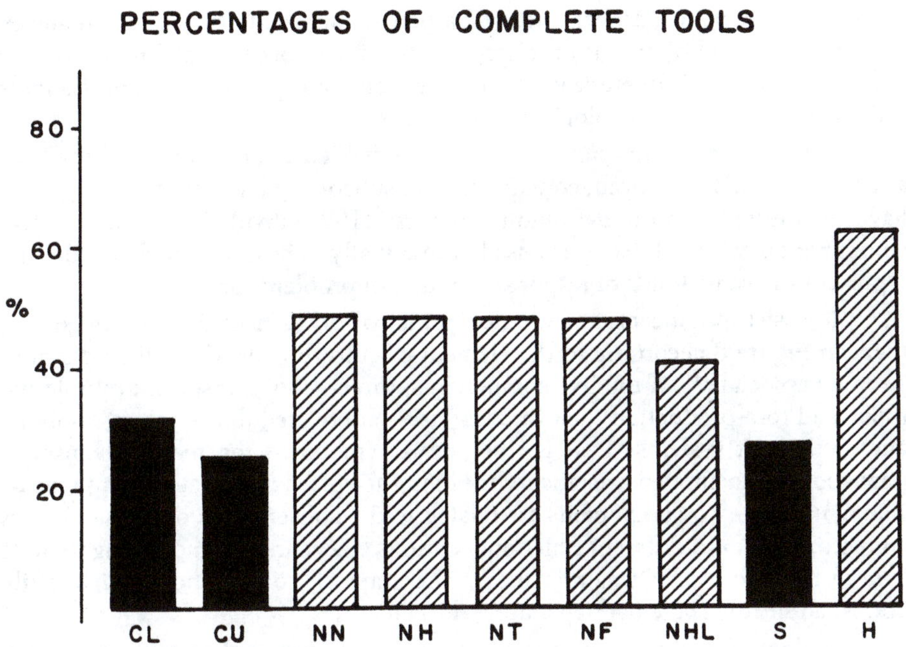


**Figure 7.** Percentages of complete tools in the chipped stone assemblages of all sampled components. Solid black columns represent sites east of the Illinois River (taken from Odell 1995).

One functional parameter does show differences, however. If people in resource-poor areas had attempted to economize, then they might be expected to have made greater use of broken edges, which in ordinary circumstances are not as suitable for most tasks as unbroken ones. I accordingly reconstructed the data from the two Archaic sites on either side of the river, finding the differences illustrated in Figure 6a. Tool users at flint-poor Campbell Hollow availed themselves of broken edges to a greater degree than did their counterparts across the river, though in no case are the results significant at the .05 level.

## RELATIONS BETWEEN CURATION AND RESPONSES TO TOOL RESOURCE AVAILABILITY

### Curation

The term "curation" has maintained a curious hold on archaeological jargon since Binford's work with the Nunamiut in the 1970s. It is a cunningly attractive concept, because it connotes special attention paid to specific items of a culture's material repertoire. Tool users made their implements specially for the occasion, transported them to another site, resharpened them extensively, or whatever. However, the *kind* of attention referred to is critical, because different adaptive strategies employed different kinds of "curation," but this level of specificity is seldom achieved in archaeological treatises.

I have argued in this paper that the concept of curation, as originally defined and subsequently employed, contains several subconcepts which, taken together, have constituted a sort of definition. Bamforth (1986) divided the term into five components, which I have assessed individually. These components do not represent the same kinds of responses to human problems or needs.

To assist in conceptualizing their principal differences, I have categorized them in Figure 8 according to the element of culture with which they are most strongly associated. Of the five components, three—tool transport, multiple-use tools, and tool production in advance of use—are related to the group's mobility and settlement systems. That is, they either embodied the moves themselves (transport) or facilitated existence while living at a place or contemplating a move. The remaining two components are related to the conservation of tools either by converting them to different functional entities (recycling) or modifying them to suit current exigencies (maintenance). These are, of course, the *principal* affiliations, admitting some overlap among the categories.

Settlement and mobility organization are unquestionably important determinants of curational practices, and were seminal considerations in the initial formulation of the concept of curation (Binford 1973, 1977, 1979; Shott 1986; Kuhn 1992). They are *not,* however, the *only* determinants, and part of the ambiguity of the term stems from the fact that, by relating to at least two major cultural systems, it has been asked to do too much. It is no coincidence that, in commenting on the inherent complexity of curation, Bamforth (1986) chose recycling and maintenance to make his point. These are the components least

| RESPONSE COMPLEX | INDIVIDUAL BEHAVIORS | | |
|---|---|---|---|
| | Mobility/Settlement | Tool Conservation | Tool Extravagance |
| 1. Curation | 1a. Production of tools in advance of use<br>1b. Design of tools for multiple uses<br>1c. Transport from one location to another | 1d. Tool recycling<br>1e. Tool maintenance | |
| 2. Scarcity-induced economizing activity | | 2a. Tool recycling<br>2b. Tool maintenance | 2c. Core bashing<br>2d. Bipolar reduction<br>2e. Extreme tool breakage<br>2f. Extreme tool exhaustion<br>2g. Extreme use intensity |

Figure 8. Individual behaviors associated with curation and scarcity-induced economizing activity.

exclusively associated with settlement and mobility organization. As demonstrated in this paper, they are also the qualities that are *least theoretically justifiable* within the lithic data set—particularly "the presence of retouch on a used piece as a simple indication of maintenance and the rare discard of an unbroken tool as an indication of recycling" (Bamforth 1986:41). It is ironic that, despite crippling flaws in his choice of variables, Bamforth's principal point concerning the complexity of curation remains valid.

If the term "curation" is to be useful in the future, its scope will have to be restricted. Based on considerations presented here, the most parsimonious usage would retain those elements associated with mobility and settlement, and discard the ones associated with tool conservation. While employment in this manner would remove some of the current ambiguities, the salvageability of the term remains to be seen.

Restriction of the meaning of curation to issues of settlement and mobility organization would not automatically clarify the specific dynamics of the system; it would only provide a label for the mobility. As a case in point, Binford (1977, 1979) initially applied the concept to the logistically organized Nunamiut; Shott (1986:43) later applied it to American Plains-dwelling Plainview peoples of the early Holocene period, who were arguably more residentially organized than the Nunamiut. Both authors were discussing mobility, though the mobility in either case was probably organized quite differently; and both attributed a high level of curation to the group each was studying. I am not saying that either author was wrong, but simply that they were talking about different phenomena. Similarly, the results of my Illinois Valley study demonstrate a totally different level of curation depending on whether the term is defined as the preparation of tools in advance of use, in which case curation *increases* with time; or as the design of tools for multiple uses, in which case it *decreases* (according to the specific variables used). Labelling a cultural practice as "curation" is not going to solve any problems concerning the organization of technology. Mobility is mobility no matter how it is organized, and in current usage of the term, different kinds of mobility may be called curational.

## Responses to Tool Resource Availability

It is indisputable that the availability of tool raw material affected the ways that implements were produced and utilized (Bamforth 1986; Odell 1989; Marks et al. 1991). Insofar as this behavior is conservational, it is identical to the recycling and maintenance considered characteristic of tool curation (Figure 8). To distinguish the complex of responses to tool resource availability from curational forces, the former has been labelled "scarcity-induced economizing activities."

Behavior that developed as a result of tool raw material shortages was not just conservational, however—it was also extravagant. As demonstrated in the

Campbell Hollow assemblages, these people bashed their cores to sub-atomic particles, suffered extreme tool breakage, and utilized portions of implements that they would otherwise have considered too marginal to employ. In my experience these traits are unique to scarcity-induced economizing activities, and can be employed to distinguish this behavior from curation. Although it is rather obvious, it is also significant that scarcity-induced economizing activities possess components less directly related to mobility organization and settlement parameters than activities recognized as curation.

The relationship between curation, however defined, and scarcity-induced economizing behavior will never be totally clear, as tool conservation is a constituent of both. This fact is illustrated by the use of cores at the Smiling Dan site (Figure 5), which was probably a result of both sets of forces. Core frequency was as low as it was because there was no readily available chert on that side of the Illinois River, inducing tool users to exhaust their cores more completely. It was as high as it was because, as a relatively large base camp, people probably exerted the extra effort required to stockpile some stone and keep it on hand.

Despite the similarities in behavioral response, this study demonstrates that it is operationally possible to separate at least some of these effects. In analyzing curation, the results of both typo-technological (the biface type) and functional (hafting through use-wear analysis) variables have provided useful information. In analyzing the larger question of hunter-gatherer mobility organization, variables such as heat alteration, retouched and utilized polar co-ordinates/tool, functional units/tool, and different tasks/tool have also been employed (Odell 1995). For discerning the existence of scarcity-induced economizing behavior, on the other hand, variables emphasizing completeness and fragmentation have proven most helpful. Cores are particularly useful, because bipolar cores, if recognized, indicate conservational strategies, and negative evidence suggests that cores were turned into more usable flakes. Tool breakage clearly emerges as a viable indicator of economizing strategies, because of either the production of new forms from tools that were exhausted or no longer needed, or increased fragmentation from more intensively utilized implements. The only specifically functional variable that was able to discriminate this behavior also involved fragmentation—the use of broken edges.

I have been fortunate in possessing a data set that includes a wide variety of site types and chronological periods, a situation that has facilitated an analysis of forager mobility and the issue of curation. Equally fortunate is the specific geographical placement of sites in the data set, including locales both with and without available chert within 10 km of each other. This situation is not very common in the history of lithic research, and has provided a unique window on a wide range of forces and constraints within the same data set.

## ACKNOWLEDGMENTS

The data reported here were accumulated during several years spent with the Center for American Archeology, and their accumulation was funded by the Illinois Department of Transportation. I acknowledge the directors of the FAP 408 Project and the Principal Investigators of the sites studied here for their cooperation in the research in which I was engaged. With respect to this paper, I owe a debt of gratitude to the conference participants, who unabashedly drew attention to logical flaws in my original argument; and particularly to Mike Shott and Brian Hayden, who made specific comments on the manuscript. Because of the efforts of these people, this paper is better than previous drafts, though its conclusions undoubtedly remain unacceptable to some. Figures 1–7 are the work of Frieda Odell-Vereecken, while Figure 8 was drawn up by Dale Phelps.

## REFERENCES

Ahler, S. A. 1971. *Projectile Point Form and Function at Rodgers Shelter, Missouri.* Missouri Archaeological Society, Research Series 8.

Ahler, S. A., and R. B. McMillan. 1976. Material Culture at Rodgers Shelter: A Reflection of Past Human Activities. In *Prehistoric Man and His Environments: A Case Study in the Ozark Highland,* edited by W. R. Wood and R. B. McMillan, pp. 163–199. Academic Press, New York.

Ammerman, A., and M. Feldman. 1974. On the "Making" of an Assemblage of Stone Tools. *American Antiquity* 39:610–616.

Bamforth, D. B. 1986. Technological Efficiency and Tool Curation. *American Antiquity* 51:38–50.

Binford, L. R. 1973. Interassemblage Variability—the Mousterian and the "Functional" Argument. In *The Explanation of Culture Change: Models in Prehistory,* edited by C. Renfrew, pp. 227–254. Duckworth, London.

Binford, L. R. 1977. Forty-seven Trips: A Case Study in the Character of Archaeological Formation Processes. In *Stone Tools as Cultural Markers: Change, Evolution and Complexity,* edited by R. V. Wright, pp. 24–36. Australian Institute of Aboriginal Studies, Canberra.

Binford, L. R. 1979. Organization and Formation Processes: Looking at Curated Technologies. *Journal of Anthropological Research* 35:255–273.

Binford, L. R. 1980. Willow Smoke and Dogs' Tails: Hunter-Gatherer Settlement Systems and Archaeological Site Formation. *American Antiquity* 45:4–20.

Bleed, P. 1986. The Optimal Design of Hunting Weapons: Maintainability or Reliability. *American Antiquity* 51:737–747.

Callahan, E. 1987. *An Evaluation of the Lithic Technology in Middle Sweden during the Mesolithic and Neolithic.* Societas Archaeologica Upsaliensis, AUN 8, Uppsala, Sweden.

Chatters, J. C. 1987. Hunter-Gatherer Adaptations and Assemblage Structure. *Journal of Anthropological Archaeology* 6:336–375.

Cole, F.-C., and T. Deuel. 1937. *Rediscovering Illinois.* University of Chicago Press, Chicago.

Cook, T. G. 1986. A Dispersed Harvesting Economy: The Titterington Phase. In *Foraging, Collecting, and Harvesting,* edited by S. Neusius, pp. 175–200. Center for Archaeological Investigations, SIU Carbondale, Occasional Paper no. 6.

Crabtree, D. E. 1972. *An Introduction to Flintworking*. Occasional Papers of the Idaho State University Museum, Pocatello, no. 28.

Farnsworth, K. B., and J. B. Walthall. 1983. In the Path of Progress: Development of Illinois Highway Archeology, and the FAP 408 Project. *American Archaeology* 3:169–181.

Francis, J. 1991. Lithic Resources on the Northwestern High Plains: Problems and Perspectives in Analysis and Interpretation. In *Raw Material Economies among Prehistoric Hunter-Gatherers*, edited by A. Montet-White and S. Holen, pp. 305–319. University of Kansas, Publications in Anthropology, no. 19.

Gramly, R. Michael. 1980. Raw Materials Source Areas and "Curated" Tool Assemblages. *American Antiquity* 45:823–833.

Hayden, B. 1987. From Chopper to Celt: The Evolution of Resharpening Techniques. *Lithic Technology* 16:33–43.

Henry, D. O. 1989. Correlations between Reduction Strategies and Settlement Patterns. In *Alternative Approaches to Lithic Analysis*, edited by D. Henry and G. Odell, pp. 139–155. Archeological Papers of the American Anthropological Association, no. 1.

Holen, S. R. 1991. Bison Hunting Territories and Lithic Acquisition among the Pawnee: An Ethnohistoric and Archaeological Study. In *Raw Material Economies among Prehistoric Hunter-Gatherers*, edited by A. Montet-White and S. Holen, pp. 399–411. University of Kansas, Publications in Anthropology, no. 19.

Jeske, R., and R. Lurie. 1993. The Archaeological Visibility of Bipolar Technology: An Example from the Koster Site. *Midcontinental Journal of Archaeology* 18:131–160.

Johnson, J. K. 1989. The Utility of Reduction Trajectory Modeling as a Framework for Regional Analysis. In *Alternative Approaches to Lithic Analysis*, edited by D. Henry and G. Odell, pp. 119–138. Archeological Papers of the American Anthropological Association, no. 1.

Keeley, L. H. 1982. Hafting and Retooling: Effects on the Archeological Record. *American Antiquity* 47:798–809.

Kelly, R. L. 1983. Hunter-Gatherer Mobility Strategies. *Journal of Anthropological Research* 39:277–306.

Kelly, R. L. 1988. The Three Sides of a Biface. *American Antiquity* 53:717–734.

Koldehoff, B. 1987. The Cahokia Flake Tool Industry: Socioeconomic Implications for Late Prehistory in the Central Mississippi Valley. In *The Organization of Core Technology*, edited by J. Johnson and C. Morrow, pp. 151–185. Westview Press, Boulder.

Kornfeld, M., K. Akoshima, and G. C. Frison. 1990. Stone Tool Caching on the North American Plains: Implications of the McKean Site Tool Kit. *Journal of Field Archaeology* 17:301–309.

Kuhn, S. L. 1991. "Unpacking" Reduction: Lithic Raw Material Economy in the Mousterian of West-Central Italy. *Journal of Anthropological Archaeology* 10:76–106.

Kuhn, S. L. 1992. On Planning and Curated Technologies in the Middle Paleolithic. *Journal of Anthropological Research* 48:185–214.

Luedtke, B. E., and J. T. Meyers. 1984. Trace Element Variation in Burlington Chert: A Case Study. In *Prehistoric Chert Exploitation: Studies from the Midcontinent*, edited by B. Butler and E. May, pp. 287-298. Center for Archaeological Investigations, SIU Carbondale, Occasional Paper no. 2.

Lurie, R. 1989. Lithic Technology and Mobility Strategies: The Koster Site Middle Archaic. In *Time, Energy and Stone Tools*, edited by R. Torrence, pp. 46–56. Cambridge University Press, Cambridge.

Mallouf, R. J. 1981. *A Case Study of Plow Damage to Chert Artifacts: The Brookeen Creek Cache, Hill County, Texas*. Texas Historical Commission, Office of the State Archaeologist, Report no. 33.

Marks, A. E., J. Shockler, and J. Zilhao. 1991. Raw Material Usage in the Paleolithic. The Effects of Local Availability on Selection and Economy. In *Raw Material Economies among Prehistoric*

*Hunter-Gatherers*, edited by A. Montet-White and S. Holen, pp. 127–139. University of Kansas, Publications in Anthropology, no. 19.

Moss, E. H., and M. H. Newcomer. 1982. Reconstruction of Tool Use at Pincevent: Microwear and Experiments. *Studia Praehistorica Belgica* 2:289–312.

Myers, A. 1989. Reliable and Maintainable Technological Strategies in the Mesolithic of Mainland Britain. In *Time, Energy and Stone Tools*, edited by R. Torrence, pp. 78–91. Cambridge University Press, Cambridge.

Nelson, M. C. 1991. The Study of Technological Organization. In *Archaeological Method and Theory*, vol. 3, edited by M. Schiffer, pp. 57–100. University of Arizona Press, Tucson.

Odell, G. H. 1981a. The Mechanics of Use-Breakage of Stone Tools: Some Testable Hypotheses. *Journal of Field Archaeology* 8:197–209.

Odell, G. H. 1981b. The Morphological Express at Function Junction: Searching for Meaning in Lithic Tool Types. *Journal of Anthropological Research* 37:319–342.

Odell, G. H. 1984. Chert Resource Availability in the Lower Illinois Valley: A Transect Sample. In *Prehistoric Chert Exploitation: Studies from the Midcontinent*, edited by B. Butler and E. May, pp. 45–67. Center for Archaeological Investigations, SIU Carbondale, Occasional Paper 2.

Odell, G. H. 1989. Fitting Analytical Techniques to Prehistoric Problems with Lithic Data. In *Alternative Approaches to Lithic Analysis*, edited by D. Henry and G. Odell, pp. 159–182. Archeological Papers of the American Anthropological Association, no. 1.

Odell, G. H. 1994a. Prehistoric Hafting and Mobility in the North American Midcontinent: Examples from Illinois. *Journal of Anthropological Archaeology* 13:51–73.

Odell, G. H. 1994b. The Role of Stone Bladelets in Middle Woodland Society. *American Antiquity* 59:102–120.

Odell, G. H. 1995. *Stone Tools and Mobility in the Illinois Valley: From Hunting-Gathering Camps to Agricultural Villages.* International Monographs in Prehistory, Ann Arbor, Michigan.

Odell, G. H., and F. Odell-Vereecken. 1980. Verifying the Reliability of Lithic Use-Wear Assessments by "Blind Tests": The Low-Power Approach. *Journal of Field Archaeology* 7:87–120.

Oswalt, W. 1976. *An Anthropological Analysis of Food-Getting Technology.* John Wiley and Sons, New York.

Owen, L. R., and G. Unrath. 1989. Microtraces d'usure dues à la préhension. *l'Anthropologie* 93:673–688.

Reher, C. A. 1991. Large Scale Lithic Quarries and Regional Transport Systems on the High Plains of Eastern Wyoming: Spanish Diggings Revisited. In *Raw Material Economies among Prehistoric Hunter-Gatherers*, edited by A. Montet-White and S. Holen, pp. 231–284. University of Kansas, Publications in Anthropology, no. 19.

Roper, D. C. 1979. Breakage Patterns of Central Illinois Woodland Projectile Points. *Plains Anthropologist* 24:113–121.

Rubey, W. W. 1952. *Geology and Mineral Resources of the Hardin and Brussels Quadrangles (in Illinois).* U.S. Geological Survey, Washington, D.C., Professional Paper no. 218.

Schiffer, M. B. 1976. *Behavioral Archaeology.* Academic Press, New York.

Shea, J. J. 1988. Spear Points from the Middle Paleolithic of the Levant. *Journal of Field Archaeology* 15:441–450.

Shott, M. 1986. Technological Organization and Settlement Mobility: An Ethnographic Examination. *Journal of Anthropological Research* 42:15–51.

Shott, M. 1989a. Diversity, Organization, and Behavior in the Material Record: Ethnographic and Archaeological Examples. *Current Anthropology* 30:283–301.

Shott, M. 1989b. On Tool-Class Use Lives and the Formation of Archaeological Assemblages. *American Antiquity* 54:9–30.

Stordeur, D. (editor). 1987. *La main et l'outil: manches et emmanchements préhistoriques*. Maison de l'Orient Mediterranéen, Lyon.

Sullivan, A. P., III, and K. C. Rozen. 1985. Debitage Analysis and Archaeological Interpretation. *American Antiquity* 50:755–779.

Thomas, D. H. 1983. *The Archaeology of Monitor Valley. 1. Epistemology.* Anthropological Papers of the American Museum of Natural History, vol. 58, part 1, New York.

Torrence, R. 1983. Time Budgeting and Hunter-Gatherer Technology. In *Hunter-Gatherer Economy in Prehistory: A European Perspective,* edited by G. Bailey, pp. 11–22. Cambridge University Press, Cambridge.

Torrence, R. 1989. Retooling: Towards a Behavioral Theory of Stone Tools. In *Time, Energy and Stone Tools,* edited by R. Torrence, pp. 57–66. Cambridge University Press, Cambridge.

Webb, W. S. 1974. *Indian Knoll.* University of Tennessee Press, Knoxville.

Wiant, M. D., and H. Hassen. 1985. The Role of Lithic Resource Availability and Accessibility in the Organization of Technology. In *Lithic Resource Procurement: Proceedings from the Second Conference on Prehistoric Chert Exploitation,* edited by S. Vehik, pp. 101–114. Center for Archaeological Investigations, SIU Carbondale, Occasional Paper 4.

Wiant, M. D., and C. R. McGimsey (editors). 1986. *Woodland Period Occupations of the Napoelon Hollow Site in the Lower Illinois Valley.* Center for American Archeology, Research Series, vol. 6.

Winters, H. D. 1969. *The Riverton Culture, a Second Millenium Occupation in the Central Wabash Valley.* Illinois State Museum, Reports of Investigations, no. 13.

Wylie, H. G. 1975. Tool Microwear and Functional Types from Hogup Cave, Utah. *Tebiwa* 17:1–31.

*Chapter* 3

# *Is Curation a Useful Heuristic?*

STEPHEN E. NASH

## ABSTRACT

The heuristic value of the concept of "curation" is examined at two levels. First, Binford's assertion that Middle Paleolithic stone tool assemblages are "largely non-curated" is tested in an analysis of nominal data from five stratigraphically distinct assemblages from Tabun Cave. The data suggest that behavior commonly described as "curation" was, in fact, present during the Middle Paleolithic. This leads to a more theoretical examination of what exactly "curation" has come to mean in archaeological interpretation. A review of the literature suggests that the "curation" concept has been used to describe and explain a great deal of morphological, technological, and assemblage-level variability with little or no standardization achieved in its usage. In fact, it is evident that "curation" now means vastly different things to different archaeologists. Archaeologists should, at the very least, make explicit their use of the term and concept; and archaeology might be better served if some standardization in usage could be achieved.

## INTRODUCTION

Binford's (1973; 1976; 1979:269) assertion that Middle Paleolithic stone tool assemblages are "largely non-curated" has been challenged but rarely tested (though see Marks 1988). In this paper I examine Binford's assertion on the basis of data from the chronostratigraphically significant Middle Paleolithic site of

**STEPHEN E. NASH** • University of Arizona, Tucson, Arizona 85721.

*Stone Tools: Theoretical Insights into Human Prehistory*, edited by George H. Odell. Plenum Press, New York, 1996.

Tabun, near Mount Carmel, Israel (Jelinek 1975, 1982). The data suggest that behavior commonly described as curation in the archaeological literature did indeed affect the formation of assemblages at Tabun; therefore, Middle Paleolithic assemblages at this site are curated (*senso* Binford 1973). This result, when considered in conjunction with conclusions offered by Marks (1988), suggest that the curated/expedient dichotomy is not useful for explaining differences between Middle and Upper Paleolithic assemblages.

This conclusion leads logically into a more general consideration of the interpretive value of curation. It is argued that, because curation has come to mean different things to different archaeologists, the concept has become diluted and has lost whatever interpretive value it once possessed. In the absence of definitive standardization of use of the term and concept, lithic analysts would do well to convey behavioral interpretations using terms and concepts other than curation.

Binford (1978) began ethnoarchaeological research among the Nunamiut Eskimos in an effort to better understand the behavioral dynamics affecting archaeological site formation. The most critical observation made by Binford was that Nunamiut hunters approach, describe, and treat items in their various tool kits in two radically different ways. He adopted the terms "curated" and "expedient" to describe these different behaviors, such that "personal gear" and "site furniture" can be described as "curated," while "situational gear" can best be described as "expedient." Curated tools thus are "produced and maintained within a technology in anticipation for future usage," while expedient tools are "produced when needed and are discarded after use" (Binford 1979:269).

Binford's (1973) original use of the term described behavior affecting individual tools, but he also referred to curated and expedient technologies. In later usage these terms have been used to describe entire assemblages (e.g., Binford 1979). One unresolved question comes immediately to mind: what exactly is curated: tools, assemblages, or technology? An additional problem stems from the fact that "curated" and "expedient" entities are often considered dichotomous endpoints on continua, describing stone tool production, maintenance, and use. Researchers often acknowledge these continua, but then establish their discussion as if "curated" and "expedient" are mutually exclusive entities. That is, the terms are usually considered dichotomous: a tool, assemblage, or technology is either curated or expedient; it is not both. A prime example of this is Binford's assertion that Upper Paleolithic assemblages are "curatorially organized," while Middle Paleolithic assemblages are "largely non-curated" and are therefore, by implication, expediently produced (Binford 1979:269; see also Marks 1988:276–277 for discussion).

The current analysis utilizes typological data from five stratigraphically distinct assemblages from Tabun to test whether behavior described elsewhere as curation is evident in Middle Paleolithic assemblages. Data were originally collected for a controlled comparison of the Tabun data with Marks' examination of curation in Middle and Upper Paleolithic assemblages in the Central Negev (Nash 1991). All the Middle Paleolithic assemblages analyzed by Marks fall "technologically and typologically within the Early Levantine Mousterian of Tabun D type" (Marks 1988:279). And while Marks (1988:276) used artifactual as well as spatial data in his evaluation of curation, Nash reasoned that the extensive stratigraphic sequence at Tabun might allow the identification of diachronic changes relevant to curation in the Middle Paleolithic assemblages examined.

Marks used resource location data and artifact assemblage characteristics to identify three general classes of sites in the Central Negev: workshop/quarries, base camps, and hunting and gathering camps. The assemblage characteristics considered by Marks include: site size, artifact density, percent cortex, Levallois blank-to-Levallois core ratio, Levallois point-to-Levallois point core ratio, percent retouched pieces, and mean core weight.

Marks' workshop/quarry sites are logically expected at lithic raw material outcrops, and are expected to have high percentages of cores and low Levallois blank-to-Levallois core ratios, assuming that the Levallois blanks were exported for use elsewhere. Base camps are expected to occur in close proximity to water and to have low percentages of (probably small) cores, because these would not be imported in large quantities. Conversely, the Levallois blank-to-Levallois core ratios at base camps are expected to be high, due to selective importation of Levallois blanks. Hunting and gathering sites are expected to fall somewhere between workshop/quarries and base camps, with artifact assemblages varying according to the particular task undertaken at the site.

A reasoned attempt was made to design Tabun data collection such that comparisons could be made with Marks' Central Negev assemblage characteristics, in the hopes that functional (in a very loose sense) similarities in the Tabun and Central Negev assemblages might be identified. Site size and artifact density data were not included in the Tabun analysis for several reasons. Proximity to lithic resources was not considered further because of the extensive lithic source area near Tabun in the Wadi Mughara. Intrasite spatial data were not considered because of the somewhat limited nature of the sample of Tabun collections available for analysis at the University of Arizona.

Data collection consisted of gathering sufficient qualitative and quantitative information to adequately describe Marks' (1988:280; Table 16.1; see Table 1 below) "assemblage characteristics" for each Tabun assemblage examined. A total of 1493 artifacts was analyzed from five arbitrarily selected Tabun assemblages. From oldest to youngest, these are: Levels 70, 66, 59, 41AN, and 35B[1]. All artifacts were measured (length, width, thickness, platform width, platform thickness) according to methods discussed by Jelinek (1975). Nominal data were recorded on the following variables pertinent to a consideration of curation: platform type, flake form, intensity of retouch, and percent exterior cortex.

Nash (1991) used partitioning $G^2$ and $X^2$ analyses to test differences between many of the Central Negev assemblages and the Tabun assemblages. Table 1 shows interesting correlations between lithic technology and site type among the Central Negev assemblages tested. For example, both the percentage of cores

Table 1. Core and Retouch Characteristics of Central Negev Assemblages

| Assemblage | Percent cores | Levallois blank to core ratio | Percent retouched | Mean core weight (grams) |
|---|---|---|---|---|
| Workshop/quarry sites | | | | |
| D2 | 27 | 1:2 | 8 | 155 |
| D42 | 14 | 1:1 | 5 | 125 |
| D44 | 13 | 1:1 | 6 | 166 |
| Base Camps | | | | |
| D15 | 7 | 16:1 | 8 | 47 |
| D35S | 5 | 19:1 | 6 | 78 |
| D35,1 | 6 | 11:1 | 5 | 78 |
| D35,3C | 3 | 96:0 | 4 | 50 |
| Ephemeral hunting or gathering sites | | | | |
| D33 | 14 | 3:1 | 2 | 254 |
| D40 | 24 | 2:1 | 3 | 190 |
| D45 | 8 | 3:1 | 4 | 88 |
| D46 | 17 | 3:1 | 5 | 156 |
| D51 | 9 | 2:1 | 7 | 109 |
| D52 | 13 | 4:1 | 6 | 44 |

After Marks 1988:280, Table 16.1; see also Munday 1976, 1977; Nash 1991.

and mean core weight are substantially lower among those sites designated as base camps than among sites designated as either workshop/quarry sites or ephemeral sites. In contrast, the Levallois blank:core ratio is much higher at base camps than among sites of either of the other two types. There appears to be more internal variability among ephemeral sites than among either base camps or workshop/quarry sites. The Tabun samples analyzed (Table 2) exhibit considerable inter-level variability on every variable, suggesting differences in site function among the levels. The nominal data analyzed below suggest that many of these differences can be accounted for by differential transport of raw materials, or of prepared cores, blanks, or tools. That is, assemblage differences may be accounted for by behavior commonly described by the "term curation."

Table 2. Core and Retouch Characteristics of Tabun Assemblages

| Assemblage | Cores percent | Levallois blank to core ratio | Percent retouched | Mean core weight (grams) |
|---|---|---|---|---|
| Level 35B | 5 | 57:0 | 7 | 73 |
| Level 41AN | 6 | 12:1 | 2 | 81 |
| Level 59 | 5 | 65:0 | 8 | 104 |
| Level 66 | 5 | 20:1 | 23 | 103 |
| Level 70 | 12 | 7:1 | 24 | 111 |

## FRAMEWORK FOR CURRENT ANALYSIS

In a consideration of curation, Bamforth (1986:39) noted that Binford's definition of curation included at least five distinct aspects of tool manufacture and use: 1) the production of implements in anticipation of use; 2) the design of implements for multiple uses; 3) the transport of implements from location to location; 4) maintenance; and 5) recycling. Marks (1988) also pointed out that behavior defined as curation may, in fact, occur during raw material acquisition and active or passive ("mundane") caching/storage/stockpiling processes. The framework used in the analysis of Tabun assemblages is similar to Bamforth's, but is modified according to points made by Marks. As such, curation is considered here to (have the potential to) affect behavior at four general stages in the tool production, maintenance, and use process: 1) raw material acquisition; 2) preparation and transport of cores, blanks, or tools; 3) reduction of tools during use; and 4) storage or caching. The following evaluation of curation in the Middle Paleolithic will be divided according to this framework.

### Raw Material Acquisition

While few archaeologists would describe raw material acquisition in terms of curation, raw material must, by definition, be curated if it is transported "in anticipation of future usage" (Binford 1979:269; see Toth 1985). In this context, raw material is seen as a spatially restricted resource which must be gathered and transported to the locus of use. Recent research has pointed to difficulties in the accurate interpretation of the meaning assigned to raw material acquisition by prehistoric and ethnographic groups, especially when such acquisition is "embedded" in other activities (Binford 1980; Kuhn 1991).

To examine curation at the level of raw material acquisition, it is best to examine variables that might reflect differential importation of raw material to the site in question. Such a variable is the percentage of cortex present on artifacts. One must make the dangerous assumption that artifacts within an archaeological unit are made on raw material from the same source or material population (in a statistical sense), and that the archaeological samples analyzed accurately represent the raw material and behavioral variability present at a given site. High quality raw material is plentiful within two kilometers of Tabun Cave, in the Wadi Mughara (Arthur Jelinek, personal communication 1990). This flint source seems to have been exploited consistently through time; hence we may assume that raw material variability at Tabun is relatively constant. Therefore, it is hypothesized that variability in the percentages of cortex in the five Tabun assemblages examined is due to differential importation of unprepared raw material nodules.

The following arbitrary but analytically common divisions of percent cortex were tabulated: 1) No cortex present on exterior surface; 2) < 10% cortex; 3) 10%

Table 3. Tabun assemblage cortex data (expected frequencies in parentheses)

| Assemblage | No cortex | Less than 50% | Greater than 50% | Total |
|---|---|---|---|---|
| Level 35B | 157 (158) | 66 (75) | 57 (47) | 280 |
| Level 41AN | 339 (303) | 112 (142) | 84 (90) | 535 |
| Level 59 | 92 (88) | 50 (42) | 14 (26) | 156 |
| Level 66 | 149 (166) | 98 (78) | 46 (49) | 293 |
| Level 70 | 53 (75) | 46 (35) | 34 (23) | 133 |
| Total | 790 | 372 | 235 | 1397 |

$X^2$=45.2, df=14, p<.001

- 50% cortex; 4) 50% - 90% cortex; and 5) > 90% cortex. To simplify the analysis, these data were grouped into three categories: no cortex, less than 50% cortex, and greater than 50% cortex present on the exterior surface (Table 3).

Analysis reveals the following departures from the expected amount of cortex present in each assemblage:

- Level 35B: More than the expected number of specimens with greater than 50% cortex;
- Level 41AN: Fewer than the expected number of specimens with less than 50% cortex;
- Level 59: Fewer than the expected number of specimens with more than 50% cortex;
- Level 66: More than the expected number of specimens with less than 50% cortex;
- Level 70: More than the expected number of specimens with cortex of any kind, fewer than the expected number of specimens with no cortex present.

No clear diachronic trends are obvious, but it seems reasonable to conclude that, if cortex presence indicates raw material importation, unprepared raw material nodules were brought to Tabun Cave during the formation of Level 70, and possibly of Level 35B. It is possible that imported raw material nodules were being reduced and exported, or simply that cortical flakes were being imported to the site, though the latter seems less intuitively reasonable.

Does this simple analysis of cortex data suggest that raw material acquisition and importation occurred at Tabun "in anticipation of future usage" (Binford 1979:269)? Yes. Is such behavior best described as curation? It is not, especially when raw material is present and readily available two kilometers from the site. Indeed, the concept of curation is only marginally useful in instances where acquisition and transport of exceptional or exotic raw materials are evident (e.g., Otte 1992). Even then, one must consider other factors, including the embeddedness of the acquisition strategy (Binford 1980), and archaeologically invisible social, religious, and political constraints on raw material acquisition, that may have affected deposition of exotic material at a given site (Gould, Koster, and Sontz 1971; Gallagher 1977; Pokotylo and Hanks 1989). Such variables are difficult enough to consider in ethnographic research; given the current debate on the presence, absence, and nature of Middle Paleolithic symbolism (e.g., Chase and Dibble 1987), such considerations are even more difficult to evaluate.

## Transport of Cores, Blanks, or Tools

Bamforth (1986:38) noted that curated technologies contain tools that are manufactured in anticipation of future use and which are transported from locality to locality. Even at the most basic level, cores and blanks may be transported as easily as finished tools. If it can be demonstrated that prepared cores, blanks, or tools were produced, maintained, or transported in anticipation of future use at Tabun, then curation must by definition have been present during the Middle Paleolithic, at least at this site. Stone tool assemblage content depends to some degree on site function, which should be related at some level to a given site's proximity to important resources. Such factors are held constant for the Tabun assemblages because the site itself is a rockshelter, though it has been demonstrated that the site was occupied through a number of different paleoenvironmental regimes that may have affected resource availability (Jelinek 1982).

In an attempt to identify differential transport of cores, blanks, or tools in the Tabun assemblages, several variables can be considered: percentage of cores (and mean core weight), platform type, and flake form. The percentage of cores in an assemblage is considered a rough measure of core abundance. Platform type and flake form are examined to understand relative differences in core preparation that occurred during the formation of each assemblage, the assumption being that prepared core technologies indicate curation, while the lack of extensive core preparation indicates expediency.

The Tabun assemblages are notable for their low percentage of cores, especially when compared to the Central Negev workshop/quarry and ephemeral sites (Tables 1 and 2): only in Level 70 do cores constitute more than 10% of the analyzed assemblage. No assumption of ideal or normal percentage of cores in a given assemblage is made, but if core importation occurred at Tabun, it happened during the formation of Level 70. This coincides with the cortical data from Level 70 considered above, which suggests greater raw material importation at this time than during the formation of the other Tabun assemblages. Metric data on cores were not considered in the analysis because of problems associated with raw material outcrop variability and comparability with Marks' data. Mean core weight, which assumes uniform material density, is offered instead as an indicator of core size, when it was removed from systemic context.

Table 4. Tabun assemblage platform type data (expected frequencies in parentheses)

| Assemblage | Plain | Prepared | Cortical | Total |
|---|---|---|---|---|
| Level 35B | 84 (68) | 119 (129) | 23 (29) | 226 |
| Level 41AN | 76 (97) | 182 (184) | 65 (42) | 323 |
| Level 59 | 13 (38) | 103 (72) | 10 (16) | 126 |
| Level 66 | 92 (79) | 146 (150) | 26 (34) | 264 |
| Level 70 | 56 (39) | 58 (73) | 14 (17) | 128 |
| Total | 321 | 608 | 138 | 1067 |

$X^2$=70.9, df=14, p<.001

Platform type (Table 4) and flake form (Table 5) data can be used to identify, at a basic level, whether standardized (prepared) or expedient core reduction strategies were in use during the formation of a given assemblage. It is assumed here that prepared core reduction strategies are characteristic of curated technologies, while *ad hoc* core reduction strategies are characteristic of expedient technologies. If Middle Paleolithic assemblages are "largely non-curated," there should be minimal evidence of prepared core reduction strategies.

Counts of plain, dihedral, faceted, transverse, superimposed, cortical, and chapeau de gendarme platform types were tabulated for the Tabun assemblages (Nash 1991). Platform types may be collapsed into three groups relevant to an examination of curation: cortical platforms, plain platforms, and platforms indicative of prepared cores (faceted, superimposed and, to a lesser degree, dihedral and transverse platforms). If diachronic change toward increased curation is present at Tabun, more recent assemblages should be characterized by more than the expected number of specimens indicative of prepared core reduction strategies. The data in Table 4 suggest the following:

Table 5. Tabun assemblage flake form data (expected frequencies in parentheses)

| Assemblage | Normal | Levallois | Blade | Total |
|---|---|---|---|---|
| Level 35B | 173 (150) | 37 (39) | 61 (84) | 271 |
| Level 41AN | 342 (292) | 16 (75) | 171 (162) | 529 |
| Level 59 | 47 (83) | 50 (21) | 53 (46) | 150 |
| Level 66 | 108 (150) | 61 (38) | 103 (83) | 272 |
| Level 70 | 77 (73) | 27 (19) | 28 (41) | 132 |
| Total | 747 | 191 | 416 | 1356 |

$X^2$=159.2, df=14, p<.001

- Level 35B: Contains more than the expected number of specimens with plain platforms, fewer than the expected number of specimens with prepared and cortical platforms;
- Level 41AN: Contains fewer than the expected number of specimens with plain platforms, more than the expected number of specimens with cortical platforms;
- Level 59: Contains more than the expected number of specimens with prepared core platform types, especially faceted and superimposed platforms, which are typical of Levallois flakes and Levallois points, respectively;
- Level 66: Contains more than the expected number of specimens with plain platforms, and fewer than the expected number with prepared and cortical platforms;
- Level 70: Contains more than the expected number of specimens with plain platforms, and fewer than the expected number with prepared and cortical platforms.

No diachronic trend toward the increased use of prepared cores is evident in these data, nor can we argue that older Middle Paleolithic assemblages at Tabun are characterized by expedient tool manufacture. Level 59 contains the best evidence for prepared core (curated) technologies, specifically the Levallois technique; this is not a stunning revelation.

Flake form characteristics may also be used to differentiate between prepared core and expedient tool manufacturing processes. The following flake forms were tabulated for the five Tabun assemblages: normal, angular, Levallois, first order point, second order point, first order blade, and second order blade. To simplify the analysis, these data have been grouped into Normal, Levallois, and Blade categories, and angular flakes were dropped from consideration due to their infrequent occurrence in these assemblages (Table 5). The data suggest the following:

- Level 35B: Contains fewer than the expected number of blade forms and more than the expected number of normal flakes;
- Level 41AN: Contains more than the expected number of normal flakes, though this was probably caused by data encoding procedures and is not necessarily behaviorally meaningful (see Nash 1991).[2]
- Level 59: Contains more than the expected number of Levallois pieces (particularly second order points), and slightly more than the expected number of blade forms;
- Level 66: Contains more than the expected number of blade and Levallois forms, and fewer than the expected number of normal flakes;
- Level 70: Contains slightly more than the expected number of Levallois forms and fewer than the expected number of blade forms.

It is interesting to note that Level 35B, while exhibiting a very high Levallois blank-to-Levallois core ratio (see Table 2), is characterized by fewer than the expected number of Levallois and blade forms, and by more than the expected number of plain (i.e., unprepared) platforms. These data argue for the selective importation of Levallois blanks during the formation of this assemblage.

Levels 59 and 66 at Tabun contain few or no formally prepared cores (Table 2), yet they are characterized by more than the expected quantities of Levallois pieces and blades. Two conclusions are possible. First, these forms were made elsewhere and imported to Tabun (i.e., the blanks were curated). Or second, the cores were curated and thus removed from the site prehistorically. Given the fact that no Levallois cores were present in the Level 59 assemblage (Jelinek 1982:137), the former scenario seems more likely. However, the behavioral option that actually took place is not important; evidence for either falls under current definitions of curation. Hence the core, blank, and tool data at Tabun support Marks' (1988:284) conclusion that this type of curation occurred during the Middle Paleolithic in the Levant.

These data also point to another problem: the importance of negative evidence. Binford (1973) pointed out that curated items are those least likely to be found in the archaeological record; therefore, one is often forced to use indirect measures to demonstrate curation. To wit, the low percentage of cores at Tabun may be a result of curation, rather than a demonstration that cores were not imported. Cores may simply have been removed (curated) to other sites, and thus did not enter the archaeological record at Tabun. Sackett (1982:90–94) provided an enlightening discussion of this problem.

## Tool Maintenance

Bamforth (1986:39) noted that Binford's definition of curation encompassed stone tool maintenance, which is here distinguished from recycling (Shott 1989). A simple and productive measure of tool maintenance in the Middle Paleolithic is intensity of retouch.

Given the current debate regarding the existence and use of hafting during the Middle Paleolithic (Shea 1988; Anderson-Gerfaud 1989; Holdaway 1989; Solecki 1992), this discussion of stone tool maintenance will focus on stone tool retouch as a measure of maintenance during utilization (Dibble 1984, 1987a, 1987b, 1988; Kuhn 1992). There are problems with the interpretation of retouch purely as maintenance activity (see Odell, this volume), but such data constitute the only reasonable measure of tool maintenance available for the analysis of Middle Paleolithic assemblages.

Retouch attributes for artifacts from the five Tabun levels were analyzed in the hope that diachronic change indicative of increased curation (e.g., maintenance, reuse) might become evident. Retouch types were tabulated in eight categories as defined by Dibble (1987a): no retouch, light retouch, medium retouch, heavy retouch, stepped retouch, abrupt retouch, deep retouch, and bifacial retouch. Not all retouch types were present in the analyzed assemblages, so statistical tests were conducted only on the first three possibilities—no retouch, light retouch, and medium retouch. Analysis of Table 6 reveals statistically significant differences in the occurrence and intensity of retouch within the five Tabun levels examined. These differences are summarized as follows:

- Level 35B: No statistically significant departures from the expected number of specimens in any category;
- Level 41AN: Fewer than the expected number of specimens exhibiting retouch of any kind; more than the expected number exhibiting no retouch;
- Level 59: No statistically significant departures from the expected number exhibiting retouch;
- Level 66: More than the expected numbers with retouch present, accounted for by the high frequency of scrapers in the assemblage (Nash 1991);
- Level 70: No statistically significant departures from the expected number in any category.

Again, there are no diachronic trends evident in these data, though when considered in conjunction with the data in Table 2, some inferences are possible. Note first the anomalous percentages of retouch present in Levels 66 and 70. Typological analyses (Nash 1991) reveal that Level 66 is much more heavily reduced (*senso* Dibble 1987a) than other assemblages, as indicated by the presence of numerous double and convergent scrapers. Level 70 retouched pieces exhibit much less typological patterning, but there are fewer of the heavily reduced forms. Level 41AN is again difficult to interpret, and Levels 35 and 59 are characterized by retouch frequencies comparable to some of the Central Negev sites (Table 1).

Does stone tool maintenance, as indicated by retouch presence and intensity, seem to have been a factor in the formation of Middle Paleolithic assemblages at Tabun? Without a doubt. Is such maintenance behavior best described by the term

**Table 6. Tabun assemblage retouch intensity data (expected frequencies in parentheses)**

| Assemblage | None | Light | Medium | Total |
|---|---|---|---|---|
| Level 35B | 262 (258) | 9 (12) | 6 (7) | 277 |
| Level 41AN | 525 (496) | 4 (23) | 3 (14) | 532 |
| Level 59 | 144 (139) | 4 (6) | 1 (4) | 149 |
| Level 66 | 230 (261) | 34 (12) | 16 (7) | 280 |
| Level 70 | 108 (115) | 7 (5) | 9 (3) | 124 |
| Total | 1269 | 58 | 35 | 1362 |

$X^2 = 96.43$, df = 14, $p < .001$

curation? No. Given recent contributions to the reduction intensity literature (e.g., Dibble 1987a), variability stemming from reduction can be parsimoniously described in terms other than curation, and may be interpreted simply as economizing behavior (see Odell, this volume).

### Storage and Caching

The last type of evidence for curation offered above is storage or caching of raw materials or finished tools. Caching behavior is not evident during the Middle Paleolithic, but Marks (1988) noted that *de facto* curation occurs at any archaeological site with lithic artifacts—all but the most "exhausted" lithic artifacts on the surface of a given site are "surely as usable as any buried cache" (Marks 1988:283; see also Frison 1968; Jelinek 1988). Use of curation as an explanatory concept in this sense, if it has to be used at all, is best restricted to situations in which obvious and indisputable evidence of caching is found. Given that such evidence, with all its associated symbolic baggage (Chase and Dibble 1987), is not yet known for the Middle Paleolithic, active caching is not of concern here. However, the possible long time lags between stone tool manufacture and use episodes caused by passive storage (*de facto* curation) of lithic material at any archaeological site should be noted during any analysis.

## SUMMARY

This brief analysis demonstrates that behavior commonly described or interpreted as "curation" in the archaeological literature did indeed exist during the Levantine Middle Paleolithic, as represented by the Tabun and Central Negev assemblages considered. Perhaps the strongest case may be made for Tabun Levels 35B and 59, wherein evidence suggests that Levallois blanks were imported to Tabun after being manufactured elsewhere; or Level 70, wherein evidence suggests the direct importation of cores or unmodified raw material. These results support those obtained by Marks in the Central Negev: "(it) may be said that during the Middle Paleolithic curation existed, centering around cores and blanks for tools rather than the tools themselves" (Marks 1988:284). However, it may also be said that curation occurred variously during raw material acquisition, tool maintenance, and *de facto* storage of raw material at Tabun. These data demonstrate that Binford's assertion that Middle Paleolithic assemblages are "largely non-curated" is not supported. The conceptualized dichotomy between curated and expedient technologies cannot be invoked to characterize differences between Middle and Upper Paleolithic assemblages in the Levant.

This analysis was designed to test specifically the concept of curation as an heuristic device for explaining observed differences between Middle and Upper Paleolithic stone tool assemblages. However, this is only one situation among many in the archaeological literature in which curation has been offered as an explanatory device. Some will argue that the widespread use of curation in the archaeological literature demonstrates, in and of itself, the useful and productive nature of the concept. I will argue precisely the opposite: curation has come to mean different things to different archaeologists, and the resulting varied and often ill-defined use of a rather vague concept has diluted whatever interpretive potential the term possessed in its original form. If archaeologists are interested in clear communication of behavioral interpretations, some standardization in their heuristic devices should be achieved. In the absence of such standardization, it might do lithic analysts some good to interpret assemblage variability in the absence of vague concepts such as curation.

## IS "CURATION" A USEFUL HEURISTIC?

It is important at this point to examine vernacular English use of the term "curation." Webster's New Collegiate Dictionary (G. & C. Merriam Company 1973:278) defines the following:

- curate (n): a clergyman in charge of a parish;
- curative (adj): relating to or used in the cure of diseases;
- curator (n): one that has the care and superintendence of something.

It is apparent that Binford sought to imply behavior along the lines of "care and superintendence" of tools when he used the term "curated" to describe a particular Nunamiut approach to tool manufacture and maintenance. It is also clear that, without specific modifications toward this usage, English vernacular definitions of "curated" and "curation" are not terminologically appropriate for the accurate description of activities related to the manufacture, use, and maintenance of historic or prehistoric tools.[3]

Given the English vernacular, use of the term "expedient" to describe *some* lithic tool production and use (but not, by definition, tool maintenance) seems more justified. Webster's dictionary (ibid:402–403) lists the following relevant terms:

- expedient (adj): a means of achieving a particular end; dictated by practical or prudential moves;
- expedient (n): a means to an end;
- expeditious (adj): acting with promptness or efficiency.

As adjectives, "expedient" and "expeditious" may both be appropriate for describing tools that are produced when needed and discarded immediately after use. Indeed, the ethnographic literature contains numerous references to just such behavior (e.g., Gould, Koster, and Sontz 1971; O'Connell 1987).

The concept of expedient tool manufacture and use may not account for much variability in most stone tool assemblages, but at this point its meaning is less interpretively vague than is that of curation. However, there is still no agreement on the relationship between expedient and curated: are they mutually exclusive categories or endpoints on a continuum? Can they be used simultaneously to describe various components of the same assemblage? Indeed, what does "expedient" modify—tools, assemblages, or technologies? Demonstrating "expediency" in tool manufacture and use in the archaeological record is a matter beyond the scope of the current paper, though the reader is directed to recent contributions by Parry and Kelly (1987), Johnson (1989), and Jeske (1992) for New World examples of such efforts.

In the social sciences it is difficult to coin new terms to describe newly developed interpretive concepts; therefore, researchers often adopt terms from other disciplines or from general English vernacular. Despite the fact that such terms may be well defined in their original formulation, their adoption often leads to subtle shifts or changes in definition, application, or interpretation. This is precisely what happened with the concept of curation: Binford borrowed the term, presumably from museum studies, and provided a relatively clear conceptualization of his original use of it in Nunamiut tool use strategies. Archaeologists (including Binford) have since applied the concept to a bewildering array of different research questions, including (but not limited to) the investigation of Middle and Upper Paleolithic assemblage differences, as discussed above. It subsequently enjoyed wide and varied usage in spite of early critiques:

> the phenomenon of curation [does not] seem to provide an adequate basis for explaining alleged changes in [Middle Paleolithic] inter-assemblage tool patterning. . . . [T]he curation model will have to be considerably modified or altered before it can assume an important place in the explanatory framework of assumed changes in inter-assemblage variability during the Paleolithic. (Hayden 1975:56)

The logical question arising from this situation is this: Does the concept of curation now retain any common interpretive or heuristic value in archaeological stone tool analyses? There is currently no agreement on a large number of issues surrounding curation. There is no agreement as to what is curated: tools (Binford 1973), assemblages (Binford 1979; Marks 1988), or technologies (Binford 1979; Slaughter et al. 1992). There is no agreement as to whether curation is a process that occurs at varying rates (Binford 1976; Shott 1989), or an event that either happens or does not. There is no agreement as to how (or if) curation is related to foresight and planning (Draper 1985), subsistence/settlement organization (Binford 1980), mobility (Parry and Kelly 1987; Johnson 1989; Jeske 1992), raw material constraints (Gramly 1980; Kuhn 1991), time stress (Torrence 1983), economic efficiency (Bamforth 1986), style (Sackett 1982), or stone tool use-life and the concept of utility (Shott 1989).

Provocative recent considerations of the concept examined it in terms of utility (Shott 1989) and efficiency (Bamforth 1986). Shott (1989) used a large body of ethnographic (!Kung San) data in a consideration of the effects of tool class use-life and discard processes on archaeological assemblage variability. He attempted to find "elementary properties" of tool classes that are correlated with use-life. Such analysis worked with ceramics, but was less successful within stone tool reductive technologies. To estimate use-life, Shott proposed the concept of curation:

> Curation . . . refers to the practice of maximizing the utility of tools by carrying them between successive settlements. . . .
> Theoretically, curation can be defined as the realized utility of a tool, and curation rate can be measured as the ratio of realized to potential utility. . . . For archaeological assemblages, a first approximation [of utility] is the degree of reduction undergone by items in a tool class, with which it should vary inversely. (Shott 1989:24)

In a strict sense, Binford's original definition of the term had little to do with utility as such. It referred more directly to the general efficiency and reliability of Nunamiut tools (see Bleed 1986). Utility is thus a secondary factor. Shott should be commended for providing the reader with an explicit definition of the term, but one wonders why the discussion mentions curation at all. He notes that "utility is a highly abstract concept" (Shott 1989:24); but if "curation" is measured by "a ratio of realized to potential utility" (ibid.), it remains by definition at least as abstract a concept as utility! At a theoretical level, the idea of realized versus potential utility may be useful for considerations of reduction intensity, but in most archaeological situations, quantification of this ratio is not possible. Even if it were, it is difficult to determine what is behaviorally meaningful in statements like: "asymmetric flake shavers . . . were curated at a rate of 71 percent of potential utility" (Shott 1989:27). This appears to be enumeration for enumeration's sake only.

In a critique of the general concept of curation, Bamforth noted that most archaeologists attempt to explain the term through considerations of efficiency. He concluded that "such explanations ignore the fact that curation is a complex activity and that its component parts are efficient in different ways" (Bamforth 1986:38).

If we assume that human groups attempt 1) to minimize energy expenditure (i.e., increase efficiency) and 2) to maximize utility, then where does curation fit into the explanatory scheme? Assuming for the moment that curation (in whatever form) does in fact constitute efficient behavior which increases the utility of a tool, assemblage, or technology, the archaeologist must demonstrate that the observed variabiility was caused by changes in curation and not by more immediate (primary?) factors such as raw material availability, settlement pattern, subsistence configuration, style, function, or reduction intensity. The meaning of assemblage variability in any context is still not fully understood, and when concepts which do not have generally accepted definitions are used to explain that variability, the situation only becomes more confused.

On a more positive note, important issues regarding changes toward fewer formal tool types in lithic assemblages produced by increasingly sedentary populations have been addressed with the concept of curation. In this case, it is more energetically efficient for sedentary populations to utilize expedient technologies, and for more mobile populations to curate (sic) stone tools (e.g., Jeske 1992). However, to polarize the discussion in terms of curated versus expedient technologies oversimplifies what is likely a multivariate and highly complex phenomenon (Bamforth 1986).

The concept of curation has been modified since Hayden's (1975) critique, though this modification has not been planned, negotiated, or arbitrated. Instead, it has been a passive process deriving from the failure of archaeologists to define their use of a vaguely defined concept whose potential applications in archaeological interpretation are numerous. There has thus been a proliferation of conceptually variable and ill-defined applications of the concept, and no standardization has been achieved. Given that the concept is now embedded in the literature, a standardized lexicon should be negotiated to facilitate some clarity of communication. In the absence of such standardization, we should drop the term from the archaeological literature altogether. The English language retains sufficient descriptive power to convey behavioral interpretations of assemblage variability without use of a diluted concept such as curation.

## ACKNOWLEDGMENTS

I would like to thank Arthur J. Jelinek, Anthony E. Marks, Michael J. Shott, Michael B. Schiffer, George H. Odell, Ronald H. Towner, Edward G. Nash, and an anonymous reviewer for comments that greatly improved this paper.

## NOTES

1. Level 70 is technically considered transitional between Tabun Ea and the overlying Tabun D levels, but was included for comparative purposes.
2. Level 41AN contains broken specimens that lack attributes that differentiate among the flake form categories used in the analysis. These specimens were encoded under the default category, which happened to be the "normal" flake form. Any interpretation of Level 41AN flake form data are therefore likely to be spurious. In the original analysis (Nash 1991), there was sufficient redundancy in the data to mitigate this oversight, and these supporting data are not reproduced here.
3. Magne (1989:16; see also Slaughter et al. 1992:13) argues that archaeologists should focus their analytical efforts on debitage, because "debitage largely escapes curative effects." If debitage has escaped curative effects, one hopes it remains bedridden.

## REFERENCES

Anderson-Gerfaud, P. 1990. Aspects of Behavior in the Middle Palaeolithic: Functional Analysis of Stone Tools from Southwest France. In *The Emergence of Modern Humans*, edited by P. Mellars, pp. 389–418. Edinburgh University Press, Edinburgh.

Bamforth, D. B. 1986. Technological Efficiency and Tool Curation. *American Antiquity* 51:38–50.

Binford, L. R. 1973. Interassemblage Variability: The Mousterian and the "Functional" Argument. In *The Explanation of Culture Change: Models in Prehistory*, edited by C. Renfrew, pp. 227–254. Duckworth, London.

Binford, L. R. 1976. Forty-Seven Trips: A Case Study in the Character of Some Formation Processes of the Archaeological Record. In *Contributions to Anthropology: The Interior Peoples of Northern Alaska*, edited by E. S. Hall, Jr., pp. 299–381. Paper No. 49. National Museum of Canada, Ottawa.

Binford, L. R. 1978. *Nunamiut Ethnoarchaeology*. Academic Press, New York.

Binford, L. R. 1979. Organization and Formation Processes: Looking at Curated Technologies. *Journal of Archaeological Research* 35:255–273.

Binford, L. R.1980. Willow Smoke and Dog's Tails: Hunter-Gatherer Settlement Systems and Archaeological Site Formation. *American Antiquity* 45:4–20.

Bleed, P. J. 1986. The Optimal Design of Hunting Weapons: Maintainability and Reliability. *American Antiquity* 51:737–747.

Chase, P. G., and H. L. Dibble. 1987. Middle Paleolithic Symbolism: A Review of Current Evidence and Interpretations. *Journal of Anthropological Archaeology* 6:263–296.

Dibble, H. L. 1984. Interpreting Typological Variation of Middle Paleolithic Scrapers: Function, Style, or Sequence of Reduction? *Journal of Field Archaeology* 11:431–436.

Dibble, H. L. 1987a. The Interpretation of Middle Paleolithic Scraper Morphology. *American Antiquity* 52:109–117.

Dibble, H. L. 1987b. Reduction Sequences in the Manufacture of Mousterian Implements of France. In *The Pleistocene Old World*, edited by O. Soffer, pp. 33–45. Plenum Press, New York.

Dibble, H. L. 1988. Typological Aspects of Reduction and Intensity of Utilization of Lithic Resources in the French Mousterian. In *The Upper Pleistocene Prehistory of Western Eurasia*, edited by H. Dibble and A. Montet-White, pp. 181–197. University Museum Press, Philadelphia.

Draper, N. 1985. Back to the Drawing Board: A Simplified Approach to Assemblage Variability. *World Archaeology* 17:3–17.

Frison, G. C. 1968. A Functional Analysis of Certain Chipped Stone Tools. *American Antiquity* 33:149–155.

G. & C. Merriam Company. 1973. *Webster's New Collegiate Dictionary*. Springfield, MA.

Gallagher, J. P. 1977. Contemporary Stone Tools in Ethiopia: Implications for Archaeology. *Journal of Field Archaeology* 4:407–414.

Gould, R. A., D. A. Koster, and A. H. L. Sontz. 1971. The Lithic Assemblages of the Western Desert Aborigines of Australia. *American Antiquity* 36:149–169.

Gramly, R. M. 1980. Raw Material Source Areas and "Curated" Tool Assemblages. *American Antiquity* 45:823–833.

Hayden, B. 1975. Curation: Old and New. In *Primitive Art and Technology*, edited by J. S. Raymond, B. Loveseth, C. Arnold, G. Reardon, pp. 47–59. University of Calgary, Alberta, Canada.

Holdaway, S. 1989. Were There Hafted Projectile Points in the Mousterian? *Journal of Field Archaeology* 16:79–85.

Jelinek, A. J. 1975. A Preliminary Report on Some Lower and Middle Paleolithic Industries from Tabun Cave, Mount Carmel, Israel. In *Problems in Prehistory: North Africa and the Levant,* edited by F. Wendorf and A. Marks, pp. 297–316. Southern Methodist University Press, Dallas.

Jelinek, A. J. 1982. The Tabun Cave and Paleolithic Man in the Levant. *Science* 216:1369–1375.

Jelinek, A. J. 1988. Technology, Typology, and Culture in the Middle Paleolithic. In *The Upper Pleistocene Prehistory of Western Eurasia,* edited by H. Dibble and A. Montet-White, pp. 199–212. University Museum Press, Philadelphia.

Jeske, R. J. 1992. Energetic Efficiency and Lithic Technology. *American Antiquity* 57:467–482.

Johnson, J. K. 1989. The Utility of Reduction Trajectory Modeling as a Framework for Regional Analysis. In *Alternative Approaches to Lithic Analysis,* edited by D. Henry and G. Odell, pp. 119–138. Archaeological Papers of the American Anthropological Association, no. 1.

Kuhn, S. L. 1991. "Unpacking" Reduction: Lithic Raw Material Economy in the Mousterian of West-Central Italy. *Journal of Anthropological Archaeology* 10:76–106.

Kuhn, S. L. 1992. Blank Form and Reduction as Determinants of Mousterian Scraper Morphology. *American Antiquity* 57:115–128.

Magne, M. P. R. 1989. Lithic Reduction Stages and Assemblage Formation Processes. In *Experiments in Lithic Technology,* edited by D. S. Amick, and R. P. Maudlin, pp. 15–31. BAR International Series 528, Oxford.

Marks, A. E. 1988. The Curation of Stone Tools during the Upper Pleistocene: A View from the Central Negev. In *Upper Pleistocene Prehistory of Western Eurasia,* edited by H. Dibble and A. Montet-White, pp. 275–286. University Museum Press, Philadelphia.

Munday, F. C. 1976. Intersite Variability in the Mousterian of the Central Negev. In *Prehistory and Paleoenvironments in the Central Negev, Israel, Volume I. The Avdat/Aqev Area,* edited by A. Marks, pp. 113–140. Institute for the Study of Earth and Man, Southern Methodist University, Dallas.

Munday, F. C. 1977. Nahal Aqev (D35): A Stratified, Open-Air Mousterian Occupation in the Avdat/Aqev Area. In *Prehistory and Paleoenvironments in the Central Negev, Israel, Volume II. The Avdat/Aqev Area,* edited by A. Marks, pp. 35–60. Institute for the Study of Earth and Man, Southern Methodist University, Dallas.

Nash, S. E. 1991. *Curation during the Middle Paleolithic: A Reasonable Research Focus?* Master's Thesis, Department of Anthropology, University of Arizona. University Microfilms, Ann Arbor.

O'Connell, J. F. 1987. Alyawara Site Structure and Its Archaeological Implications. *American Antiquity* 52:74–108.

Otte, M. 1992. The Significance of Variability in the European Mousterian. In *The Middle Paleolithic: Adaptation, Behavior, and Variability,* edited by H. L. Dibble and P. Mellars, pp. 45–52. University Museum Press, Philadelphia.

Parry, W. J., and R. L. Kelly. 1987. Expedient Core Technology and Sedentism. In *The Organization of Core Technology,* edited by J. Johnson and C. Morrow, pp. 285–304. Westview Press, Boulder.

Pokotylo, D. L., and C. C. Hanks. 1989. Variability in Curated Technologies: An Ethnoarchaeological Case from the MacKenzie Mountains, Northwest Territories, Canada. In *Experiments in Lithic Technology,* edited by D. Amick and R. Maudlin, pp. 49–66. BAR International Series 528, Oxford.

Sackett, J. R. 1982. Approaches to Style in Lithic Archaeology. *Journal of Anthropological Archaeology* 1:59–112.

Shea, J. J. 1988. Spear Points from the Middle Paleolithic of the Levant. *Journal of Field Archaeology* 15:441–456.

Shott, M. J. 1989. On Tool-Class Use Lives and the Formation of Archaeological Assemblages. *American Antiquity* 54:9–30.

Slaughter, M. C., L. Fratt, K. Anderson, and R. V. N. Ahlstrom. 1992. *Making and Using Stone Artifacts: A Context for Evaluating Lithic Sites in Arizona.* SWCA, Inc., Environmental Consultants, Tucson.

Solecki, R. L. 1992. More on Hafted Projectile Points in the Mousterian. *Journal of Field Archaeology* 19:207–212.

Torrence, R. 1983. Time Budgeting and Hunter-Gatherer Technology. In *Hunter-Gatherer Economy in Prehistory: A European Perspective,* edited by G. Bailey, pp. 11–22. Cambridge University Press, Cambridge.

Toth, N. 1985. The Oldowan Reassessed: A Close Look at Early Stone Artifacts. *Journal of Archaeological Science* 12:101–120.

Chapter 4

# Hunter-Gatherer Lithic Economy and Settlement Systems

## Understanding Regional Assemblage Variability in the Upper Paleolithic of Portuguese Estremadura

PAUL T. THACKER


## ABSTRACT

Intensive archaeological survey and excavation conducted in the Rio Maior region of Portuguese Estremadura has yielded a large sample of Upper Paleolithic sites. Chipped stone assemblages from technologically and typologically distinct Gravettian and Magdalenian periods exhibit both temporal and spatial variability. Valley landforms have changed little since the Early Pleistocene, which allows comparison of prehistoric land use patterns. Field survey coupled with analogy to historic gunflint workers in the nearby town of Azinheira provides insight into Late Pleistocene flint source distributions and procurement activities. Knowing these background variables is advantageous for understanding the complex dynamic between raw material selection, lithic reduction trajectories, and settlement systems. A combination of both curated and expedient technological strategies is evident in the Upper Paleolithic assemblages. By analyzing degree of planning in


**PAUL T. THACKER** • Department of Anthropology, Southern Methodist University, Dallas, Texas 75275.

*Stone Tools: Theoretical Insights into Human Prehistory*, edited by George H. Odell. Plenum Press, New York, 1996.

raw material choices and assessing economizing behavior in the reduction sequence, assem-blages are able to be interpreted as an integrated whole. This study demonstrates the importance of regional analysis not only for contributing to middle range theory, but as a method for interpreting diachronic lithic data in behaviorally significant ways.

## INTRODUCTION

Lithic assemblages result from a complex series of interrelationships involving technology, raw material selection, lithic economy, site function, and settlement/subsistence systems. Archaeologists must analyze the entire set of relationships if meaningful data concerning human behavior is to be obtained (Odell 1989; Bamforth 1990). To some degree, depth of understanding in lithic studies is correlated with the ability to identify, and thus control, one or more variables across the prehistoric landscape and in the cultural system (Perles 1992).

This paper is a first attempt at understanding diachronic lithic variability in the Upper Paleolithic of Portuguese Estremadura. The Rio Maior sample is based on intensive regional survey and excavation, with close regard to geological and archaeological site formation processes (Schiffer 1987). Flint source distributions are known, and both intersite patterns and assemblage samples are large and representative of prehistoric variability. Among Upper Paleolithic regional counterparts in southwestern Europe, the Rio Maior area is unusually appropriate for lithic analysis and settlement system studies.

Any large lithic data base consists of multiple questions, to be untangled in order. The progression of analysis presented here reflects the settlement and land use bias that a survey project cannot help but bestow on its field workers. Yet the interconnectedness of each problem will become evident, as most assemblage observations (i.e., figures in this paper) are relevant to several questions. The primary goal of this paper is to illustrate the explanatory power gained from a regional approach, coupled with analysis at all levels of lithic production (raw material procurement to discard).

## THE UPPER PALEOLITHIC OF THE RIO MAIOR REGION: LAND USE AND ASSEMBLAGE VARIABILITY

Three drainages surrounding the town of Rio Maior contain the greatest known density of open-air Upper Paleolithic sites in Portuguese Estremadura. This clustering is somewhat a function of intensity of survey, but work in other valleys has failed to locate comparable densities of sites (Marks et al. 1994). The Upper Rio Maior, Ribeira da Pa, and Penegral Valleys are shown in Figure 1.

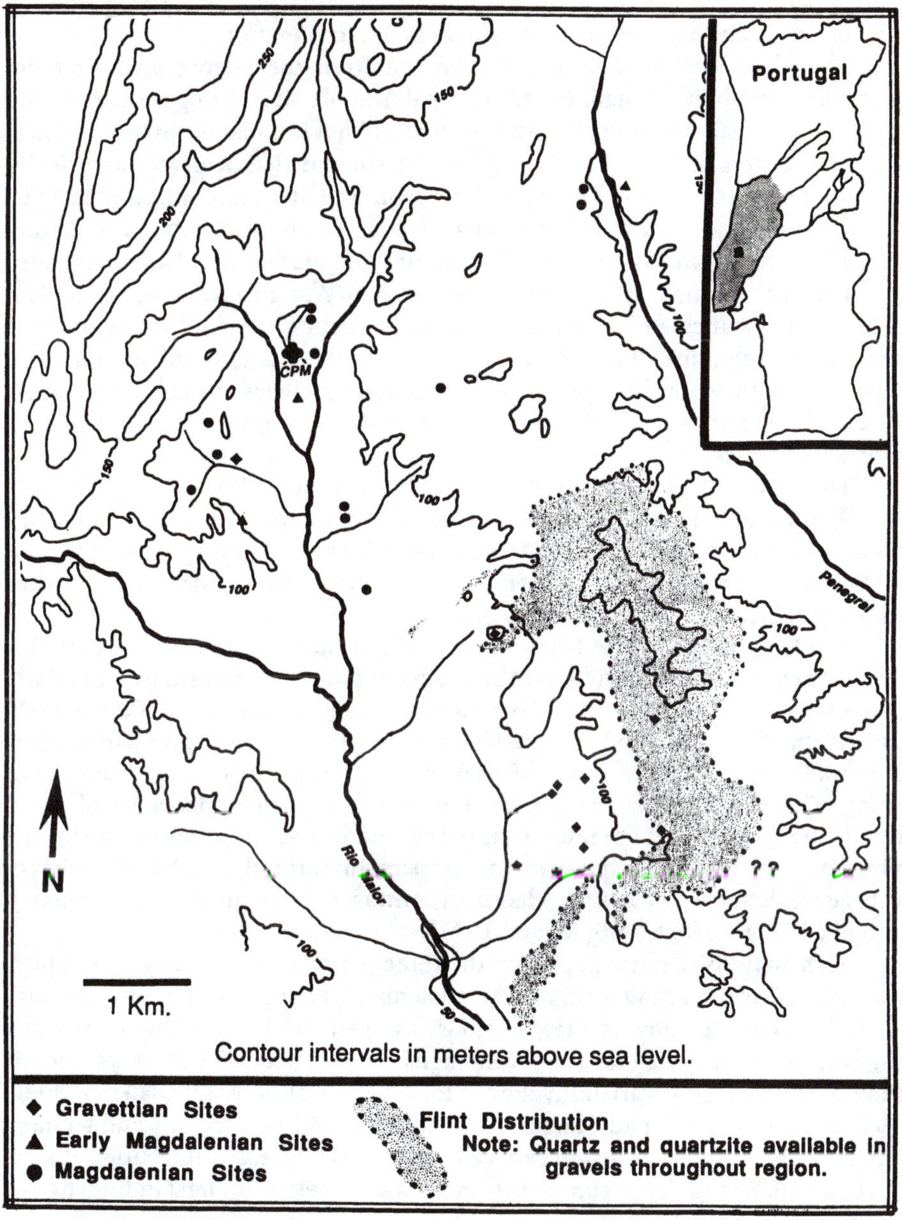


Figure 1. Site locations and flint source in Rio Maior vicinity.

Known sites are plotted by period. The site complex of Cabeco de Porto Marinho (CPM) consists of at least six stratified and dated loci containing over 20 components. A sample of two Gravettian, two Early Magdalenian, and six Magdalenian assemblages from the CPM complex are included in this study.

Only four Aurignacian sites are known from the region, and inter-site assemblage variability is high. Given the small sample size, dating problems, and high technological variability, the Aurignacian will not be included in this analysis.

The earliest Gravettian of Portuguese Estremadura has been radiometrically dated to just after 23,000 b.p. by radiocarbon and thermoluminescence (TL) techniques (Zilhao 1991 n.d.; Marks et al. 1994). The technocomplex is characterized by a fine blade and bladelet technique. The blades from Gravettian sites are narrower, thinner, and usually shorter than Aurignacian blades. Backed bladelets and truncated backed bladelets are common. Several different blade production techniques have been identified by Marks and Zilhao. Platform preparation differences have resulted in varying frequencies of thicker and larger blades. The significance of this observation may be chronological, but further work is in progress.

The Solutrean is well documented in Estremadura (Zilhao 1987, 1988 a,b, 1990a), but only one Solutrean site was found in the survey region. Future work will undoubtedly reveal more Solutrean occupations in the valleys, but it is clear that Solutrean groups either did not intensively inhabit the Rio Maior area, or they did not produce significant numbers of open-air sites.

The Magdalenian period contains the largest number of known sites in this region. The period can be typologically and technologically divided into the Early Magdalenian (ca. 16,000 to 15,000 b.p.) and the Magdalenian (ca. 12,000 to 9,000 b.p.) (Bicho 1992a,b). The Early Magdalenian assemblages contain relatively few blade tools, and a small number of backed microliths. Endscrapers outnumber burins. The Magdalenian (ca. 12,000 b.p.) witnesses the appearance of a microburin technique and common thin-nosed and ogival endscrapers. Burins are only about one-third as frequent as endscrapers. Importantly, the Magdalenian of Portuguese Estremadura has few blades, in contrast to other areas of southwestern Europe (Weniger 1989,1990; Rigaud 1991).

This analysis will use data from the three periods with the largest samples: Gravettian, Early Magdalenian, and Magdalenian. Technological continuity cannot be assumed for any of these periods, as even the Early Magdalenian and Magdalenian are separated by a chronological gap of about 3,000 years. Bicho's study of technological variability argues for a degree of Early Magdalenian/Magdalenian continuity, but additional sites and dates will be needed to understand fully the diachronic relationship between these assemblages. Connection between the Gravettian and later complexes must await better understanding of the Solutrean in Portugal.

In short, the chronostratigraphic framework for the Upper Paleolithic of Estremadura is still being constructed. The work presented here utilizes preliminary patterns derived from the Rio Maior region, as the relationship between trends in central Estremadura and the rest of Portugal is less clear. One benefit of this ambiguity is that analysis has proceeded with little theoretical or methodological biases from past Portuguese work. Furthermore, research was structured with regard to Conkey's (1987: 75) caveat against using "the implications generated by one regional analysis as starting points for another study."

## SURVEY METHODOLOGY AND REPRESENTATIVE SAMPLES

A primary obstacle in Upper Paleolithic regional analyses is locating an unbiased sample of Late Pleistocene sites (White 1983, 1985, 1987; Rigaud and Simek 1987; Wobst 1990). Field methodology during the 1991 and 1992 seasons was aimed at recovering a large sample, while understanding geological visibility.

Preliminary reconnaissance indicated that the region was geologically appropriate for systematic survey of Pleistocene landforms. Total surface survey of the Upper Rio Maior and Ribeira da Pa drainages was completed, with three 1 kilometer-wide transects surveyed in Penegral. Artifact concentrations were located, distributions mapped, and test units were excavated in both high and low density areas of each site. *In situ* assemblages from the ongoing excavations of Marks and Zilhao's National Science Foundation project were examined and sampled, and site locations were visited.

Geologic visibility and destruction of sites must be included in an assessment of the significance of site patterning. Within the valley, the only region where no Pleistocene age sites were found was the extreme northern end of the Ribeira da Pa. Here the U-shaped valley is a continuous sloping surface. Sites were either not present, deeply covered by slope wash and colluvium, or washed away. Therefore, the absence of sites may not be behaviorally significant. In contrast, the mid-valleys have suffered only local colluvial events or erosion. While a few Gravettian sites were found on the surface, most Early Upper Paleolithic levels occur as deep as two meters below the modern surface. Resulting midvalley settlement patterns are biased toward Late Upper Paleolithic visibility. This favoritism does not detract from the representativeness of the Early Upper Paleolithic sample. To date, Gravettian sites occur on well over half of the appropriately-aged landforms that have been exposed in the mid-valley. The number of sites should not be compared, but the patterning can be.

The lower valley has a much wider floodplain, and Early Pleistocene terraces have been continuously exposed since the start of the Upper Paleolithic. Archaeological visibility during survey was influenced only by local erosional or depositional events, and historical land use affecting ground cover. Site assemblages and distributions should be representative and relatively unbiased.

Roughly 80 percent of valley surfaces have been plowed, about 45 percent deep plowed. The plowing was destructive of intra-site patterning, but aided site discovery in several instances. Often deep plowing exposed artifacts, but left a meter-wide strip *in situ* between furrows.

Controlling for Late Pleistocene visibility, site distributions in Figure 1 can be evaluated. Gravettian sites cluster in the lower section of the valley. This pattern is biased by field visibility. Some Gravettian sites have been found in the mid-valley but in lower frequencies, since the levels are often more deeply buried than their Magdalenian counterparts. Taking visibility bias into account, Gravettian occupation appears to have been intensive and valley-wide.

Site location in the Magdalenian differs from the earlier Gravettian. Sites are situated exclusively in the mid-valley. This clustering is significant, since lower valley geologic visibility is higher than mid-valley. Also, the lower valley has been plowed to a greater extent than the upper or mid-valley, a situation that has exposed sites. The pattern can be interpreted as evidence for paucity of settlement. Magdalenian groups did not produce archaeologically visible occupation areas in the lower valleys.

Corresponding with settlement location shifts, the prevalence of flint as a raw material changed through time (Marks et al. 1991). Quartz and quartzite cobbles are locally available in gravel deposits throughout the valleys, and were used throughout the Upper Paleolithic. A high quality red flint occurs as redeposited cobbles in coarse sands mainly on the ridge separating the lower Rio Maior and Penegral drainages (see Figure 1 for distribution of flint source). Virtually all of the flint from Upper Paleolithic assemblages in the area was obtained from this deposit.

## FLINT SOURCES OF THE UPPER RIO MAIOR

Flint has been an important natural resource for inhabitants of the Rio Maior region throughout history. Tools chipped from the distinctive flint occur continuously from the Middle Paleolithic until 1921, when the last gunflints made in the town of Azinheira were sold (Do Paco 1959; Martins and Pereira 1992). Initial studies of the flint source indicate fluviatile redeposition of Jurassic flint cobbles in coarse sands, perhaps as late as the Early Pleistocene. Location of the parent source of cobbles has not yet been determined.

There is no prehistoric evidence of flint quarrying near Rio Maior. However, the historic record, coupled with ethnohistorical interviews conducted in 1992 in Azinheira, provides details on procurement of the raw material. Gunflint workers gathered cobbles for reduction in two ways, depending on where their property was located relative to the ridge. Persons with land on top of the ridge or away from springs dug pits into the sands, roughly one to one-and-a-half meters square, and up to five meters deep. When historically known extraction pits were revisited in 1992, they were barely evident on the surface. If such techniques were used prehistorically, there is no evidence remaining today.

The second method of flint acquisition is the better analogy for Paleolithic procurement. The Upper Rio Maior Valley has a perched water table. Springs high on the ridge between the Penegral and Rio Maior Rivers flowed during the wet months (winter), cutting into the coarse sand deposits. Flint cobbles were exposed during these erosional events and collected at the base of the narrow channels. This process certainly occurred in the Late Pleistocene, and only stopped when modern mining and eucalyptus forestry caused a recession of the water table. Thus it is not necessary to claim that flint was mined prehistorically, but rather that cobbles were obtained from the sides of these channels.

The flint distribution in Figure 1 is based on observations taken during survey. At any one point in the past, flint would have been visible on the surface in stream and spring channels in the shaded area. Predicting exact locations of exposed deposits is futile and unnecessary, given the geology of the flint deposit. The flint distribution is significant spatially, since it has a relatively distinct margin. Measurements of distance from a site to the nearest edge of the flint source may not be accurate in absolute terms for the Upper Paleolithic, but should be valid for relative comparisons. Given the lack of large-scale geologic alteration of the landscape since the Early Pleistocene, Figure 1 approximates the situation that faced Late Pleistocene hunter-gatherers.

## RAW MATERIAL PROCUREMENT AND SETTLEMENT SYSTEMS

Explaining the settlement pattern differences between Gravettian and Magdalenian periods begins with an understanding of the role of flint within each assemblage. Initial study of cores by Marks et al. (1991) demonstrated an increasing preference for flint throughout the Upper Paleolithic, culminating in the Magdalenian. Yet, survey results in 1991 and 1992 indicated that no Magdalenian sites were located within a kilometer of the flint, while Gravettian sites were frequently found on or very near the source. To explore this pattern, assemblage raw material frequencies were tabulated. Assemblage frequencies may be relatively reliable as an indicator of raw material use, especially if local quartz and quartzite were reduced in a technological strategy distinct from flint. For example, quartzite in the Late Upper Paleolithic of La Riera, Spain, was preferentially used for the production of heavier "archaic" tools, while flint was used for the finer Upper Paleolithic tools (Straus and Clark 1986; Straus 1991b). Preliminary sorting of collections from Rio Maior indicates a high frequency of heavy-duty scrapers and choppers of quartzite. Production of these expedient types of tools may inflate core frequencies, as the "core" was the desired tool.

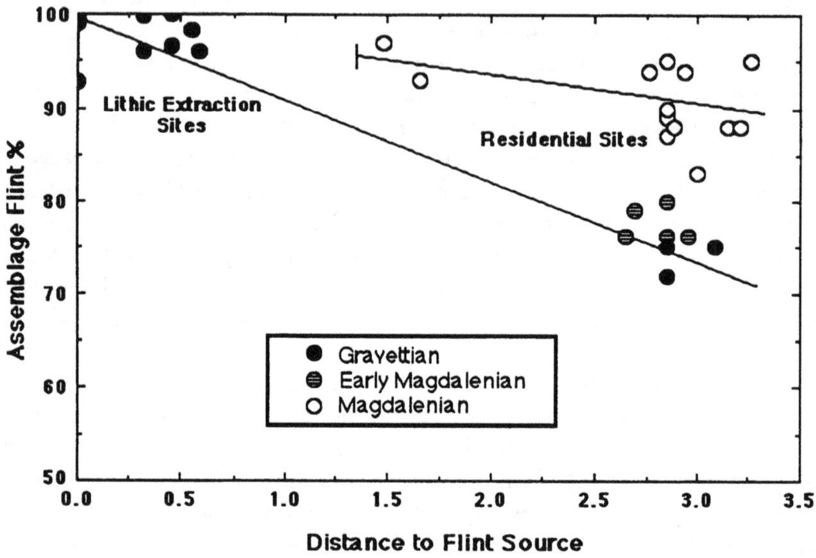


Figure 2. Assemblage flint composition vs. distance from flint source.

Figure 2 plots the relative frequency of flint within assemblages against the distance from the sites to the flint source. Several patterns emerge, which have been displayed with regression slopes through the scatterplots. Gravettian sites cluster near the flint source, as was evident visually in Figure 1. As Gravettian groups moved away from the source, the frequency of flint they used dropped off rapidly. Sites 2.5 to 3 kilometers from the source contain only about 75% flint, with quartz comprising the bulk of the other raw material. This distance–decay relationship is linear, not log-linear, and has an r-squared value of .916. Additionally, the gap between Gravettian sites in the upper left of Figure 2 and those in the lower right is not a function of site visibility.

Early Magdalenian sites are not found near the flint, and contain frequencies of flint only slightly higher than the Gravettian sites in the mid-valley. Quartz and quartzite account for the remainder of the assemblage. The Magdalenian sites are all found some distance from the flint source, but all show a high frequency of flint. The distance–decay rate for the Magdalenian is substantially lower than for the Gravettian.

Overall, the assemblage raw material data suggest that more flint is moving farther during the Magdalenian. However, numerous variables contribute to the pattern in Figure 2. Critical issues for assessing behavioral meaning within these data include differences in technology within and between periods, the technological role of quartz and quartzite, site functional differences, site reduction trajectory lengths, and procurement differences.

## UPPER PALEOLITHIC LITHIC TECHNOLOGY

Core typology has been successfully used to reveal technological differences between and within periods (Johnson and Morrow 1987; Johnson 1989). Figure 3 is a histogram of the pooled (or global) averages of flint core classes for Gravettian sites near the flint source, Gravettian sites away from the flint source, Early Magdalenian sites, and Magdalenian sites. Quartz and quartzite cores were excluded from this analysis. Cores were classified on the basis of scar dimensions into blade, bladelet, flake, and mixed cores. Bicho's (1992b, n.d.) detailed technological study of the Early Magdalenian and Magdalenian reduction sequences showed that mixed cores frequently occurred during the early stages of bladelet core preparation, as flakes were detached for shaping and setting up the core. Magdalenian periods did not demonstrate significant variability across space on the scale discussed here (< 5 km from source).

The primary diachronic pattern evident is the greater frequency of blade cores in the Gravettian, an expected pattern since blade technology should, by definition, have been most prevalent in that period. Within the Gravettian, the frequency of blade cores decreases in sites located further from the flint source. Almost all of this decreased proportion is compensated for by an increase in bladelet cores. This pattern confirms Marks et al.'s (1994:61) proposal that for the Portuguese Gravettian "there is no true distinction between the 'blades' and the 'bladelets,' in terms of size range."

Figure 4 charts the relative frequency of blades, bladelets, and cortical flakes (having >5% cortex on the dorsal surface, excluding platform) within the total unretouched flint assemblages. Again, blades are more frequent in the Gravettian than at any point in the Magdalenian. Unretouched blades are most prevalent in Gravettian sites greater than 1 km from the flint source. This observation correlates

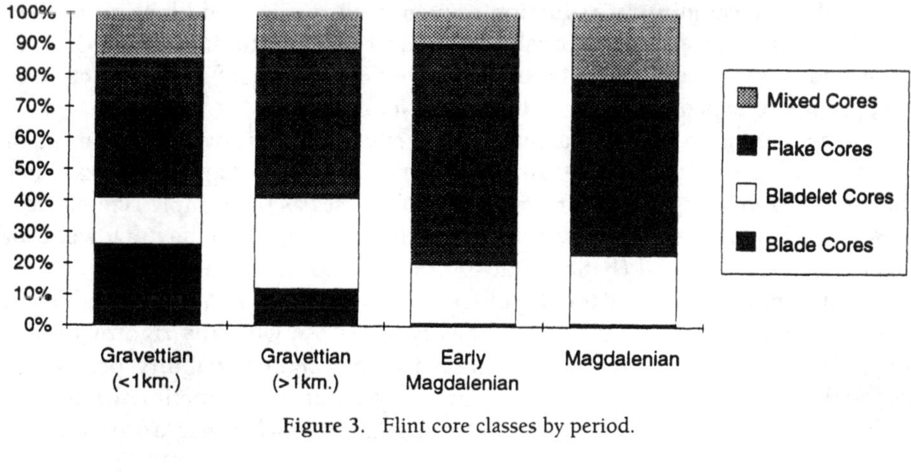


Figure 3. Flint core classes by period.

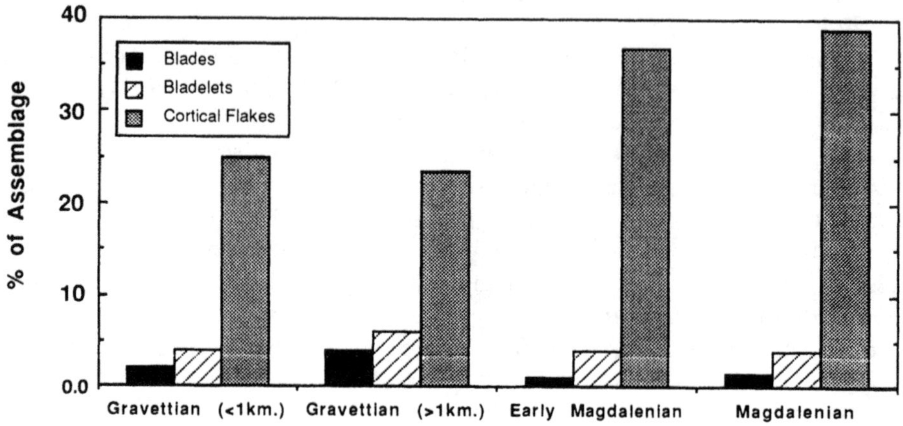


**Figure 4.** Relative frequencies of flint unretouched blades, bladelets, and cortical flakes by period.

with the reduction of blade cores into bladelet cores (Figure 3). Taken together, the Gravettian technological pattern indicates a fairly intensive utilization of flint blade cores. These cores were exhausted rapidly and disappeared from the lithic assemblage (were transformed into bladelet cores) a few kilometers from the raw material source.

If all of the Gravettian sites in the survey region were equivalent residential bases, the pattern in blade/bladelet production could only be one of differential exhaustion and discard judgements. This scenario expects groups near the flint to discard larger cores because of the abundance of high quality cobbles. Groups further away would have reduced fewer cores more intensively. But if the sites near the flint are special purpose extraction sites, as are common in the archaeological record of the local gunflint reduction sequence, then prepared blade cores may have been transported from workshop sites near the flint to multi-purpose sites in the midvalleys. The pattern of production would then be a reflection of a curated (as opposed to expedient) flint technology (Nelson 1991).

One popular method for distinguishing site function is diversity analysis of tool assemblages (Kintigh 1984; Simek and Price 1990). This approach examines the distribution of tools into a number of tool classes as the sample size of tools increases. Generally, sites with specific functions are assumed to have a lower tool diversity than residential (hence multi-purpose) areas.

The Upper Paleolithic tool typology employed by Marks and Zilhao derives from the traditional DeSonneville-Bordes type list. As with any rigorous typology, Portuguese type classes are designed to highlight variability present in regional assemblages. The typology may or may not have functional meaning with regard to actual tool use. Therefore, the large typelist was unsuitable for

**Table 1. Tool classes of the Portuguese Upper Paleolithic**

| |
|---|
| Simple Endscrapers |
| Endscrapers with Continuous Retouch around Perimeter ("on Flake") |
| Carinated Pieces |
| Flat-Nosed and Shouldered Endscrapers |
| Notches |
| Denticulates |
| Truncations |
| Perforators |
| Burin on Truncation (including Bec de Perroquet and Transverse Burin) |
| Dihedral and Busked Burins |
| Burin on Snap |
| Points |
| Backed Pieces (except Bladelets) |
| Retouched Pieces |
| Side Scrapers |
| Retouched Bladelets |
| Backed Bladelets |
| Microgravettes and Pointed Backed Bladelets |
| Multiple Mixed Burins |
| Geometrics |
| Multiple Tools |
| Splintered and Battered Pieces |
| Varia |
| Rabot and Heavy Duty Scrapers |

analysis, since diversity measures are only significant if tool classes are functionally discrete. For this study, approximately 105 types were condensed into the 24 classes of Table 1. These classes are based on gross edge shape, which should correlate with function more closely than divisions based on side of artifact retouched, shape of truncation, etc. An underlying assumption in this classification is that tool classes are mutually exclusive, and that tools are not transformed through several classes during their use lives. Initial examination of class patterning across space indicates that little reworking into new tools took place. More often a new tool was fashioned on another edge, causing it to become a multiple tool. However, overall effects of reworking on class distributions appear to be minimal.

Plots of Gravettian and Magdalenian assemblage diversity are shown in Figures 5 and 6, respectively. The Gravettian sample is small (6 sites), because of incomplete typological analysis of many of the collections. Gravettian sites near the flint source show less diversity than the sites further away, and the sites greater than a kilometer from the source show more diversity in small assemblages. These observations, while preliminary, indicate important assemblage formation differences.

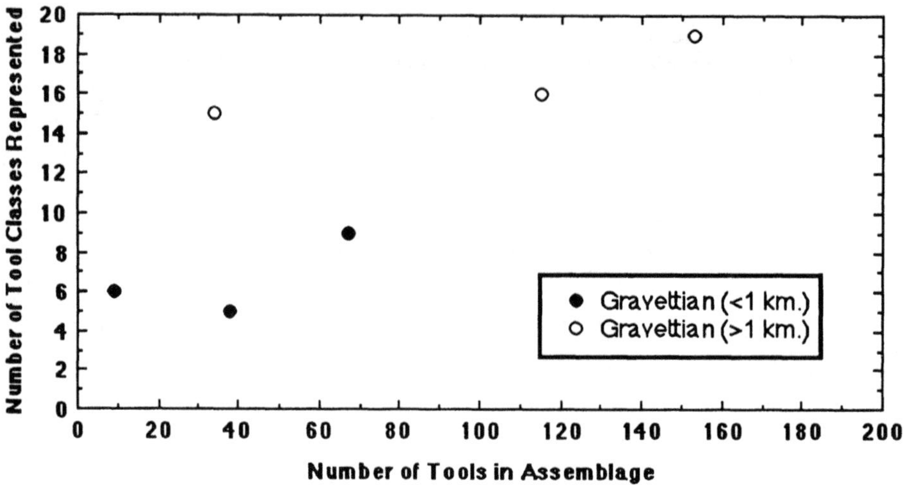


Figure 5. Gravettian tool diversity.

Magdalenian sites (Figure 6) show a classic diversity curve. Assemblage diversity varies with sample size, with no significant outliers. Interpretation of the sites as multi-purpose residential bases would fit this pattern. Coincidentally, Bordes' "100 tool rule" shows surprising insight, as both periods peak in diversity close to a 100-tool sample size.

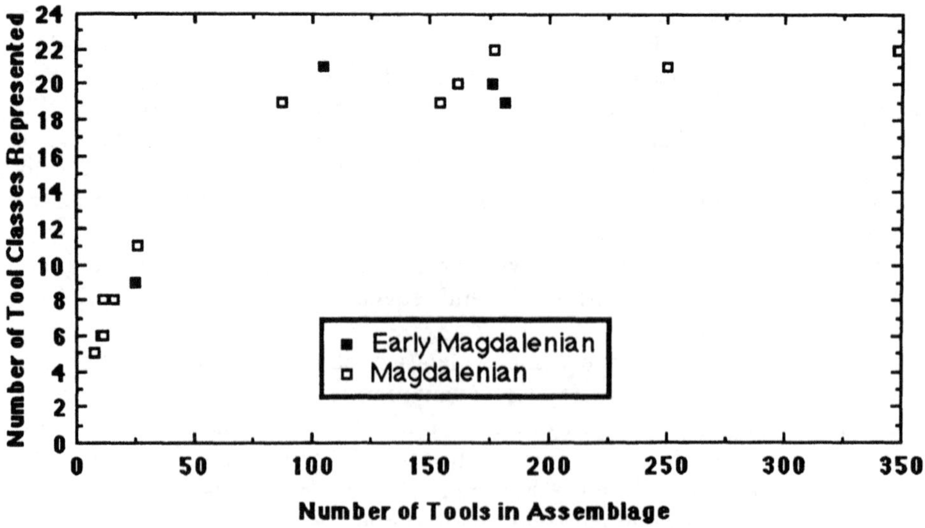


Figure 6. Early Magdalenian and Magdalenian tool diversity.

Plots of Gravettian flint tool and core frequencies (Figures 7 and 8) against total flint assemblage size strengthen a differential function hypothesis. Both retouched tools and cores occur much less frequently in assemblages close to flint deposits than away from them, once assemblage size (and, indirectly, completeness of excavation) is controlled. This pattern conforms to that expected if the sites were primary lithic reduction workshops.

The final evidence for functional site differences in the Gravettian is shown in Table 2. Tool classes from Table 1 that occur on bladelet blanks were summed and compared with the total tool collections. Less than 2 percent of tools found at sites within one kilometer of the flint were fashioned on bladelets, whereas 17 percent of tools at the midvalley Gravettian sites were bladelet tools. This observation concurs with the diversity pattern, indicating more generalized activities at sites more than a kilometer from the flint source.

Absolute site size is unknown for most locations, but preliminary estimates show little variation in site area for the midvalley Gravettian, Early Magdalenian, and Magdalenian sites. The Gravettian primary reduction sites in the lower valley tend to be larger in area, possibly reflecting continuous reoccupation of the location for flintworking.

In conclusion, tool class diversity, tool and core occurrence frequencies, and bladelet tool frequencies suggest a Gravettian settlement system involving primary lithic reduction sites close to the flint source, and residential sites in the midvalleys. Magdalenian assemblages are more generalized, as sample size accounts for most typological diversity.

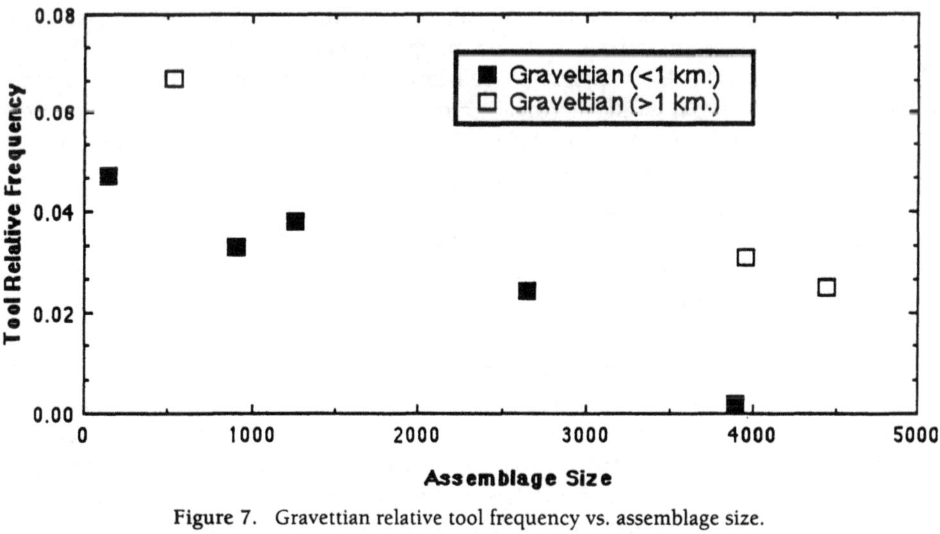


Figure 7. Gravettian relative tool frequency vs. assemblage size.

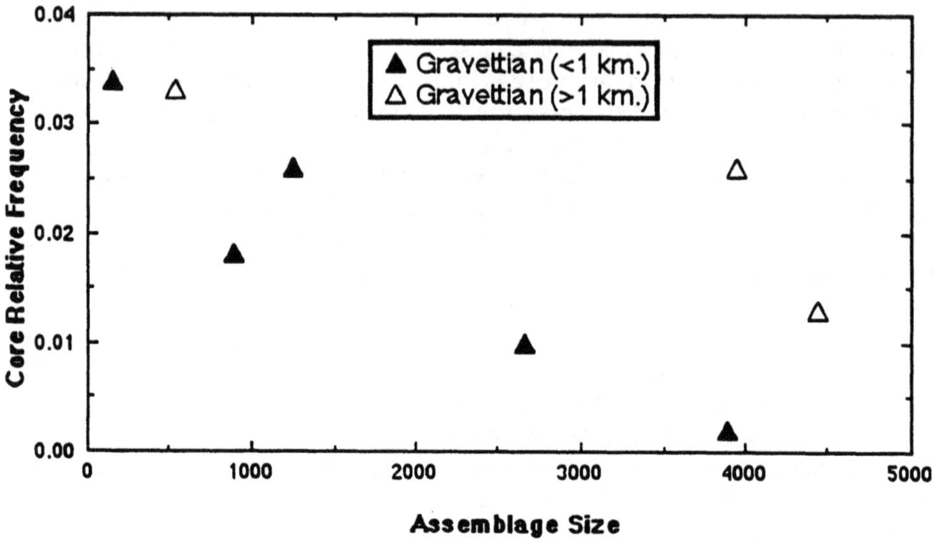


Figure 8. Gravettian relative core frequency vs. assemblage size.

## QUARTZ AND QUARTZITE AS RAW MATERIAL CHOICES

Quartz and quartzite lithic technology varies between periods. Figures 9 and 10 graph the proportions of cortical flakes,blades/bladelets, and tools/cores for these local raw materials. Comparison of these distributions with flint patterns (Figure 4) helps understand reduction strategy differences.

An expedient use of quartz or quartzite within the context of the Upper Paleolithic of Portugal would involve the removal and use of flakes from cobbles that are obtained locally. Very few or no tools/cores in the raw material would be transported from the site of production. An assemblage from an expedient technology is expected to contain a high frequency of cortical flakes, a relatively high tool and core frequency, and low frequencies of unretouched, non-cortical blades and bladelets. At this level of assemblage analysis, Nelson's (1991) terms "expedient" and "opportunistic" cannot be distinguished.

Gravettian quartz and quartzite use within a kilometer of the flint matches expected patterns of expedient reduction. Yet quartz had a more important role as distance increased from the flint source. While quartzite remained an expedient raw material, quartz was reduced by the same techniques as flint. The quartz assemblage contains relatively large quantities of unretouched blades and bladelets, and has fewer cortical flakes. Figure 9 implies that the distance–decay model hypothesized from Figure 2 is accurate. As Gravettian hunter-gatherers

**Table 2. Frequencies and raw materials of bladelet tools**

| | Relative frequency of bladelet tools (among tools) | % of Bladelet tools on flint bladelets |
|---|---|---|
| Gravettian (< 1 km) | < .02 | 100 |
| Gravettian (> 1 km) | .171 | 98 |
| Early Magdalenian | .116 | 97 |
| Magdalenian | .152 | 99.5 |

moved away from the flint, quartz replaced flint as flint blade and flake cores were exhausted. However, flint was still desired for bladelet tool production, as will be demonstrated later.

Quartzite was worked extensively, but expediently, at Early Magdalenian residential sites (Bicho 1992b, n.d.), with large flakes being the apparent desired blank. Bicho's work detailed variation in quartz and quartzite core forms throughout the Magdalenian. The main difference between Magdalenian flint reduction and that of other raw materials lies in bladelet technology. Bladelets and bladelet tools occurred almost exclusively on flint during the Magdalenian (Table 2).

Diachronically, reduction of quartz was most frequent in the Gravettian, as a locally available alternative to flint. Quartzite occurred most significantly in the

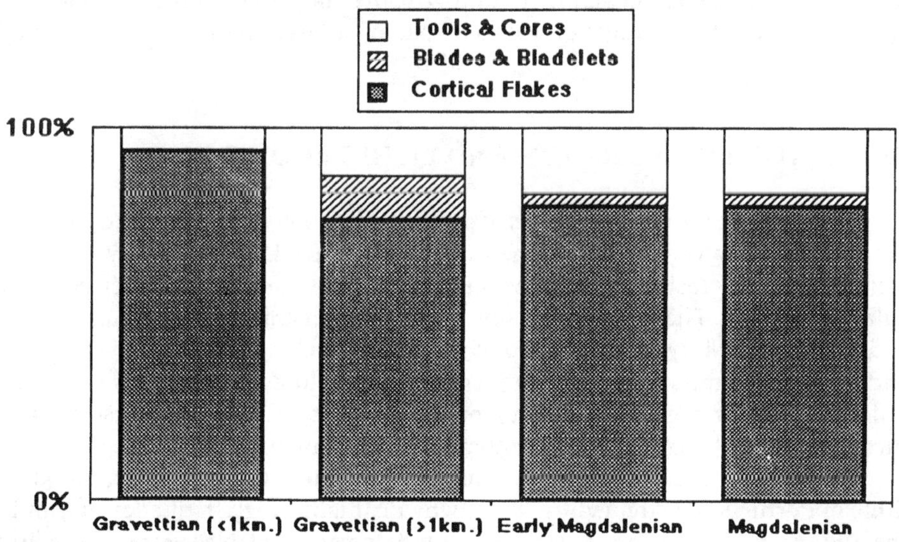


**Figure 9.** Quartz assemblage ratios by period for tools and cores, blades and bladelets, and cortical flakes.

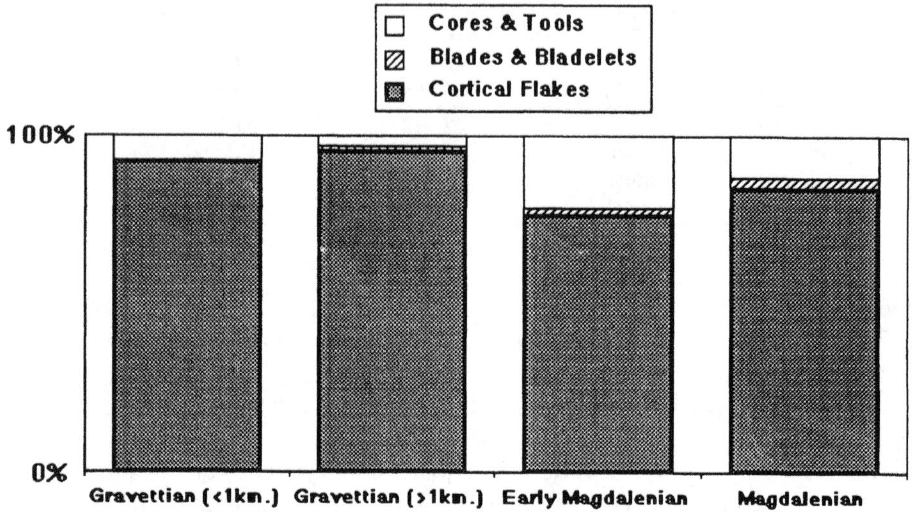


**Figure 10.** Quartzite assemblage ratios by period for tools and cores, blades and bladelets, and cortical flakes.

Early Magdalenian, as an expedient source of large tools and flake blanks. Magdalenian sites contain very small frequencies of both quartz and quartzite, but were reduced in a manner similar to flint. The overall pattern of the Magdalenian is of a very high flint frequency, without spatial variation on this micro-regional scale.

## RAW MATERIAL SELECTION AND LITHIC ECONOMY

Correlating the above patterns of raw material choices and site function does not explain human adaptation in the Upper Paleolithic. Lithic economy, mobility, technology, and subsistence systems are interrelated segments of hunter-gatherer cultural systems. The first part of this article demonstrated that the diachronic and spatial variability in lithic assemblages of the Rio Maior area were significant and not idiosyncratic. The following section will explore the consequences of the differing lithic strategies for human groups, as well as attempt to understand why the technology changed from Gravettian to Magdalenian periods.

Three questions stem from raw material variability during the Upper Paleolithic of Portuguese Estremadura. First, why maintain a high frequency of flint in the lithic reduction process if other raw materials were available? Second, was flint more intensely economized in the Magdalenian? Third, how was flint transported to sites located some distance from the raw material source? Answers to these

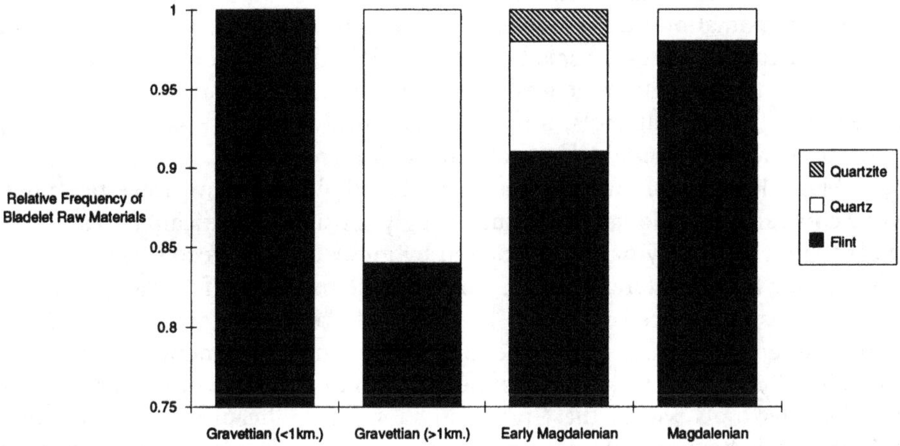


**Figure 11.** Relative frequencies of bladelet raw materials.

questions shed light on the explanation of the lithic component of the adaptive system.

The flint found in the source area is of high quality, despite being somewhat brittle for knapping purposes (J. Tixier, personal communication 1991). Studies of Upper Paleolithic assemblages in Spain and Germany, as well as Paleoindian assemblages in the New World, indicate that the finer fracture qualities of flint make it preferred over quartzite and other rock types for the manufacture of bladelets and other fine tools (Goodyear 1989; Straus 1991b). This hypothesis was tested by pooling all unretouched bladelets by period and recording raw material proportions (Figure 11). Gravettian sites have the lowest flint bladelet percentage, with increased proportions of flint in Early Magdalenian and Magdalenian levels. Table 2 shows that bladelet tools were almost always fashioned on flint, regardless of period. The significance of this pattern lies in the technological flexibility of flint. Technologies, such as the Rio Maior Magdalenian, that utilize a microburin technique to manufacture geometrics, required a high quality raw material. In this situation, raw material choice and technology are not independent.

The high frequency of flint bladelet tools in the Gravettian was also caused by this relationship. Quartz was a substitute for flint for most other parts of the lithic system, but not for bladelet production. Quartz bladelets were either desired "as is," meaning that they were used without retouching, or they were debitage produced during the manufacture of large core-scrapers or carinated scrapers. This observation implies that as distance from flint source increased in the Gravettian, assemblage flint frequencies should have continued to decay, except for flint bladelet tool frequencies. Unfortunately, this prediction must await further excavations.

Flint economy and intensity of use are difficult variables to measure. A simple approximation can be reached using weight comparisons, when distance to source area differences are small. Heavier tools and cores should indicate either a less efficient technology or less intensive use. Table 3 compiles the pooled average weights of flint tools, flint cores, and flint cortical flakes. Several very heavy outliers were excluded from the Gravettian core sample.

Magdalenian tools and cores are significantly lighter than Gravettian ones, while cortical flake weights remain surprisingly constant. Significant tool and core weight differences may have been caused by intensity of use/economy of reduction, or may reflect different starting points in the trajectory (for example, if raw material was introduced to a site in different sizes). The drop in weight of Gravettian cores between sites close to the flint source (94 grams) and those in the mid-valley (39 grams) supports the scenario of transport of prepared cores. Non-bladelet tools within the Gravettian show insignificant weight variability, indicating production and use-lives of tools on curated cores similar to those closer to the flint source. Magdalenian cores are lighter (46 and 31 grams) than their Gravettian primary reduction site counterparts, because they were either produced on smaller cobbles or were discarded after more intense use.

Cortical flake weights and frequencies can narrow these possibilities. Flint cortical flakes are more prevalent in the Magdalenian assemblages than in either set of Gravettian sites (Figure 4). Combine this observation with equivalent cortical flake weights for all periods, and the postulate that flintknappers commenced with different sizes of flint cobbles becomes unlikely. If Magdalenian groups were procuring smaller cobbles, then either cortical flake frequencies or weights would have decreased. These data suggest a longer reduction trajectory at the Magdalenian sites than at either of the Gravettian clusters. Flint cobbles were introduced to midvalley Magdalenian sites without preparation. Lighter tool weights for the Magdalenian confirm either that reduction was more intense during the Magdalenian, or that technological differences resulted in more efficient economy of raw material.

**Table 3. Average Flint, Tool, Core, and Cortical Flake Weights by Period**

| | Flint tools (excluding bladelet tools) | | Flint cores | | Flint cortical flakes | |
|---|---|---|---|---|---|---|
| Period | Mean weight | Sample size | Mean weight | Sample size | Mean weight | Sample size |
| Gravettian (< 1 km) | 13.9 | 137 | 94.1 | 102 | 10.1 | 826 |
| Gravettian (> 1 km) | 14.1 | 35 | 38.5 | 17 | 11.4 | 159 |
| Early Magdalenian | 10.3 | 27 | 45.7 | 180 | 11.7 | 67 |
| Magdalenian | 11.6 | 73 | 30.9 | 478 | 9.6 | 527 |

## CONCLUSIONS: PLANNING AND CONSERVATION OF RAW MATERIAL

Despite the almost ideal situation of a known flint source, representative site locations, and large assemblage samples, the Rio Maior data set demonstrates the complexity of lithic technological organization and its interface with settlement systems. The conclusions reached by the preceding series of analyses do not cleanly fit most models of hunter-gatherer technological organization. Diachronic lithic strategy differences occurred in spite of several common selective pressures. For example, flint is consistently the medium of choice for bladelet tool manufacturing throughout the Upper Paleolithic. Yet flint frequencies (within assemblages) change through time independently of the emphasis on bladelet tool production. Furthermore, different raw materials were reduced according to different strategies within the same period.

In the Paleolithic of the Rio Maior, planning or anticipation of raw material needs often affected technological organization. Three types of organizational behavior, based on planning/anticipation, have been distinguished: curated technology, expedient technology, and opportunistic behavior (Nelson 1991). The term "curation" (*sensu* Nelson) is not limited to tool transport and reworking, but includes carrying prepared cores to a site. Expedient technologies are differentiated from curated ones by on-site raw material availability and lower time stress (Torrence 1983). Expediency is anticipated by hunter-gatherers occupying the site. Either groups plan activities in areas of known raw material, or visit locales of cached materials. Opportunistic strategies are unplanned and situational. Local raw material is utilized, but manufacturing is for immediate use. Nelson (1991:81) proposed that, in some cases, opportunistic reduction would not include a prepared core technique, since the site would not be occupied for an appreciable length of time. These three categories, distinguished by degree of anticipation and raw material transport, are useful for interpretation of the Portuguese Upper Paleolithic.

In addition to planning, lithic strategies in the survey region vary in the degree to which raw material was conserved. Conservation of raw material is often confused with curation, but the two are not necessarily linked (see Odell, this volume). In fact, the Gravettian and Magdalenian sites from Portuguese Estremadura demonstrate this distinction.

Gravettian specialized flint extraction sites were used to prepare blade cores. At these workshops, an occasional quartz or quartzite cobble was flaked. However, the planned activities at this location revolved around flint procurement. Quartz and quartzite reduction at sites near flint sources therefore constituted opportunistic behavior.

Flint blade cores from extraction sites were transported in anticipation of use ("curated") to Gravettian mid-valley residential bases. These cores were reduced quickly (becoming bladelet cores) and were replaced by local quartzes in all assemblage aspects except bladelet tool manufacture. Thus at Gravettian sites more than a kilometer from flint sources, quartz (and occasionally quartzite) was utilized in an expedient strategy. Prepared quartz cores and debitage mirroring the flint assemblage support a degree of anticipated use and lower time stress at these sites. Overall, the Gravettian strategy was both "curated" and expedient. Some flint was transported for bladelet tools, but most residential activities could be performed using local quartz and quartzite. This two-part strategy concurs with Nelson's assertion that "it is crucial that curation and expediency not be perceived as mutually exclusive systems" (Nelson 1991:65).

Flint reduction trajectories were truncated at both Gravettian site types, as would be expected when specialized procurement sites were employed. Sites at the source contain few finished tools, while early stages of reduction are under-represented at the more distant sites. Flint is not particularly economized in Gravettian technological organization. High quality raw material was exploited when available, but otherwise was only needed for bladelet tool blanks. A lower concern for the conservation of flint (relative to the Magdalenian) results in a steep distance-decay slope.

Early Magdalenian and Magdalenian strategies vary only in degree, and both are distinct from the earlier Gravettian data set. Survey methodology and high geologic visibility for the lower valley suggests that the absence of Magdalenian sites within a kilometer of the flint source was a real behavioral pattern. Whole cobbles were brought to mid-valley residential sites from streambeds more than a kilometer away. Either flint procurement took place in the course of other daily activities in the source area, or specialized procurement trips (taking less than a day) were organized.

Magdalenian groups reduced flint cobbles in a more economical/intensive manner than did Gravettian counterparts. Cores and tools were smaller upon discard, and flint use does not decline with distance as rapidly as in the Gravettian. Magdalenian reduction strategies produced smaller tools and included a micro-burin technique, both characteristics that make flint a more favorable medium than quartz or quartzite. Magdalenian hunter-gatherers chose to conserve raw material in compensation for procurement distance.

Quartz and quartzite reduction was generally expedient during the Early Magdalenian and Magdalenian. While evidence is strongest for planned use of quartzite during the Early Magdalenian, Bicho's (1992b) work with core organization demonstrates the existence of both quartz and quartzite reduction strategies that were more planned than opportunistic throughout the Magdalenian.

The Upper Paleolithic of Portuguese Estremadura demonstrates much of the confusion surrounding the term "curation." Both degree of anticipation and conservation of raw material must be explored within a technological strategy. Gravettian groups anticipated a need for flint (bladelet tools), and transported prepared cores, but did not feel pressure to conserve large amounts of flint.

Expedient strategies using local quartzes and quartzites sufficed for most lithic tool needs. In contrast, Magdalenian groups possessed a technology that efficiently reduced flint, but also required a high-quality medium. Expedient strategies did not supplement transported flint in lithic production, possibly due to time stress. A lithic strategy that anticipated future use and conserved raw material resulted. This scenario explains the occurrence of more flint farther from the source in Magdalenian assemblages.

Ultimately, these different strategies reflect other variables, such as paleoenvironmental change and/or population increase. The climate of Portuguese Estremadura was much more temperate during the Last Glacial Maximum than the Paleolithic centers of Spain and France (Zilhao 1987,1990b; Straus 1991a,1992; Bicho 1992b). No large herd migrations took place in Estremadura, in contrast to many regions of Paleolithic Europe and Late Pleistocene North America. Glacial conditions did impact the region by compressing species into a homogeneous mosaic of microzones. A broad mix of species was continuously available in the sheltered lower regions, even in glacial times. Climatic changes affected the expansion of the forest zones, but never eliminated them. Given the lack of faunal preservation from Upper Paleolithic sites near Rio Maior, corroborating evidence for the lithic data presented here is not available. Environmental change was probably one of degree, not radical replacement.

Although the survey region reported here is probably smaller than the average hunter-gatherer range during the Late Pleistocene, this likelihood does not diminish the trends and hypotheses presented. By locating sites with distinctive assemblage characteristics caused by proximity to flint sources (Gravettian extraction sites), assemblages from residential bases have been distinguished. Residential bases throughout Estremadura would have had access to equivalent environmental zones, given the relatively homogeneous landforms. No specialized hunting stations, butchering sites, or other site types have been discriminated for Estremadura. Certainly large portions of the settlement system have been archaeologically invisible or geologically destroyed. Residential bases outside the survey region are expected to follow patterns visible in the survey region just a few kilometers from the flint source.

In the current state of Portuguese Paleolithic research, there is little evidence for seasonally varied group activities, rounds, or adaptations. On a larger scale, coastal excursions or extreme inland seasonal movements are possible. Should sites representing these movements be found, the patterns described in this paper would remain valid for central and eastern Estremadura, but would be supplemented by evidence for a change in macro-regional land use due to seasons, drought, or some other paleoenvironmental variable. Additional work is planned for valleys further away from the flint source than those surveyed thus far. As more distant sites are found and assemblages excavated, the relationships and hypotheses presented here will be tested.

Understanding Portuguese Upper Paleolithic stone assemblage variability requires examining site locational data, raw material procurement trends, typological diversity, and technological organization. The interface of these observations with settlement systems requires an integrated approach, attempting to control for as many variables as possible. Lithic assemblages from Gravettian and Magdalenian periods demonstrate both expedient and curated technological strategies. The role of lithic production within the broader context of hunter-gatherer adaptation is understood by examining anticipated use of raw materials, and conservation/intensity of flint reduction during the Late Pleistocene in Portuguese Estremadura. The complexity evident in the patterns discussed here underscores the value of multiple, theory-informed analyses at all levels of stone tool production.

## ACKNOWLEDGMENTS

The foundation of this paper is the work of Anthony Marks, Joao Zilhao, and Nuno Bicho. Their comments and discussion have immeasureably improved my analysis. Carlos Pereira, Miguel Martins, and the Camara Municipal de Rio Maior deserve thanks for their contributions, particularly their translation and interview help with residents of Azinheira. Reid Ferring, Antonio Carvalho, Jeff Shokler, Joao Ladeira, Katina Lillios, Elizabeth Pintar, Francisco Almeida, and Brooks Elwood all were a source of needed criticism and support. I appreciate comments on a draft of this paper by Michael Shott, Frederic Sellet, Steven Rosen, Mike Rondeau, and Brian Hayden. A National Science Foundation Dissertation Improvement Grant and an Institute for the Study of Earth and Man (SMU) Seed Grant funded the survey project reported here. Finally, I thank George Odell for taking interest in my work, and encouraging my participation in the Tulsa Conference on Theory in Lithic Analysis.

## REFERENCES

Bamforth, D. B. 1990. Settlement, Raw Material, and Lithic Procurement in the Central Mohave Desert. *Journal of Anthropological Archaeology* 9:70–104.

Bicho, N. F. 1992a. Magdalenian Flint Technology at the Site of Cabeco de Porto Marinho, Rio Maior, Portugal. In *Proceedings of the VI International Flint Symposium*, Madrid.

Bicho, N. F. 1992b. *Technological Change in the Final Upper Paleolithic of Rio Maior, Portuguese Estremadura*. Ph.D. Dissertation, Department of Anthropology, Southern Methodist University.

Bicho, N. F. n.d. The Role of Quartz and Quartzite in the Magdalenian of Cabeco de Porto Marinho, Rio Maior, Portugal. In *The Role of Quartzite and Other Non-Flint Raw Materials in the Iberian Paleolithic*, edited by N. Maloney, L. Raposa, and M. Santonja. BAR International Series, Oxford.

Conkey, M. W. 1987. Interpretive Problems in Hunter-Gatherer Regional Studies: Some Thoughts on the European Upper Paleolithic. In *The Pleistocene Old World: Regional Perspectives,* edited by O. Soffer, pp. 63–78. Plenum Press, New York.

Do Paco, A. 1959. *Notas Arqueologicas da Regiao de Alcobertas (Rio Maior).* I Congresso Nacional de Arqueologia, Lisbon.

Goodyear, A. C. 1989. A Hypothesis for the Use of Cryptocrystalline Raw Materials among Paleoindian Groups of North America. In *Eastern Paleoindian Lithic Resource Use,* edited by C. Ellis and J. Lothrop, pp. 1–9. Westview Press, Boulder.

Johnson, J. K. 1989. The Utility of Production Trajectory Modeling as a Framework for Regional Analysis. In *Alternative Approaches to Lithic Analysis,* edited by D. Henry and G. Odell, pp. 119–138. Archaeological Papers of the American Anthropological Association, vol. 1.

Johnson, J. K., and C. A. Morrow (editors). 1987. *The Organization of Core Technology.* Westview, Boulder.

Kintigh, K. 1984. Measuring Archaeological Diversity by Comparison with Simulated Assemblages. *American Antiquity* 49:44–54.

Marks, A. E., N. Bicho, J. Zilhao, and C. R. Ferring. 1994. Upper Pleistocene Prehistory in Portuguese Estremadura: Results of Preliminary Research. *Journal of Field Archaeology* 21:53–68.

Marks, A. E., J. Shokler, and J. Zilhao. 1991. Raw Material Usage in the Paleolithic: The Effects of Local Availability on Selection and Economy. In *Raw Material Economies among Prehistoric Hunter-Gatherers,* edited by A. Montet-White and S. Holen, pp. 127–139. University of Kansas, Publications in Anthropology 19, Lawrence.

Martins, M., and C. Pereira. 1992. Da exploracao de silex na regiao de Rio Maior. *Arqueologia Industrial* II:2, Lisbon.

Nelson, M. C. 1991. The Study of Technological Organization. *Archaeological Method and Theory* 3:57–100.

Odell, G. H. 1989. Fitting Analytical Techniques to Prehistoric Problems with Lithic Data. In *Alternative Approaches to Lithic Analysis,* edited by D. Henry and G. Odell, pp. 159–182. Archeological Papers of the American Anthropological Association, no. 1.

Perles, C. 1992. In Search of Lithic Strategies. In *Representations in Archaeology,* edited by J.-C. Gardin and C. Peebles, pp. 222–247. Indiana University Press, Bloomington.

Rigaud, J.-P. 1991. *Le Magdalenien en Europe.* ERAUL 38, Liège.

Rigaud, J.-P. and J. F. Simek. 1987. "Arms Too Short to Box with God": Problems and Prospects for Paleolithic Prehistory in Dordogne, France. In *The Pleistocene Old World: Regional Perspectives,* edited by O. Soffer, pp. 47–62. Plenum Press, New York.

Schiffer, M. B. 1987. *Formation Processes of the Archaeological Record.* University of New Mexico Press, Albuquerque.

Simek, J. F., and H. A. Price. 1990. Chronological Change in Perigord Lithic Assemblage Diversity. In *The Emergence of Modern Humans,* edited by P. Mellars, pp. 243–261. Cornell University Press, Ithaca.

Straus, L. G. 1991a. Southwestern Europe at the Last Glacial Maximum. *Current Anthropology* 32:189–199.

Straus, L. G. 1991b. The Role of Raw Materials in Upper Paleolithic and Mesolithic Stone Artifact Assemblage Variability in Southwest Europe. In *Raw Material Economies among Prehistoric Hunter-Gatherers,* edited by A. Montet-White and S. Holen, pp. 169–186. University of Kansas Publications in Anthropology 19, Lawrence.

Straus, L. G. 1992. *Iberia before the Iberians: The Stone Age Prehistory of Cantabrian Spain.* University of New Mexico Press, Albuquerque.

Straus, L. G., and G. A. Clark. 1986. *La Riera Cave: Stone Age Hunter-Gatherer Adaptations in Northern Spain*. Anthropological Research Papers 36. Arizona State University Press, Tempe.

Torrence, R. 1983. Time Budgeting and Hunter-Gatherer Technology. In *Hunter-Gatherer Economy in Prehistory: A European Perspective*, edited by G. Bailey, pp. 11–22. Cambridge University Press, Cambridge.

Weniger, G.-C. 1989. The Magdalenian in Western Central Europe: Settlement Pattern and Regionality. *Journal of World Prehistory* 3:323–372.

Weniger, G.-C. 1990. Germany at 18,000 BP. In *The World at 18,000 BP (Volume 1: High Latitudes)*, edited by O. Soffer and C. Gamble, pp. 170–192. Unwin Hyman, Boston.

White, R. 1983. Changing Land-Use Patterns across the Middle/Upper Paleolithic Transition: The Complex Case of the Perigord. In *The Mousterian Legacy: Human Biocultural Change in the Upper Pleistocene*, edited by E. Trinkaus, pp. 113–121. BAR International Series 164, Oxford.

White, R. 1985. *Upper Paleolithic Land Use in the Perigord: A Topographic Approach to Subsistence and Settlement*. BAR International Series 253, Oxford.

White, R. 1987. Glimpses of Long-Term Shifts in Late Paleolithic Land Use in the Perigord. In *The Pleistocene Old World: Regional Perspectives*, edited by O. Soffer, pp. 263–278. Plenum Press, New York.

Wobst, H. M. 1990. Minitime and Megaspace in the Palaeolithic at 18K and Otherwise. In *The World at 18,000 BP (Volume 1: High Latitudes)*, edited by O. Soffer and C. Gamble, pp. 331–343. Unwin Hyman, Boston.

Zilhao, J. 1987. O Solutrense da Estremadura Portuguesa. Uma proposta de interpretacao paleoantropologica. *Trabalhos de Arqueologia* 4. Lisbon.

Zilhao, J. 1988a. Plaquette gravée du Solutéen supérieur de la Gruta do Caldeirao (Tomar, Portugal). *Bulletin de la Societe Prehistorique Francaise* 85:106–109.

Zilhao, J. 1988b. Nouvelles datations absolues pour la préhistoire ancienne du Portugal. *Bulletin de la Société Préhistorique Francaise* 85:247–250.

Zilhao, J. 1990a. The Portuguese Estremadura at 18,000 BP: The Solutrean. In *The World at 18,000 BP (Volume 1: High Latitudes)*, edited by O. Soffer and C. Gamble, pp. 109–125. Unwin Hyman, Boston.

Zilhao, J. 1990b. Le Solutréen du Portugal: environnement, chronologie, industries, peuplement, origines. In *Les Industries à Pointes Foliacées du Paléolithique Supérieur Européen*, edited by J. Kozlowski, pp. 486–501. ERAUL 42, Liege.

Zilhao, J. 1991. Aurignacien et Gravettien au Portugal. Paper to be published in the Proceedings of the XII UISPP Congress, Bratislava.

Zilhao, J. n.d. The Upper Paleolithic of Portugal: Past Research and Current Perspectives. In *Recent Research on the European Paleolithic*, edited by R. E. Webb. BAR International Series, Oxford. In press.

# Part III

# Stone Tools and Complex Societies

Lithic analysis has traditionally been the domain of hunter-gatherer studies, because the introduction of metallurgy in late prehistoric times initiated the demise of the stone tool in several parts of the world. Prehistoric cultural groups that existed subsequent to this introduction were regarded by archaeologists as metal-using and therefore uninteresting from a lithic perspective. This attitude has prevailed until recently, but stone tool-based archaeological analyses have begun to appear in such complex society-yielding regions as the Mediterranean (Whallon 1978; Cauvin 1983), the Maya heartland (Rovner 1975; Shafer 1983; Clark 1987; Lewenstein 1987; Hester and Shafer 1991), and the American Southeast during Mississippian times (Johnson 1987; Koldehoff 1987).

Perhaps one reason for interest in the stone tools of American complex societies is that metal tools were never mass produced for a proletarian market in these regions. Hence stone continued to be the favored medium for implement production right up to the colonization of these lands by Europeans. Yet even in these areas, archaeologists have traditionally preferred to analyze materials other than stone. This preference is a result partly of the existence of culturally sensitive artifacts of nonlithic origin such as pottery, and partly of the devolution of core technology that universally appears to have been the result of restrictions in mobility (Parry and Kelly 1987).

In regions in which the products of metallurgy sifted down to the lower classes (e.g., Europe and the Near East), scholars of the Chalcolithic and Bronze Ages have studiously avoided this issue and have written very little at all about stone. Among those who have considered this medium, the perception of a sudden and precipitous replacement of stone tools by metal ones has persisted. Until now, that is, for Steve Rosen's work for this volume shows that the replacement of stone by metal was both slower than previously thought, and more selective. Employing data from Israel, Rosen has established that the replacement took 3000 years to complete and proceeded in stages, each stage governed at least as much by outside factors such as expanded trade routes and economic restructuring as by the intrinsic advantages of metal. The fact that people clung to their time-honored lithic technologies until there existed powerful reasons to relinquish them fits nicely with traditional concepts of cultural conservatism, but until Rosen's study, the piecemeal tool replacement that conformed to these regularities had never been established.

The next two chapters explore the role of lithic technologies within complex societies in which metals were known, but stone continued to be important. Jay Johnson's study of Mayan implements focuses not on providing detailed descriptions of Mayan assemblages, but on investigating certain elements of the culture's complex social organization. These include its centralization, social control, and manufacture of specialized ritual objects. Johnson draws some interesting parallels between Maya tool production sites such as Colha in Belize and late Middle Archaic Benton period sites in northwest Alabama and northeast Mississippi. Benton sites exhibit characteristics similar to Maya production sites, but their cultural organization was structurally different, and their society provided an entirely different context for these technologies.

Michael Nassaney's chapter is principally concerned with the consequences of social ranking among denizens of the pre-Mississippian Plum Bayou culture of central Arkansas. Considering procurement patterns of raw material for chipped stone tools, Nassaney examines several sites for evidence of elite control of lithic resources. Although centralized management of chert and novaculite sources appears not to have been successfully accomplished, there is some evidence for control of quartz crystal. This finding has interesting ramifications for the succeeding Mississippian period and for the study of emerging elites in general.

## REFERENCES

Cauvin, M.-C. (editor). 1983. *Traces d'utilisation sur les outils néolithiques du Proche Orient.* Travaux de la Maison de l'Orient, no. 5, Lyon.

Clark, J. 1987. Politics, Prismatic Blades, and Mesoamerican Civilization. In *The Organization of Core Technology,* edited by J. Johnson and C. Morrow, pp. 259–284. Westview Press, Boulder.

Hester, T. R., and H. J. Shafer (editors). 1991. *Maya Stone Tools.* Prehistory Press, Madison.

Johnson, J. K. 1987. Cahokia Core Technology in Mississippi: The View from the South. In *The Organization of Core Technology,* edited by J. Johnson and C. Morrow, pp. 187–205. Westview Press, Boulder.

Koldehoff, B. 1987. The Cahokia Flake Tool Industry: Socioeconomic Implications for Late Prehistory in the Central Mississippi Valley. In *The Organization of Core Technology,* edited by J. Johnson and C. Morrow, pp. 151–185. Westview Press, Boulder.

Lewenstein, S. M. 1987. *Stone Tool Use at Cerros: The Ethnoarchaeological and Use-Wear Evidence.* University of Texas Press, Austin.

Parry, W. J., and R. L. Kelly. 1987. Expedient Core Technology and Sedentism. In *The Organization of Core Technology,* edited by J. Johnson and C. Morrow, pp. 285–304. Westview Press, Boulder.

Rovner, I. 1975. *Lithic Sequences from the Maya Lowlands.* PhD dissertation, Department of Anthropology, University of Wisconsin at Madison.

Shafer, H. J. 1983. The Lithic Artifacts of the Pulltrouser Area: Settlement and Fields. In *Pulltrouser Swamp: Ancient Maya Habitat, Agriculture, and Settlement in Northern Belize,* edited by B. Turner II and P. Harrison, pp. 212–245. University of Texas Press, Austin.

Whallon, R. J. 1978. Threshing Sedge Flints: A Distinctive Pattern of Wear. *Paléorient* 4:319–324.

*Chapter 5*

# The Decline and Fall of Flint

STEVEN A. ROSEN

## ABSTRACT

The replacement of chipped stone technologies by metallurgy cannot be viewed as a simple linear process, nor as the complementary rise and fall of competing technologies. Both technologies are complex arrays of distinct sub-technologies, each of which has its own developmental trajectory. The "replacement" of chipped stone tools by metal equivalents occurred episodically, and perceptions of the process as linear are more the result of our own foreknowledge of the importance of metallurgy to later societies than of their own absolute importance to earlier ones. The replacement process should not be seen as one based merely on the greater utility of metal tools tempered by the time it took to develop requisite technologies. As in most histories of technology, other social and economic factors played primary roles.

## INTRODUCTION

The importance of metallurgy to our own civilization is such that its significance is often simply assumed for ancient people. The long evolution in metallurgical technology from simply hammering native copper, to smelting and alloying into bronze, and finally to the introduction of iron and steel, tends to be seen as directional, and perhaps in some sense, almost inevitable. The explanations for the adoption of metallurgy are often considered self-evident. The inherent

**STEVEN A. ROSEN** • Archaeology Department, Ben-Gurion University of the Negev, Beer Sheva 84 105, Israel.

*Stone Tools: Theoretical Insights into Human Prehistory*, edited by George H. Odell. Plenum Press, New York, 1996.

advantages of metal over flint are assumed, and if the process of replacement of lithic technology by metallurgy took over 3000 years, this is usually considered to be a consequence of the length of time required to make the requisite metallurgical discoveries and developments (e.g., Childe 1951:95–99, but see p. 99 for qualification).

More recently, the origins and development of metallurgy have been examined from more sophisticated technological and socio-political perspectives (e.g., Rothenberg et al. 1978; Waldbaum 1978; Muhly 1980; Heskel 1983; Shalev and Northover 1987; Levy and Shalev 1989). However, there has been little attempt to understand the role of the flint implements being replaced. In fact, the conception that flint artifacts represented a dying or decaying technology, as opposed to the "metallic wave of the future," still occasionally surfaces; although more often, the lithic artifacts are simply ignored. This lack of balance may be at least partially attributable to the often assumed primary role of metallurgy in the rise of urban and state society, as expressed through long-distance trade and craft specialization (e.g., Childe 1951:95–99; Redman 1978:270; Kempinsky 1983, 1989; Zaccagnini 1983; Renfrew 1984; Ilan and Sebanne 1989).

A second factor in this "skewed" perspective is the general orientation of Near Eastern and Levantine historical archaeology. Researchers working in Palestine during the early stages of archaeological inquiry recognized the significance of flint artifacts (e.g., Petrie 1891:49–50; Macalister 1912:121–128; Neuville 1930, 1934/5; Crowfoot 1936, 1937; Guy and Engberg 1938:70). However, as Levantine historical archaeology concentrated on more textual and historically related questions (cf. Glock 1985) and less on ethnography and anthropology, the importance of lithic analysis declined. As a consequence, even the existence of flint artifacts in historical contexts has often gone unrecognized. In the past two decades this has changed considerably, and archaeologists have concentrated on a wider range of types of evidence than in the past. Thus not only stone tools, but also faunal and floral assemblages, microartifacts, chemical analysis, etc., have begun to provide a much fuller picture of the archaeology of the Levant.

Metallurgy and lithic technology are generally viewed as (contrasting) integrated technologies. Although such an approach is justified to a degree, it can also serve to mask significant complexities. From another perspective, both metallurgy and lithic technology can and should be broken down into the different types of technologies used in the manufacture of different tools. That such a division may be significant in metallurgy is evident in the contrasting modes of Chalcolithic copper utensil manufacture (lost wax versus simple mold), seemingly corresponding to distinct orders of material culture and economic organization (e.g., Rosen 1986a, n.d.; Shalev and Northover 1987; Levy and Shalev 1989; Shalev et al. 1992). With respect to lithic artifacts, there are distinct modes of manufacture for different tool types which, aside from the use of flint as raw material, have little in common with each other (Rosen 1989a). Even the shared aspect of raw material is questionable, since different tool types often require different types of flint.

Thus the replacement process is not simply the exchange of one technology and raw material for another. There is no simple "rise" of metallurgy accompanied by a correspondingly simple "fall" of flint. Both modes of manufacture are arrays of different raw materials and technologies; they required a wide range of skills to produce a similarly wide range of tools.

Finally, each technological system includes a set of factors, each of which must have played a role in the ultimate replacement process. These include relative functional efficiency of tools from different raw materials for different tasks; relative efficiency of manufacture of different tool types; relative availability of raw material; and a range of social and economic factors, such as relative social or symbolic value of different raw materials, relative degrees of manufacturing specialization, capital expenses, and interdependence of the stages of manufacture. Given such a picture, the replacement of flint by metals cannot be viewed as a simple linear process, nor as the rise and fall of competing technologies.

## THE DECLINE OF FLINT: A DESCRIPTION

The use of chipped stone tools in the Near East is known to recent times in the form of rectangles for flintlock guns (Oakley 1975:26) and threshing sledge teeth (Bordaz 1965, 1969; Whallon 1978; Ataman 1992). In spite of technological continuity from ancient times, it is clear that in prehistoric and early historic times there was significantly greater use of chipped stone technologies. The period of decline in the Levant corresponds to the span between the later Neolithic (ca. 5000 B.C.) and the middle of the Iron Age (ca. 800 B.C.), by which time chipped stone assemblages seem to have virtually disappeared, with the exception of threshing teeth (which have not been unambiguously identified in the archaeological record, either).

The decline can be traced archaeologically on two levels: an absolute numerical decrease in flint tools and waste, and a decline in typological diversity, with sickle blades and associated waste comprising an ever-increasing proportion of lithic assemblages until they, too, disappear around the 9th-8th centuries B.C. This undoubtedly reflects a reduction in the variety of tasks in which flint tools were used. These two trends are not independent.

On a more profound level, the long decline in the use of chipped stone was accompanied by increasing specialization in manufacture and associated changes in the modes of distribution. Not only were there trends toward fewer tools and fewer uses, but also toward fewer people in the manufacturing process, even when the tools themselves, like sickles, may have been used by a relatively high proportion of the population.[1] These economic changes, relating more to the evolution of social complexity than to technological developments, act as an independent factor in the lithic system, and thus may have influenced the adoption of metallurgy.

## QUANTITATIVE DECLINE

Quantitative decline of stone tools is evidenced by Figures 1–5 and Tables 1–3. However, the details of the decline are also of interest, since it is only there that the specific mechanisms of change can be traced. Unfortunately, detailed definition of the quantitative trends is difficult.

The most significant problem has been the lack of archaeological recognition of the issue and the resulting scarcity of good data. In recent years, improved collection has provided more reliable samples. However, we are still left with serious problems of comparing assemblages from strata and sites which may not be strictly comparable. Thus, for example, collection methods, which should be relatively easy to standardize (and for all intents and purposes, *have* been standardized in prehistoric archaeology), vary from the collection of pretty pieces with the discard of virtually all waste, to the rigorous sieving of all sediments and collection of all artifacts. Methodological statements are also often missing in publications, so that attempts to account for collection biases are impossible.

Other problems include variability between and within sites (including such aspects as population density and areal functions), and post-depositional factors causing loss, displacement, and mixing of assemblages. This is especially common in historic city sites, the subject of major earth moving projects such as the construction of city walls, palaces, etc. The great size of these sites is also to be considered. No site is ever fully excavated, and comparisons must minimally consider differences in area or volume excavated, data which are only rarely available.

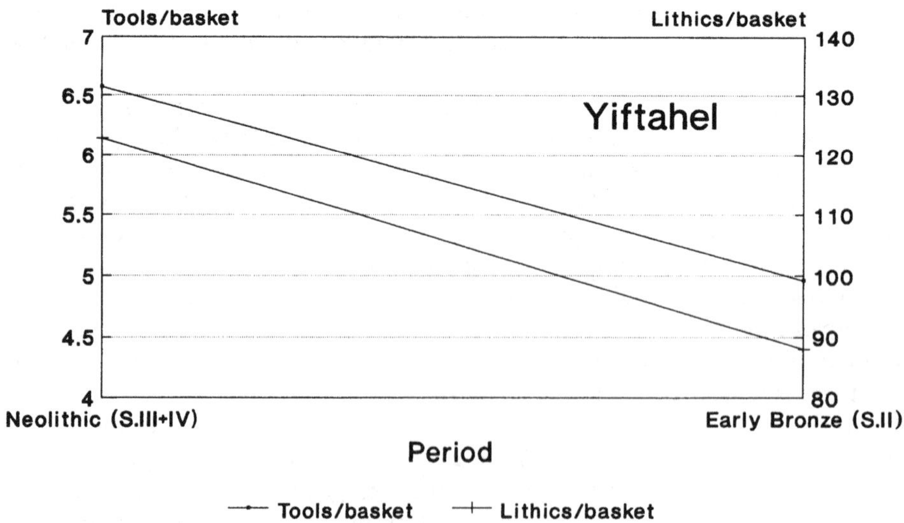


Figure 1. Lithic frequency decline in absolute numbers from Yiftahel (per basket).

In short, detailed absolute quantitative comparisons between sites (as opposed to relative proportions) are probably not possible. However, chronological analysis can still be undertaken if restricted to intra-site comparison, i.e., comparison between strata of a single site. Thus, at least problems of differences in collection bias are minimized. Furthermore, the issue of differences in extent of excavation between strata can be accounted for by reference to the excavation units used by the archaeologist at the site. Presumably these do not change between strata. With respect to Figures 1–5, all lithic assemblages (with the exception of those from Ir David - Jerusalem) were standardized with respect to artifacts per basket. A basket in Levantine archaeology is a relatively arbitrary provenience unit which in theory is not a standardized volume or area. In practice it does provide a rough gauge of intensity of excavation (Gilead n.d.). Analysis was limited to baskets containing lithic artifacts, so that baskets without them were not included in the calculations. At Ir David it was possible to calculate the proportional extent of excavation for different periods, allowing data to be standardized by multiplication to 100%.[2]

Given these qualifications, the graphs presented in Figures 1–5 should be viewed as reflecting trends.[3] They suggest three phase changes in the decline of chipped stone technology. The first occurred in the Late Neolithic-Early Bronze Age, as reflected in the graphs from Yiftahel, Sataf, and Qasis (Figures 1–3). The Qasis graph (Figure 3) shows a steep decline from the Neolithic to the Early Bronze Age, followed by a levelling off. The steepness of the slope at Sataf (Figure 2, Table

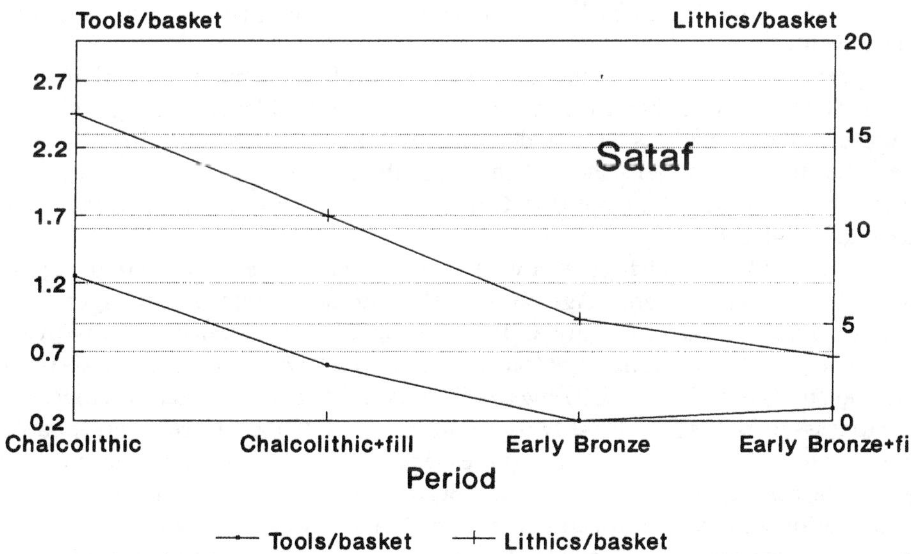


Figure 2. Lithic frequency decline in absolute numbers from Sataf (per basket).

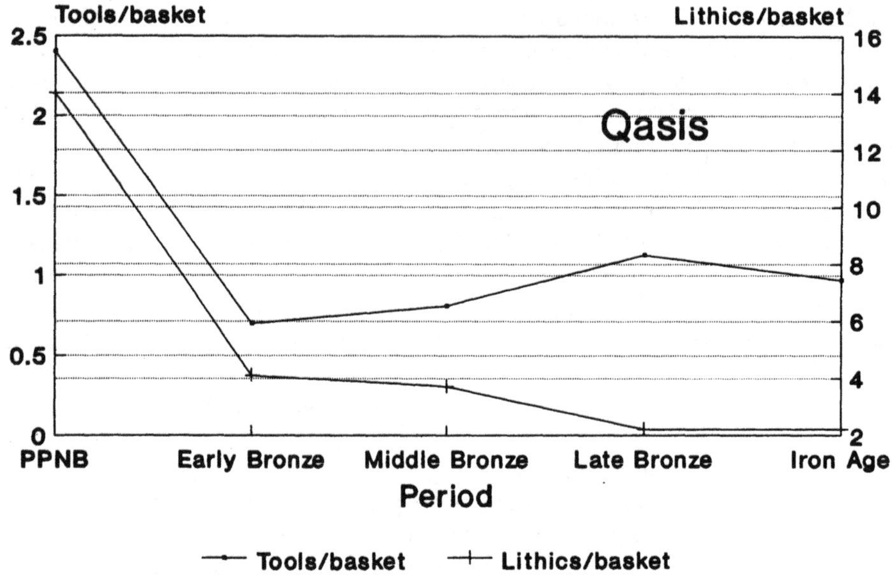


**Figure 3.** Lithic frequency decline in absolute numbers from Qasis (per basket).

1) suggests that much of this decline occurred from the Chalcolithic to the Early Bronze Age, a transition which is not noted for any obvious advances in metallurgical technology.

Beyond the graphs, subjective evaluation of Neolithic and Chalcolithic assemblages, impossible to quantify due to the problems outlined above, suggests that there was also a decline from the Neolithic to Chalcolithic, perhaps in the terminal stages of the Neolithic. Unfortunately, no good data from sites spanning the Neolithic-Chalcolithic transition are yet available for detailed analysis.

A second phase change, to a yet lower level of lithic exploitation, occurred following the Early Bronze Age, around the Middle or Late Bronze Ages. The materials from Ir David (Figure 4, Table 1) suggest that this decline took place from Early to Middle Bronze Age (see also endnote 2), whereas the materials from Batashi and Qasis (Figures 5, 3) suggest the possibility of a later decline. However, the tool assemblages from these two sites are more ambiguous. Intrusions probably play a large role in these assemblages, especially in light of the small size of the assemblages. Regardless, by the Late Bronze-Iron Age, it appears that lithic exploitation was on a significantly lower level than it was during the Early Bronze Age. The anomaly of the Qasis tool proportions is probably a function of small sample size (Table 1).

**Table 1. Lithic Frequencies Showing Decline in Absolute Numbers from Selected Sites**

| Site | Total baskets | Total tools | Total lithics | Tools/ basket | Lithics/ basket |
|---|---|---|---|---|---|
| Yiftahel[a] | | | | | |
| Neolithic (Str. III+IV) | 100 | 657 | 12275 | 6.57 | 122.75 |
| Early Bronze (Str. II) | 47 | 233 | 4149 | 4.96 | 88.00 |
| Sataf | | | | | |
| Chalcolithic | 40 | 50 | 643 | 1.25 | 16.08 |
| Mixed Chalcolithic + Classical Fill | 77 | 47 | 823 | .60 | 10.69 |
| Early Bronze I | 35 | 7 | 183 | .20 | 5.23 |
| Mixed Early Bronze+ Classical Fill | 35 | 10 | 116 | .29 | 3.31 |
| Qasis[b] | | | | | |
| PPNB | 5 | 12 | 70 | 2.40 | 14.00 |
| Early Bronze | 349 | 243 | 1446 | .70 | 4.10 |
| Middle Bronze | 77 | 62 | 281 | .81 | 3.70 |
| Late Bronze | 38 | 43 | 83 | 1.13 | 2.20 |
| Iron Age | 34 | 33 | 77 | .97 | 2.30 |
| Batashi | | | | | |
| Middle Bronze | 23 | 18 | 155 | .78 | 6.74 |
| Late Bronze | 101 | 61 | 181 | .60 | 1.79 |
| Iron I | 40 | 33 | 68 | .83 | 1.70 |
| Iron II | 70 | 32 | 110 | .46 | 1.57 |

[a]At Yiftahel the tool totals do not include retouched flakes, due to the disturbed nature of the site and the difficulty in distinguishing between post-depositional damage and intentional retouch of flakes. Numbers from Yiftahel in this table are from unpublished research by author on excavations conducted by E. Braun.
[b]In order to account for differences caused by differential sieving of strata at Qasis, the figures do not include chips, the <2 cm category and the class most significantly affected by sieving.

The final phase constituted the near disappearance of flint artifacts, and is therefore not seen in the graphs. It is most readily apparent in the excavations at the Iron Age strata at Batashi, where by stratum II only 10 flint artifacts were collected (Tables 2–3), in spite of the fact that this stratum was as extensively exposed as those beneath it. One problem here is whether the total absence of reports on lithic assemblages from the Persian period (following the Iron Age) is the result of the virtual absence of artifacts, or an absence of archaeological recognition. Personal experience strongly suggests the former.[4]

Finally, it is important to note that there does not appear to be evidence for intra-period decline, especially in the Chalcolithic and Early Bronze Ages. Thus single period sites in which stratigraphic separation of assemblages is possible (e.g., *Chalcolithic* - Shiqmim [Levy and Rosen 1987], Horvat Beter [Yeivin 1959]; *Early Bronze Age* - Yarmouth [Rosen 1988], Arad [Shick 1978], Bab edh Dhra

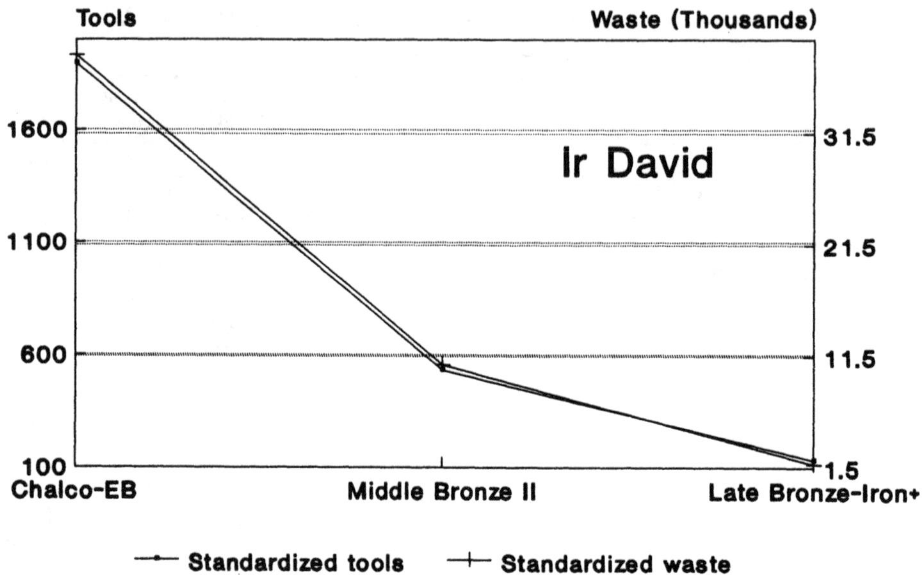


Figure 4. Lithic frequency decline in absolute numbers from Ir David (percentage by area excavated).

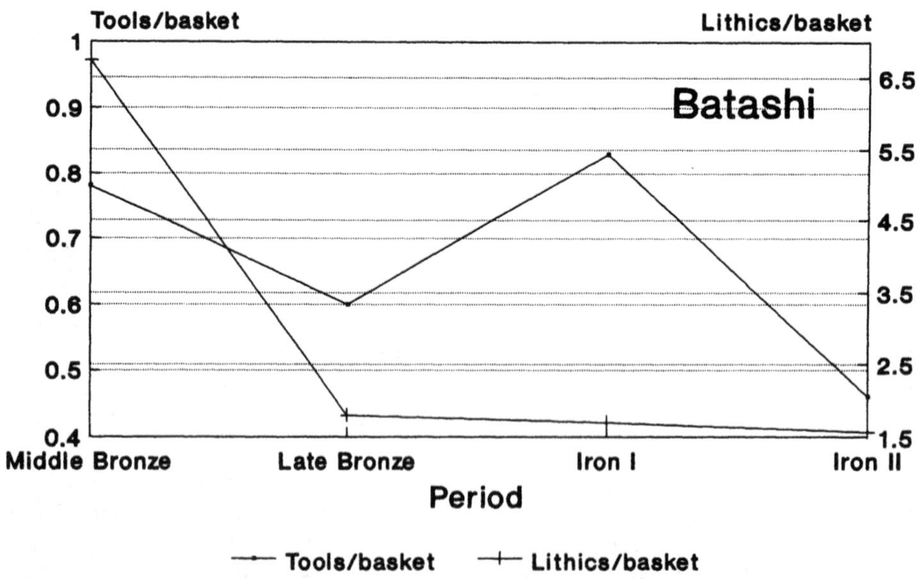


Figure 5. Lithic frequency decline in absolute numbers from Batashi (per basket).

**Table 2. Lithic Frequencies from Tell Batashi, Including Only Artifacts Assignable to Unambiguous Strata**

| Period | Stratum | Waste | Non-sickle tools | Sickle blades | Total tools | Total |
|---|---|---|---|---|---|---|
| Middle Bronze IIc | XI | 126 | 11 | 4 | 15 | 141 |
| Middle Bronze II—Late Bronze | X | 11 | 1 | 2 | 3 | 14 |
| Late Bronze | IX | 9 | 2 | 2 | 4 | 13 |
| Late Bronze | VIII | 14 | 1 | 7 | 8 | 22 |
| Late Bronze | VII | 12 | 7 | 19 | 26 | 38 |
| Late Bronze | VI | 6 | 1 | 10 | 11 | 17 |
| Iron I | V | 35 | 8 | 25 | 33 | 68 |
| Iron II | IV | 43 | 3 | 18 | 21 | 64 |
| Iron II | III | 8 | 1 | 2 | 3 | 11 |
| Iron II | II | 7 | 0 | 3 | 3 | 10 |
| Persian | I | 3 | 0 | 0 | 0 | 3 |

[McConaughy 1979]) do not show discernible patterns of decline in lithic exploitation. Although materials from older excavations such as Jericho, excavated in the 1950's (Crowfoot Payne 1983), may show apparent patterns of intra-period decline, low numbers and biased recovery techniques render these collections of limited quantitative value.

The data on intra-period decline for the post-Early Bronze Age is more problematic, especially in light of the low number of artifacts in the first place. In most cases it has been necessary to lump strata in order to get reasonable numbers for analysis, and problems of intrusion are more severe. We may be dealing with a constant downward trend, and not a period of stability followed by a final decline.

**Table 3. Lithic Frequencies from Tell Batashi, by General Period, Including Artifacts with Ambiguous Stratigraphic Association But not Ambiguous Period Association**

| Period | Waste | Non-sickle tools | Sickle blades | Total tools | Total lithics |
|---|---|---|---|---|---|
| Middle Bronze IIc | 126 | 11 | 4 | 15 | 141 |
| Middle Bronze II—Late Bronze | 11 | 1 | 2 | 3 | 14 |
| Late Bronze | 120 | 13 | 48 | 61 | 181 |
| Iron I | 35 | 8 | 25 | 33 | 68 |
| Iron I-II | 14 | 1 | 1 | 2 | 16 |
| Iron II | 78 | 7 | 25 | 32 | 110 |
| Persian | 3 | 0 | 0 | 0 | 3 |

## TYPOLOGICAL DECLINE

Typological decline is somewhat easier to establish than quantitative decline, since the important issues are the proportional decline and presence/absence of specific types. Comparisons between sites and strata are thus less sensitive to differences in areal extent of excavation, population density, and length of occupation. The key issue is the proportional representativeness of the collection, i.e., non-selected recovery. To a degree, in historical period sites in the Levant, this is reflected by the general quantities of waste and debitage present in the assemblage. As a general rule, sites with moderate quantities of debitage imply unbiased collection.

One important qualifying factor is the possibility of significant differences in the function of sites and loci. For example, with respect to functional variability between sites, there is a clear decrease in sickle percentages from north to south in the Early Bronze Age, undoubtedly reflecting general levels of intensity of agricultural exploitation (Rosen 1989a). For this reason, Central and Southern Negev assemblages have been excluded from the analysis. With respect to intra-site variation, the assemblage from Gezer (Rosen 1986b) seems to be a set of caches from a specialized sickle blade production workshop and is not directly comparable to assemblages derived from general occupation strata. That from Tell Dan also seems to derive from a specialized industrial area. Nevertheless, given the large areal extent of most excavations and the mixing occurring in fills, usually this seems not to be a problem.

Table 4 presents the proportions of selected tool classes, from sites ranging from the Pre-Pottery Neolithic B through the Iron Age. Figure 6 presents a summary of the decline and disappearance of tool types; and Figures 7–11 present the proportional decline of specific types based on the data from Table 4.

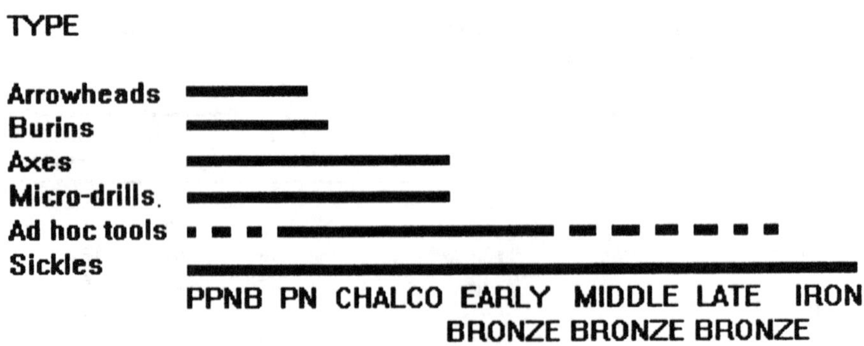


Figure 6. Summary of lithic decline by type.

Table 4. Lithic Frequencies of Selected Tool Classes, by Site and Period

| | Burin | % | Arrow | % | Sickle | % | Axe | % | Other | % | Total |
|---|---|---|---|---|---|---|---|---|---|---|---|
| Yiftahel–PPNB | 167 | 16.2 | 195 | 18.9 | 200 | 19.4 | 60 | 5.8 | 409 | 39.7 | 1031 |
| Jericho–PPNB | 913 | 31.8 | 405 | 14.1 | 999 | 34.8 | 40 | 1.4 | 514 | 17.9 | 2871 |
| Abu Gosh–PPNB | 207 | 12.5 | 337 | 20.4 | 764 | 46.2 | 88 | 5.3 | 258 | 15.6 | 1654 |
| Beisamun–PPNB | 70 | 20.2 | 67 | 19.3 | 46 | 13.3 | 51 | 14.7 | 113 | 32.6 | 347 |
| Beidha–PPNB | 165 | 5.8 | 901 | 31.6 | 379 | 13.3 | 71 | 2.5 | 1335 | 46.8 | 2851 |
| Ain Ghazal–PPNB | 1333 | 37.6 | 209 | 5.9 | 313 | 8.8 | 208 | 5.9 | 1486 | 41.9 | 3549 |
| Sheik Ali–PPNB | 20 | 4.6 | 24 | 5.5 | 132 | 30.3 | 10 | 2.3 | 250 | 57.3 | 436 |
| Sha'ar Hagolan–PN | 18 | 3.6 | 16 | 3.2 | 70 | 13.8 | 17 | 3.4 | 385 | 76.1 | 506 |
| Ain Ghazal–PN | 644 | 29.8 | 113 | 5.2 | 51 | 2.4 | 142 | 6.6 | 1213 | 56.1 | 2163 |
| Sheik Ali–PN | 9 | 6.6 | 0 | 0.0 | 12 | 8.8 | 3 | 2.2 | 112 | 82.4 | 136 |
| Nahal Zehora–PN | 47 | 18.0 | 0 | 0.0 | 20 | 7.7 | 20 | 7.7 | 174 | 66.7 | 261 |
| Tsaf–PN | 5 | 5.1 | 0 | 0.0 | 6 | 6.1 | 10 | 10.1 | 78 | 78.8 | 99 |
| Grar–Chalco | 4 | 0.4 | 0 | 0.0 | 166 | 16.9 | 48 | 4.9 | 763 | 77.8 | 981 |
| Sataf–Chalco | 0 | 0.0 | 1 | 1.0 | 25 | 25.8 | 3 | 3.1 | 68 | 70.1 | 97 |
| Charbush–Chalco | 0 | 0.0 | 0 | 0.0 | 14 | 9.0 | 0 | 0.0 | 142 | 91.0 | 156 |
| Shiqmim–Chalco | 0 | 0.0 | 0 | 0.0 | 39 | 2.3 | 86 | 5.1 | 1561 | 92.6 | 1686 |
| Horvat Beter–Chalco | 0 | 0.0 | 0 | 0.0 | 15 | 3.9 | 14 | 3.7 | 354 | 92.4 | 383 |
| Ir David–Chalco/EB | 0 | 0.0 | 0 | 0.0 | 14 | 9.9 | 3 | 2.1 | 125 | 88.0 | 142 |
| Yiftahel–EB | 9 | 3.9 | 1 | 0.4 | 10 | 4.3 | 11 | 4.7 | 202 | 86.7 | 233 |
| Qasis–EB | 3 | 1.2 | 2 | 0.8 | 97 | 39.9 | 3 | 1.2 | 138 | 56.8 | 243 |
| Har Tuv–EB | 6 | 3.8 | 0 | 0.0 | 35 | 22.4 | 0 | 0.0 | 115 | 73.7 | 156 |
| Beit Yerah–EB | 0 | 0.0 | 0 | 0.0 | 77 | 49.0 | 1 | 0.6 | 79 | 50.3 | 157 |
| Hesi–EB | 2 | 0.6 | 0 | 0.0 | 77 | 21.3 | 1 | 0.3 | 282 | 77.9 | 362 |
| Bab edh Dhra–EB | 4 | 0.6 | 0 | 0.0 | 60 | 9.6 | 0 | 0.0 | 561 | 89.8 | 625 |
| Yarmuth–EB | 0 | 0.0 | 0 | 0.0 | 25 | 33.3 | 0 | 0.0 | 50 | 66.7 | 75 |
| Shadud–EB | 1 | 0.3 | 0 | 0.0 | 39 | 9.8 | 1 | 0.3 | 358 | 89.7 | 399 |
| Sha'ar Hagolan–MBI | 1 | 1.1 | 0 | 0.0 | 35 | 37.6 | 0 | 0.0 | 57 | 61.3 | 93 |
| Ir David–EB/MB | 0 | 0.0 | 0 | 0.0 | 27 | 24.8 | 2 | 1.8 | 80 | 73.4 | 109 |
| Malha[a]–MB1–2 | 0 | 0.0 | 0 | 0.0 | 77 | 22.4 | 11 | 3.2 | 255 | 74.3 | 343 |
| Batashi–MB | 0 | 0.0 | 0 | 0.0 | 6 | 33.3 | 0 | 0.0 | 12 | 66.7 | 18 |
| Qasis–MB | 0 | 0.0 | 0 | 0.0 | 29 | 46.8 | 1 | 1.6 | 32 | 51.6 | 62 |
| Qasis–LB | 0 | 0.0 | 0 | 0.0 | 26 | 60.5 | 0 | 0.0 | 17 | 39.5 | 43 |
| Deir el Balah–LB-Iron | 0 | 0.0 | 0 | 0.0 | 210 | 71.7 | 0 | 0.0 | 83 | 28.3 | 293 |
| Wawiyat–LB-Iron | 0 | 0.0 | 0 | 0.0 | 63 | 73.3 | 0 | 0.0 | 23 | 26.7 | 86 |
| Batashi–LB | 0 | 0.0 | 0 | 0.0 | 48 | 78.7 | 0 | 0.0 | 13 | 21.3 | 61 |
| Qasis–Iron | 0 | 0.0 | 0 | 0.0 | 18 | 54.5 | 0 | 0.0 | 15 | 45.5 | 33 |
| Hesi–Iron | 0 | 0.0 | 0 | 0.0 | 110 | 31.2 | 0 | 0.0 | 243 | 68.8 | 353 |
| Batashi–Iron | 0 | 0.0 | 0 | 0.0 | 51 | 76.1 | 0 | 0.0 | 16 | 23.9 | 67 |
| Ir David–Iron | 0 | 0.0 | 0 | 0.0 | 33 | 42.3 | 0 | 0.0 | 45 | 57.7 | 78 |
| Qiri–Iron | 0 | 0.0 | 0 | 0.0 | 17 | 44.7 | 0 | 0.0 | 21 | 55.3 | 38 |

[a]Note the presence of an anomalously high number of typologically Chalcolithic axes in MB loci at Malha, most likely intrusive. Data sources: Yiftahel PPNB–Garfinkle 1987:73; Yiftahel EB–Rosen unpublished; Jericho PPNB–Crowfoot Payne 1983 (summarized in Garfinkle 1987:73); Abu Gosh PPNB–Lechevallier 1978 (summarized in Garfinkle 1987:73); Beisamun PPNB–Lechevallier 1978 (summarized in Garfinkle 1987:73); Beidha PPNB–Mortensen 1970; Ain Ghazal PPNB and PN–Rollefson et al. 1992; Sheikh Ali PPNB and PN–Garfinkle 1992:Tables 40 and 69; Sha'ar Hagolan PN–Garfinkle 1992:Table 56; Nahal Zehora PN–Gopher and Orelle 1989; Tsaf PN–Gopher 1988-9; Grar Chalco–Gilead 1989; Shiqmim Chalco–Levy and Rosen 1987; Horvat Beter Chalco–Rosen 1987a; Bab edh Dhra EB–McConaughy 1979; Yarmuth EB–Rosen 1988; Shadud EB–Rosen 1985; Qiri Iron–Rosen 1987b. All other data from unpublished research by the author.

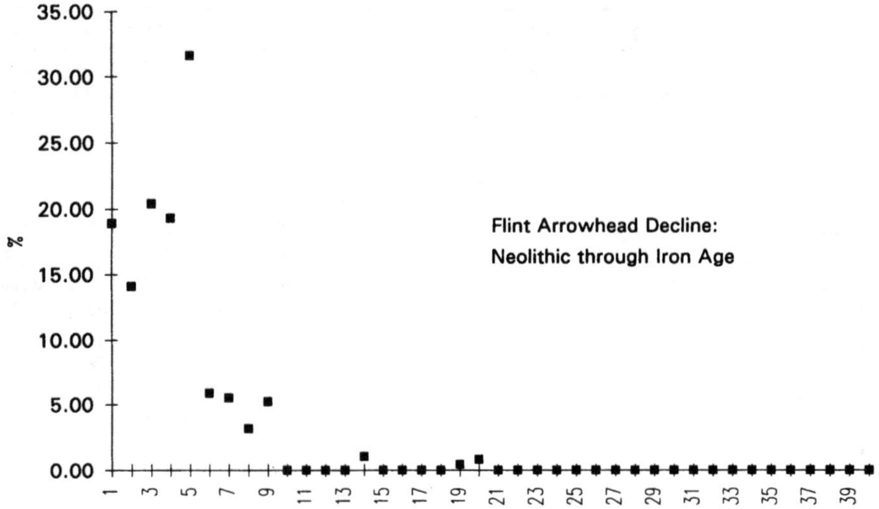


**Figure 7.** Decline of arrowheads by site, grouped according to period. Data and site numbers as in Table 4.

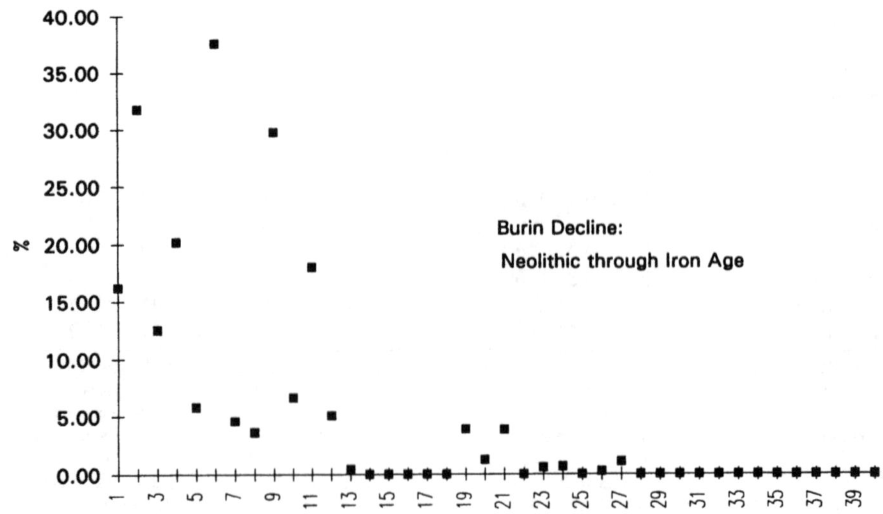


**Figure 8.** Decline of burins by site, grouped according to period. Date and site numbers as in Table 4.

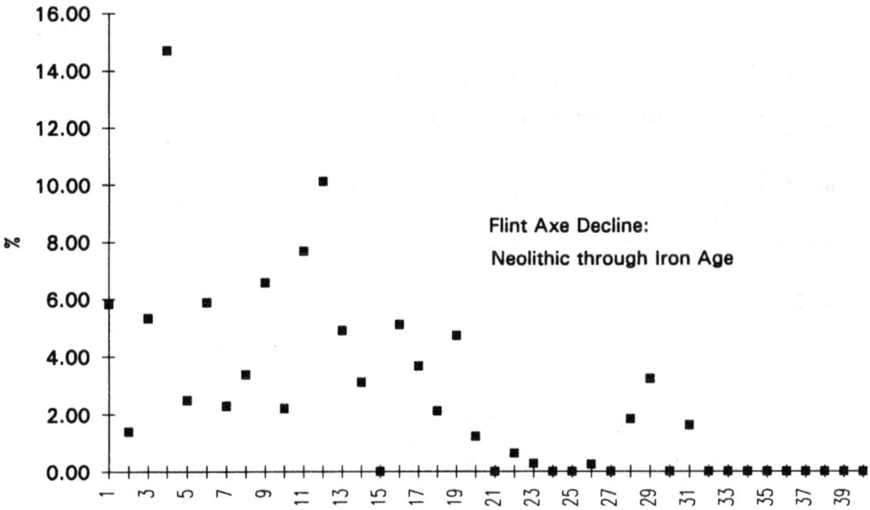


Figure 9. Decline of axes by site, grouped according to period. Data and site numbers as in Table 4.

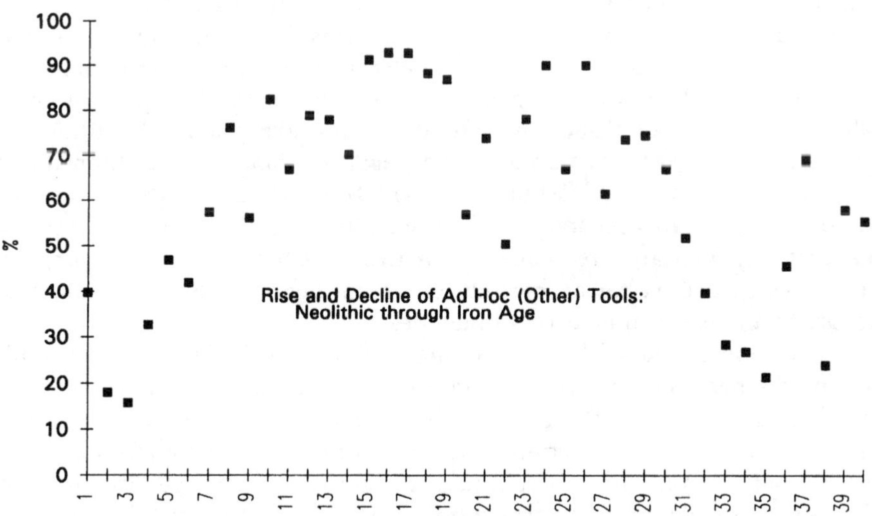


Figure 10. Rise and decline of ad hoc (other) tools by site, grouped according to period. Data and site numbers as in Table 4.

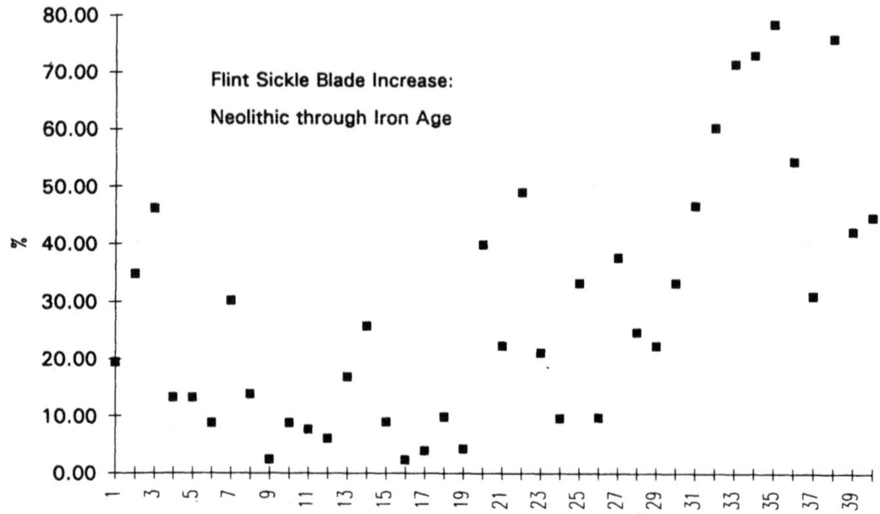


**Figure 11.** Rise of sickle percentages by site, grouped according to period. Data and site numbers as in Table 4.

One may distinguish three basic stages in the typological decline of chipped stone industries: a pre-Early Bronze Age stage, a Middle-Late Bronze Age stage, and a final, Late Iron Age stage. As with the quantitative data, there is little evidence for intra-period decline in the Chalcolithic and Early Bronze Ages, and the typological composition of the industries seems to remain stable within these periods. Data are not precise enough for determination in the later periods.

The pre-Early Bronze Age phase (Figure 6) shows the disappearance of the following general tool classes: arrowheads, burins, axes,[5] and microlithic drills (Figure 12). The exceptions occur in the Negev and Sinai, where chipped stone arrowheads were in use at least until the Early Bronze II. Some special tools were introduced during this general period, most notably the tabular (fan-) scraper, but quantitatively these are rare in most northern sites. Choppers and chopping tools are also typical of the Chalcolithic industries of the Beersheva Basin, although they are rare in, or absent from, earlier industries.

On a more detailed level, this long period from the end of the Neolithic through the beginning of the Early Bronze Age, spanning some 2000 years, can be subdivided into two phases. Arrowheads and burins drop out of the lithic repertoire some time in the terminal Neolithic, or in the transition from Neolithic to Chalcolithic. Axes and microlithic drills are present in both Neolithic and Chalcolithic assemblages, but absent from Early Bronze Age assemblages.[6]

In conjunction with these typological changes, some tool classes exhibit a trend toward less standardization and regularity in form. Tools such as borers,

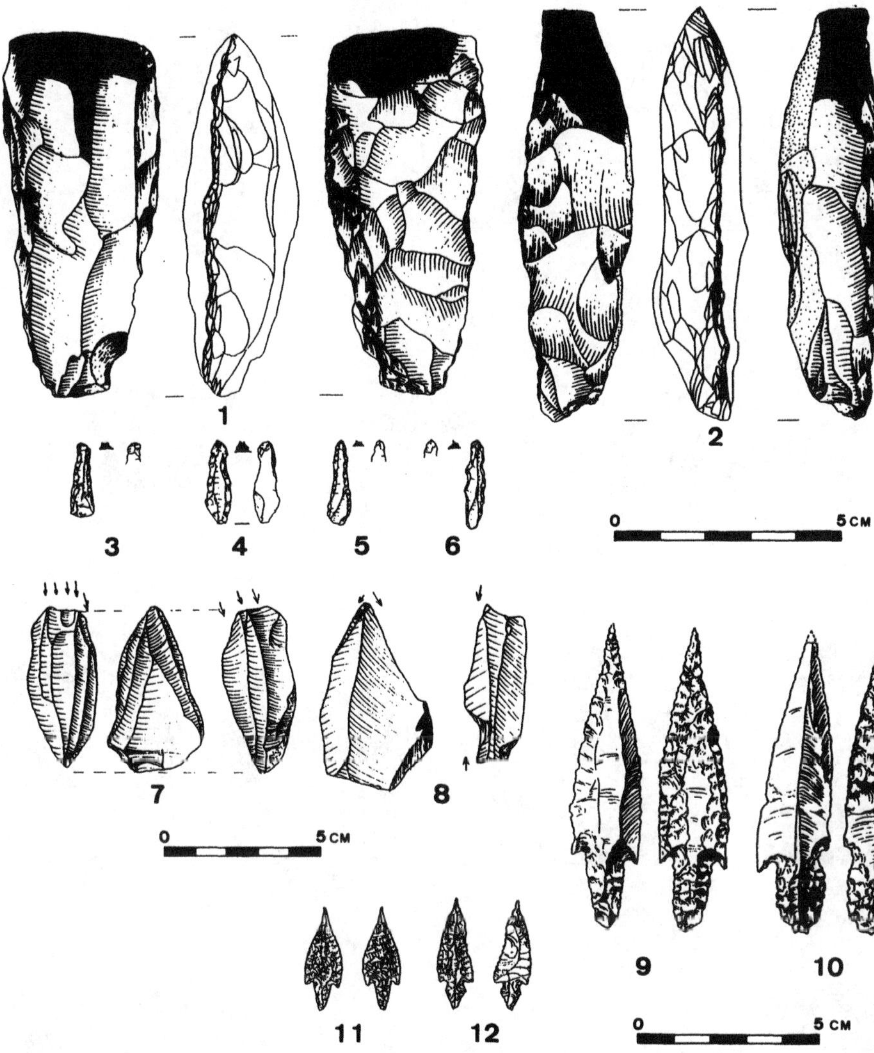


**Figure 12.** Axes (upper scale), microlithic drills (upper scale), burins (central scale), arrowheads (lower scale). Sources: axes: Sataf (unpublished); microlithic drills: Nahal Nizzana 103 (Burian and Friedman 1987:Fig. 7:9–12); burins: Sha'ar Hagolan (Stekelis 1972: Pl. 22:6–8); arrowheads: Jericho (Crowfoot Payne 1983:Fig. 333:1–2, 309:10–11).

scrapers, backed blades, and knives, more or less morphologically standardized in the Neolithic, "degenerated into what I have termed "ad hoc" tools by the Early Bronze Age (Figure 13). These are tools which are often difficult to classify into traditional formal tool types, although they may bear some of the traditional

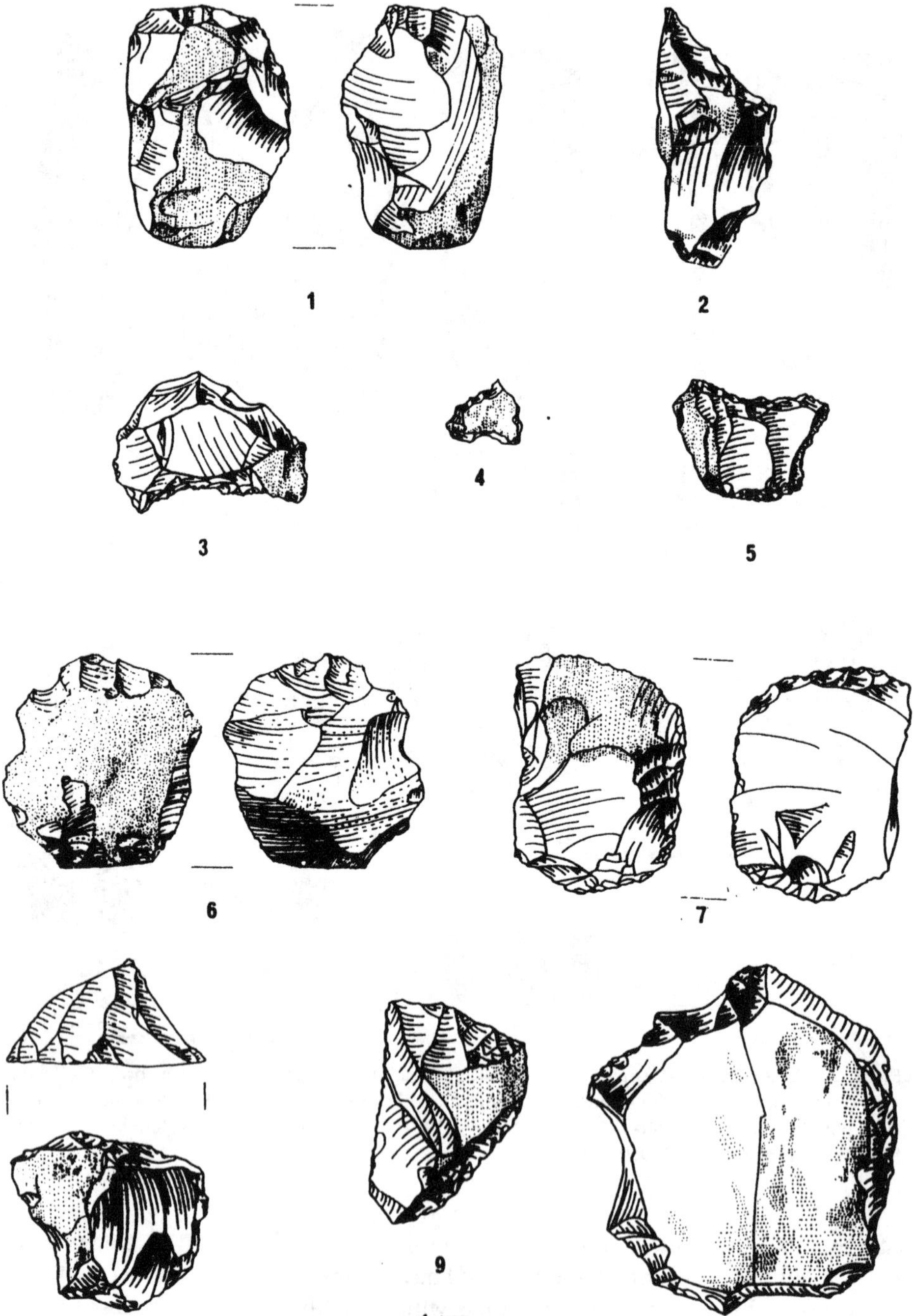


Figure 13. Ad hoc tools: Yarmouth Early Bronze Age (Rosen 1988:Pl. 50).

attributes. This process of destandardization is also evident in the increasing number of retouched flakes, notches, and denticulates found in Chalcolithic and especially Early Bronze Age assemblages. In terms of technology, this is also reflected in the increasing proportion of expedient flake cores from the Neolithic through the Early Bronze Age (cf. Parry and Kelly 1987). These expedient cores are the only ones found at most Early Bronze Age sites, excepting the rare specialized workshops and loci.

The second phase, the Early Bronze to Middle/Late Bronze Age transition, shows the decline of virtually all formal elements with the exception of sickle blades. Tabular (fan-) scrapers disappeared, apparently in the Middle Bronze I (also called Intermediate Bronze, Early Bronze IV, and EB-MB). The general ad hoc class seems to have continued, but at a much lower level of use. Thus, by the Late Bronze and Iron Ages, sickle blades (Figure 14) usually constitute between 50% and 90% of any assemblage (Figure 6, Table 4).

The final phase shows the disappearance of flint sickle blades, ca. 800 B.C. The data from Tell Batashi (Tables 2 and 3) are the only ones available to provide any precision for this final decline.

## THE DECLINE OF FLINT: AN EXPLANATION

Given our preconceptions concerning the replacement of flint by metal, perhaps the most striking conclusion to be drawn from the above data is that a significant quantitative and typological decline in lithic use took place before the introduction of metallurgy. The disappearance of flint arrowheads was a consequence of the decline in the importance of hunting in late Neolithic and Chalcolithic times, and neither the Chalcolithic nor the Early Bronze Age show metal arrowheads. In fact, with the exception of the desert regions, there is little evidence for use of the bow and arrow. The typological disappearance of arrowheads must have been accompanied by a quantitative decline as well, since during most of the Neolithic, arrowheads always constituted a significant proportion of any tool assemblage, sometimes as high as 40%. Such high proportions left similarly high quantities of waste. The simple elimination of the category could account for a noticeable decline in flint mass.

As noted above, burins are also rare-to-absent in post-Neolithic assemblages. Assuming their function as bone and woodworking implements, it is difficult to account for their disappearance. There are no obvious metal replacements in the Chalcolithic, although by the Early Bronze Age, saws are known, as at Kfar Monash (Hestrin and Tadmor 1963). It is more likely that they were simply replaced by other elements in the lithic repertoire, such as borers and retouched/utilized flakes.

The disappearance of axes after the Chalcolithic may indeed be a consequence of their replacement by copper equivalents, as I have suggested elsewhere

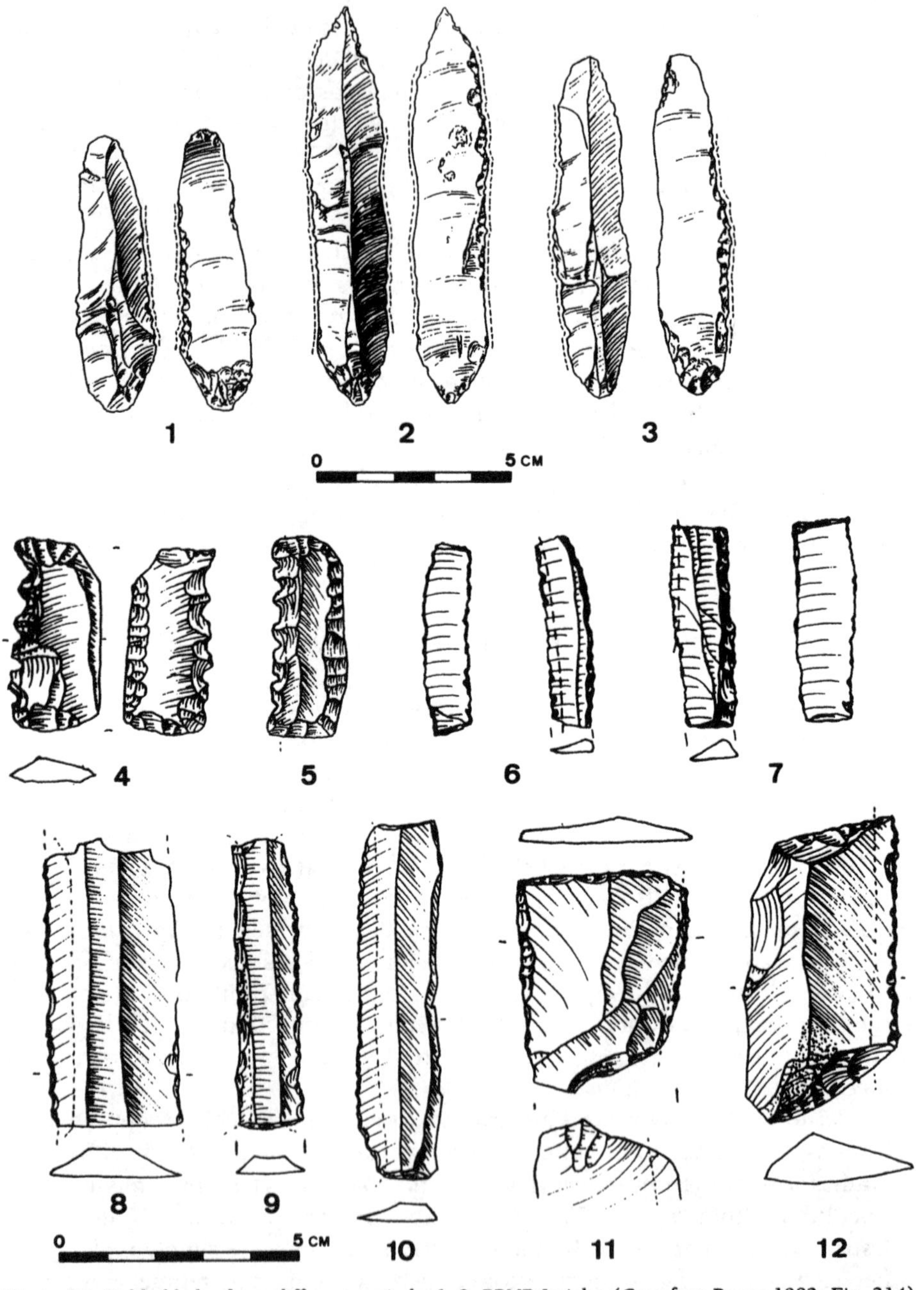


**Figure 14.** Sickle blades from different periods. 1–3: PPNB Jericho (Crowfoot Payne 1983: Fig. 314) (upper scale; all others, lower scale); 4–5: Pottery Neolithic from mixed fills at Hesi (unpublished); 6–7: Chalcolithic Shiqmim (Levy and Rosen 1987:Fig. 10.1:4–5); 8–10: Early Bronze Age Hesi (unpublished); 11–12: Iron Age Hesi (unpublished).

(Rosen 1984). However, the issue is more complex than simple replacement. Although stone axes were common in the Chalcolithic and absent from the Early Bronze Age, copper axes were present in both periods, raising the question as to why the stone implements were not replaced in the Chalcolithic, with the introduction of the metal tools. Of note here is the apparent rapidity of the replacement. To the best of current knowledge, chipped stone axes were present in the material culture of the Chalcolithic until the very end of the period, and yet were absent from the earliest layers of the Early Bronze Age. The transition between these periods was not a long one, usually estimated at less than 100 years.

Furthermore, assuming that axes were used for woodworking (e.g., Semenov 1964:126-134; Keeley 1983), the softness of the pure copper used for manufacturing the axes (Key 1980; Potaszkin and Bar-Avi 1980; Shalev and Northover 1987; Shalev et al. 1992) suggests that they were probably not much more efficient than their stone equivalents (e.g., Coles 1973:20–21). Thus mere utilitarian efficiency seems not to adequately explain this technological transition.

In this context several points should be made with respect to Chalcolithic and Early Bronze Age metallurgy. First, there appear to be significant contrasts between the two periods. Even discounting the clearly cultic Nahal Mishmar hoard (Bar-Adon 1980; per contra Moorey 1988), a high proportion of Chalcolithic artifacts apparently served a ritual function. In contrast, the copper artifacts recovered from Early Bronze Age Arad were exclusively utilitarian (Ilan and Sebbane 1989). Furthermore, the 67 axes and scores of copper awls found at Arad outnumber the entire corpus of copper implements (as opposed to slags and ores) found at Chalcolithic village sites. This situation is clearly related to the development of extensive trade networks in the Early Bronze Age, especially with South Sinai, the southern Negev, and perhaps Feinan, known copper sources (Amiran et al. 1973; Beit Arieh 1986; Rothenberg and Glass 1992).

There is also a technological contrast between the periods. The Chalcolithic shows both sophisticated lost wax cultic implements cast from a copper antimony alloy (Shalev and Northover 1991; Shalev et al. 1992; per contra Key's [1980] initial determination of an arsenic alloy) and simple mold, pure copper objects. In the Early Bronze Age, a technique using pure copper seems to have been the only one in use. This technique is significantly less difficult than lost wax alloy casting, and is more easily mass produced. It is probably also less time-consuming (read "cheaper") than the chipping and polishing of flint axes, each of which must, in essence, be handcrafted. In the case of breakage, although flint axes can often (but not always) be repaired with a transverse blow across the working edge, their effectiveness is reduced and they must eventually be discarded. There is good evidence that copper axes could be, and indeed were, recycled into the system (e.g., Shalev and Northover 1987; Ilan and Sebbane 1989), preserving a major part of the capital investment.

Thus the replacement of flint axes by metal ones may have been related to the establishment of steady supplies and trade routes, perhaps established specifically for the copper trade. The rapidity of the process may have been a result of differential value. Thus, if the copper was being mined or collected by the local inhabitants in the desert, and they were exchanging it for goods imported from the north, exchange values may have rendered the copper ores, and perhaps even the tools themselves, relatively inexpensive.

Ethnographically it is possible to draw an analogy with the disruption of the conch shell - flint axe trade in aboriginal Australia. The rapid replacement of flint axes by steel appears to have been a consequence of the breaking of the monopoly on axe production with the introduction of European steel axes, and a reduction in their value as a result. In short, it was no longer worthwhile to produce flint axes, since the price the Europeans asked for the steel ones was less than that required to maintain the production of flint axes (Sharp 1952). Of added note here is the general economic disruption resulting from the introduction of European steel axes.

The fact that extensive trade systems were established during the Early Bronze Age, and that the Chalcolithic systems were clearly of a lesser order of complexity, suggests that the copper trade itself was not the prime mover in establishing these systems. In fact, the copper trade system is a consequence of Early Bronze Age social complexity, not vice versa. The replacement of flint axes stems more from the evolution of Early Bronze Age social complexity and the concomitant establishment of extensive trade routes than it does from advantages derived from the copper tools themselves.

The fourth tool type to disappear by the Early Bronze Age was the microlithic drill (e.g., Macdonald 1932:8; Roshwalb 1981:166–7; Burian and Friedman 1987), which was utilized for bead production. The relative abundance of copper awls, especially at Arad in the Early Bronze Age (Ilan and Sebbane 1989), suggests that these metal tools might have replaced the flint drills (Rosen 1984). As with axes, they are less subject to breakage and more efficiently manufactured. However, re-evaluation again suggests a more complex picture. It is unclear whether copper drills can be used for drilling beads, given that the copper is often softer than the bead material, whereas flint is harder. In terms of chronology, as with the flint axes, copper awls are present in the Chalcolithic as well as the Early Bronze Age (disregarding the question of slight typological differences [Ilan and Sebbane 1989]), which leaves the question of timing open. Experimental archaeology and microwear analysis could help resolve some of these issues.

Furthermore, from a sampling perspective, flint drills are known in very large quantities from fewer than a half dozen very specialized sites (Rosen 1987), and only in very small quantities, if at all, from most sites. Bead production was a specialized activity, and the absence of drills in Early Bronze Age sites may simply be the result of sampling bias. Given the specialized nature of bead manufacture, the difference in general spatial distribution between the flint drills in the Chalcolithic, showing a very clustered distribution, and the copper awls of the

Early Bronze Age, showing a much more evenly spaced distribution (e.g., Ilan and Sebbane 1987) also suggests that they may be functionally distinct.

Thus it is currently not possible to determine if the perceived disappearance of microlithic drills is genuine, and if so, the result of metal replacements. Certainly, these tools do disappear at some point after the Chalcolithic (see endnote 6); given the continued use of beads of all types, they were presumably replaced by metal drills.[7]

The second phase in the decline of chipped stone assemblages occurred in the post-Early Bronze Age, some time between the beginning of the Middle Bronze II (ca. 1900 B.C.) and the Late Bronze Age (ca. 1500–1200 B.C.). This transition shows the decline and near disappearance of the whole range of ad hoc tools, which comprised the bulk of most Early Bronze Age assemblages. By the end of this period, sickles were essentially the only representative of chipped stone technologies remaining in significant use.

This was also the first period associated with the common availability of true bronze, and with a major increase in the general accessiblity of metals. This "innovation" is undoubtedly related to the development of the great Near Eastern trade networks of the second millennium B.C. (Muhly 1980). On the simplest level it is clear that bronze tools did indeed replace flint. However, as in the earlier phase, this replacement process has as much to do with trade and specialization as it does with inherent properties of the metals themselves.

As an industry, the ad hoc or expedient tools comprising the bulk of Early Bronze Age assemblages show little evidence for standardization or specialization in manufacture, contrasting significantly with the sickle industry. Cores and waste from these tools totally dominate the debris from Early Bronze Age sites, ample evidence that they were produced on-site (Rosen 1989a). Raw materials were almost always local, according well with the idea of local production. The ubiquity of these tools and their waste, the simplicity of the technology, and the seeming absence of real spatial concentrations, indicates that the ad hoc assemblage was fundamentally a domestic industry, produced expediently for immediate use by the users.

The key point here is that, whereas the flint tools used for domestic purposes (the ad hoc assemblage) comprised an "expedient" industry, the bronze replacements were the products of specialized production. Aside from any advantages to be derived from the use of metals, either in production efficiency or use, there is a clear element of evolving economic specialization, essentially the decline of the domestic economy, in Sahlins' (1972) terms. It is notable that the one rural site from the Middle Bronze II available for study, Malha, on the outskirts of modern Jerusalem, shows significantly greater quantities of ad hoc tools than other sites (Table 4; notwithstanding problems of intrusions, as at most of the later sites). The same is true of the desert Middle Bronze I pastoral sites (e.g., Gilead 1973), in spite of the clear involvement in the metal trade evident in these sites (Cohen

1986:295–298). It is not surprising that social and economic specialization would be less evident in the rural and pastoral periphery.

The continued use of flint sickles also requires explanation. Steensberg (1943; also Coles 1973:34–39) demonstrated that differences in harvesting efficiency between bronze and flint sickles are minimal, at least with respect to new tools. Although his experiments did not incorporate the issue of edge erosion and resharpening, the ease with which one may serrate a flint blade or sharpen a bronze sickle suggests that this was perhaps not a major issue. Regardless, the flint sickle assemblages show a wide range of edge retouch. The blanks, without gloss and presumably before use, are most often fresh, with little retouch. Various degrees of serration seem to reflect stages in the use and resharpening of the sickles.

Given the absence of significant functional differences between bronze and flint sickles, the apparently greater efficiency in manufacture of bronze sickles should have resulted in the replacement of the flint. However, flint sickles were already the product of specialized production and exchange. As such, the advantages derived from the possibly greater production efficiency of the bronze tools was minimized. Furthermore, given the presence of high quality flint throughout Israel, it is likely that, as a raw material, flint was much less expensive.

The final phase in the decline of chipped stone artifacts occurred during the Iron Age, with the disappearance of flint sickles. Unlike the previous replacement processes, the introduction of iron sickles seems to represent a genuine functional improvement. Steensberg (1943) documents the significant improvement in harvesting efficiency of iron sickles over flint and bronze. An interesting aspect here is that flint sickles are probably more efficiently manufactured than their iron counterparts. Since early iron smelting did not reach temperatures high enough to produce molten iron, iron tools had to be worked into shape with hammer and tongs, and could not be cast. Thus, the iron tools were apparently of such greater utility that they replaced the flint, in spite of lesser efficiency in manufacture.

The timing of the replacement is also of note. Waldbaum (1978:41–42, 57) indicated that, during the 11th century B.C., metal assemblages were still dominated by bronze, but by the 10th century B.C., iron tools already outnumbered bronze. This discrepancy was most pronounced for weapons, but is evident in other realms as well. Muhly (1980) suggested that the rise of iron metallurgy is in fact related to the decline of Bronze Age civilization toward the end of the second millennium B.C., the breakdown in international trade networks, and the resulting need to develop a replacement for the increasingly restricted access to copper and tin sources.

Flint sickles, not represented in Waldbaum's work, were ultimately replaced only in the 9th or 8th centuries B.C. This is not surprising. There is a pattern to the introduction of iron, such that it probably was initiated as a higher status metal (and probably more expensive as well), used primarily for weapons and perhaps for tools for which bronze and flint were simply inappropriate. The final replace-

Table 5. Ir David Tool and Assemblage Totals Calibrated for Effects of Intrusion and Percentage of Excavation (See Endnote 2)

| | Chalco/EB | MBII | LB, Iron Per/Hell |
|---|---|---|---|
| Relative population density | low | high | high |
| Percent excavated area | 7.5 | 20.5 | 72 |
| Estimated length of occupation (years) | 300 | 100 | 500 |
| Total tabular scrapers + axes/adzes | 9 | 4 | 2 |
| Index of intrusion | — | 0.31 | 0.13 |
| Total tools | 142 | 109 | 95 |
| Total lithics | 2894 | 2206 | 1294 |
| Tools calibrated for intrusions | — | 46 | 56 |
| Total lithics calibrated for intrusions | — | 937 | 518 |
| Tool calibration percent excavation | 1893 | 229 | 78 |
| Total lithics cal. percent excavation | 38,587 | 4570 | 719 |

ment of flint sickles occurred only after iron was readily available, even on the domestic level.

It is notable here that the continued use of flint in threshing sledges (Bordaz 1965; Whallon 1973) does not constitute an exception to the process of decline outlined above. Threshing sledge teeth are, on the one hand, the products of efficient specialized manufacture, and on the other, are associated with the rural periphery. They are also probably not less efficient than iron for the purpose of separating the wheat from the chaff.

## CONCLUSIONS

The decline and virtual disappearance of chipped stone technology in the Levant over the course of some 3000 years is clearly related to the rise of an alternative technology—metallurgy. However, the rise of metallurgy was not dictated by its inherent advantages over lithic technology in a strict sense, nor by perceived needs to improve existing technologies. Rather, the gradual replacement of flint by metal can be seen as a long series of by-products of other social, cultural, political, and technological events. Some of these can be related to increasing demographic and social complexity, others to expanding trade, and others to the nature of craft specialization and changes in economic organization. The role of relative efficiency of manufacture is clearly of importance. Functional advantages of metal tools over flint counterparts seem to have played only a minimal role in the replacement process.

Thus the first noticeable decline in lithic exploitation, in the terminal Neolithic, is not related to metallurgy at all, and the first 1000 years of metallurgy, in the Chalcolithic, are not accompanied by any decline in lithic technology. The first replacement, of flint axes by copper ones, seems to have been a result of expanded trade routes and exchange systems and their ease of manufacture, rather than properties of the respective tools themselves. The second stage in the replacement process shows the decline of homemade products, the ad hoc flint industry, and its replacement by specialist-produced bronze tools. Again it is difficult to trace any obvious specific advantages to the metal tools themselves, and the replacement process seems to reflect more the imposition of a new economic organization, a trend from the household economy to one based on specialization and exchange. It is no accident that flint sickles remain in use during this period since they are both functionally efficient relative to bronze, and were already the product of specialized production.

The final decline occurred only with the ready availability of iron, and is the only example in which metal has a demonstrable advantage over flint. In this case, in fact, the flint tools were probably more easily produced than metal ones.

An important issue here is that the initial uses to which the metals were put seem not to have actually displaced the flint. Thus the primarily cultic functions of Chalcolithic metallurgy did not displace flint technology. Copper axe replacement of flint is a later occurrence, in essence a by-product of a technology initially developed for other purposes.

The same is true for the weapons-dominated bronze industry of the Middle and Late Bronze Ages. By this period flint arrowheads had been out of use for several thousand years (in the Levant—Egypt is a somewhat different case). And flint was never used for body armor or swords. Thus, again in the initial stages of bronze production, there was little functional overlap between the metal and flint tools. However, once the industry was in place, a wide range of implements could be produced, effectively replacing the less specialized flint industry. The greater flexibility of metal as a raw material, and the wider range of uses to which it may be put, are also factors. One smith can produce a far wider range of tools than a flint knapper, essentially providing better opportunities for the support of craft specialists.

Finally, a similar picture obtains for the Iron Age. Iron technology was not developed as a replacement for flint, but for bronze, and the ultimate decline of flint sickles was in a sense a by-product of the decline of bronze and its replacement by iron. That is, the replacement of flint was incidental to a different process altogether.

Thus what seems in hindsight to be a clear trend, the decline of flint and the rise of metallurgy, is in reality a set of discrete and occasionally almost unrelated skips and jumps. There is no linear inverse correlation between the rise of one and the decline of the other. In fact, metallurgy created entirely new sets of functions which had no parallels in the lithic repertoire. And in at least one case, that of the decline of arrowheads and hunting, lithic functions were simply abandoned, due to factors totally unrelated to metallurgy.

These ideas suggest that trends in technological change, which appear to us from the perspective of modern technology as improvements or as logical developments, need not have been so. In Spratt's (1982) terms, the critical paths of these technologies were very complex, and it is not all clear that a linear or utilitarian approach is really appropriate for examination of technological replacement involving more than a single episode of innovation (cf. also Schiffer and Skibo 1987). Complex, long-term, undirected technological change, in many ways like biological evolution, can appear to be progressive and linear. This illusion can mask far more important cultural developments.

## ACKNOWLEDGMENTS

I am grateful to George Odell for inviting me to participate in this conference. Some of the data collection on which this paper is based was conducted while I was a National Endowment for the Humanities Fellow at the Albright Institute in 1985. Needless to say, data collection from the large number of sites would have been impossible without the active help of numerous archaeologists who were not only good enough to allow me access to their collections, but also generous enough to significantly modify their collection strategies to better fit my goals. I am grateful to George Odell, Mike Shott, Brian Hayden and Avi Gopher for reviewing an earlier version of this manuscript. I would also like to thank all the participants in the conference for making it one of the most stimulating that I have ever attended.

## NOTES

1. The evidence for increasing specialization in manufacture derives from major contrasts in the distribution of different core, waste, and tool types, and in the discovery of a few primary and secondary workshop sites for blade and/or sickle manufacture. Some types (ad hoc elements) were clearly manufactured on site, while other types were unquestionably imported. For data and discussions, see Rosen (1986, 1987a, 1988, 1989a,b).
2. The problem of intrusions and mixing is also acute, especially considering that the bulk of most assemblages derives from fill loci. It is possible to attempt a calibration of degree of mixing, based on intrusive index fossils (also see Hesse [1986] for discussion of a similar problem with respect to bones). Assuming that artifacts mix more or less randomly with no selection process involved, if an index of intrusion can be calculated from a known quantity, an index fossil, it can then be applied to non-diagnostic artifacts as well. The assumption is that intrusions derive from the stratum beneath the one being excavated. Thus the original donor stratum consisted of the material in it, plus the material intruding into the above stratum. The formula for this calibration is as follows:

   $A_0 = A_1 - \underline{I} \times N_{1-I}$

   where $A_0$ = the original unmixed assemblage, to be used for further study

$A_1$ = the assemblage as recovered from the excavation, including intrusions
N = the size of the donor stratum (the lower stratum)
I = the index of intrusion, calculated as the number of index fossils in $A_0$ divided by the number of index fossils in the donor stratum plus the number in $A_0$

The formula was first developed in consultation with Ilan Sharon. It is also possible to compute several intrusion indices, based on different index fossils, and to average the results. Figure 10 and Table 5 show the results of such an analysis performed on the materials from Ir David. In essence the decline from Early Bronze to Middle Bronze Age is accentuated, but there is little other change. A similar phenomenon occurs at Qasis.

3. The differences in scales between sites in Figures 1–5 are caused by differences in types of fill and matrix (e.g., alluvial, colluvial, construction, floors, destruction, etc.), in intensity of occupations, and in collection strategies. Use of a single scale for all would render some of the graphs unreadable. Data are taken from Table 1. Approximate dates for periodization: Pre-Pottery Neolithic B: 7400–6000 B.C. (uncalibrated $C^{14}$); Pottery Neolithic A: 6000–5000 B.C. (uncalibrated $C^{14}$); Pottery Neolithic B: 5000–4000 B.C. (uncalibrated $C^{14}$); Chalcolithic: 4500–3500 B.C. (calibrated $C^{14}$); Early Bronze Age: 3500–2300 B.C. (calibrated $C^{14}$); Middle Bronze I = Intermediate Bronze: 2300–1950 B.C. (calibrated $C^{14}$ and calendric dates); Middle Bronze II: 1950–1550 B.C. (calendric dates); Late Bronze: 1550–1200 B.C. (calendric dates); Iron Age: 1200–600 B.C. (calendric dates).

4. The "Persian" Period assemblage from Tell Hesi (Bennett et al. 1989) is almost undoubtedly an amalgam of artifacts from earlier periods, brought together by the vagaries of Persian pit digging. Numerous intrusions are present in the illustrations, and seem not to be recognized as such. Thus on p. 236 the authors claim that no lithic artifacts from the Pottery Neolithic were present in the Persian period assemblage, but two pages previously (Fig. 186:2–4) they illustrate classic examples of PN arrowheads (which are recognized as intrusions, but not as Pottery Neolithic!). It is not possible to apply the formula noted above to this issue, since typologically one cannot distinguish between Iron Age sickles and alleged Persian period sickles, the only diagnostic artifacts on which one might perform a calibration.

5. A distinction must be made between polished stone votive axes, more akin to lapidary than chipped stone technology, and the axe-adze family common in the Chalcolithic. Although chipped stone axes sometimes show polish on the working edge, the body of the piece is always left in its chipped state, whereas votive axes are fully ground. The two types also differ in raw material, technology of manufacture, and morphology. They contrast in function as well, the votive axes usually assumed to be ritual in function, in contrast to the utilitarian nature of the second class. Votive axes are found in later periods.

   The axe family in the Chalcolithic is usually divided into three general classes—axes, adzes, and chisels—often with several subcategories according to size or variation in morphology (e.g., picks, small axes, transverse axes) (Cauvin 1968; Lee 1973; Roshwalb 1981:45–52; Levy and Rosen 1987). A distinction between the axe and the adze can be made according to the cross-section shape of the working edge. Axes show a symmetrical cross-section, often lens-shaped but sometimes rectilinear, with the working edge located in the middle, whereas adzes show an asymmetric cross-section, with the working edge skewed to the lower side of the implement. These differences are usually assumed to reflect differences in method of hafting and in function, such that axes were used for chopping, while adzes were used for chipping (e.g., Semenov 1964:126–134; Lee 1973). Chisels are long and narrow, and are easily distinguished from both of the above. Presumably they were also functionally different, serving as chisels or punches.

6. Microlithic drills have been recovered from a limited number of sites, and most of these are inappropriate for quantitative analysis. Therefore, this category has not been included in Table

4. A summary of the data on these tools is presented in Rosen 1987a. Excavations conducted by the author in the summer of 1993 (after the first version of this paper was presented) at the Camel Site, an Early Bronze I pastoral encampment in the south Central Negev, recovered some 30 microlithic drills, the first time these have been recovered from chronologically Early Bronze Age contexts. The site is early within the period, and culturally seems to fall outside the traditional Early Bronze Age urban system of the Mediterranean zone. The interplay between center and periphery in terms of technological change is of interest here, especially in light of the recovery of two copper awls as well, but data are still too few to properly address the question. Regardless, the general argument is not affected.

7. Drill bits can also be made from bits of shell or bone (Hodges 1976:106), but are not any more efficient than flint. Thus, given that flint drills had been employed for such a long time, it is unclear why they might have been replaced by shell or bone. In any case, there is no evidence for such an occurrence.

## REFERENCES

Amiran, R., I. Beit Arieh, and J. Glass. 1973. The Interrelationship between Arad and Sites in the Southern Sinai in the Early Bronze II. *Israel Exploration Journal* 23:193–198.

Ataman, K. 1992. Threshing Sledges and Archaeology. In *Prehistoire de l'agriculture: Nouvelles approches experimentales et ethnographiques,* pp. 305–319. CNRS, Monographies du CRA 6, Paris.

Bar Adon, P. 1980. *The Cave of the Treasure.* Israel Exploration Society, Jerusalem.

Beit Arieh, I. 1986. Two Cultures in South Sinai in the Early Bronze Age. *Bulletin of the American Schools of Oriental Research* 263:27–54.

Bennett, W. J., Jr., J. B. Sollberger, and A. F. Gettys. 1989. Flint Tools. In *Tell el Hesi. The Persian Period (Stratum V),* ed. W. J. Bennett, Jr., and J. A. Blakely, pp. 231–256. Eisenbrauns, Winona Lake, Indiana.

Bordaz, J. 1965. The Threshing Sledge - Ancient Turkish Grain Separating Method Still Proves Efficient. *Natural History* 74:216-229.

Bordaz, J. 1969. Flint Flaking in Turkey. *Natural History* 78:73–77.

Burian, F., and E. Friedman. 1987. Chalcolithic Borer Industry at Site 103 - Nahal Nizzana. *Mitekufat Haeven: Journal of the Israel Prehistoric Society* 20:160–172.

Cauvin, J. 1968. *Fouilles de Byblos IV: Les outillages neolithiques de Byblos et du littoral libanais.* Maisonneuve, Paris.

Childe, V. G. 1951. *Man Makes Himself.* New American Library, New York.

Cohen, R. 1986. *The Settlement of the Central Negev in the Light of Archaeology and Literary Sources during the 4th - 1st Millennia.* Ph.D. dissertation, Hebrew University, Jerusalem (in Hebrew with English abstract).

Coles, J. 1973. *Archaeology by Experiment.* Scribners, New York.

Crowfoot, J. 1936. Notes on the Flint Implements of Jericho 1935. *Liverpool Annals of Archaeology and Anthropology* 22:174-184.

Crowfoot, J. 1937. Notes on the Flint Implements of Jericho 1936. *Liverpool Annals of Archaeology and Anthropology* 24:35–52.

Crowfoot Payne, J. 1983. The Flint Industries of Jericho. In *Excavations at Jericho V,* by K. A. Kenyon and T. A. Holland, pp. 622–758. Oxford Press, Oxford.

Gilead, D. 1973. Flint Industry of the Bronze Age from Har Yeruham and Tell Nagila. In *Excavations and Studies*, edited by Y. Aharoni, pp. 133–143. Tel Aviv University, Tel Aviv (in Hebrew with English abstract).

Gilead, I. n.d. *Investigations at the Chalcolithic Site of Grar, Northern Negev.* Beersheva VI - Ben-Gurion University Press, Beersheva. In press.

Glock, A. E. 1985. Tradition and Change in Two Archaeologies. *American Antiquity* 50:464–477.

Guy, P. L. O., and R. M. Engberg. 1938. *Megiddo Tombs.* University of Chicago Press, Chicago.

Heskel, D. 1983. A Model for the Adoption of Metallurgy in the Ancient Middle East. *Current Anthropology* 24:362–366.

Hesse, B. 1986. Animal Use at Tel Miqne-Ekron in the Bronze Age and Iron Age. *Bulletin of the American Schools of Oriental Research* 264:17–28.

Hestrin, R., and M. Tadmor. 1963. A Hoard of Tools and Weapons from Kfar Monash. *Israel Exploration Journal* 13:265–288.

Hodges, H. 1976. *Artifacts: An Introduction to Early Materials and Technology.* John Baker, London.

Ilan, O., and M. Sebbane. 1989. Copper Metallurgy, Trade, and the Urbanization of Southern Canaan in the Chalcolithic and Early Bronze Age. In *L'Urbanisation de la Palestine a l'age du bronze ancien*, edited by P. de Miroschedji, pp. 139–162. BAR International Series 527, Oxford.

Keeley, L. H. 1983. Microscopic Examination of Adzes. In *Excavations at Jericho V*, by K. M. Kenyon and P. A. Holland, p. 759. Oxford University Press, Oxford.

Kempinsky, A. 1983. Early Bronze Age Urbanization of Palestine: Some Topics in a Debate. *Israel Exploration Journal* 33:235–241.

Kempinsky, A. 1989. Urbanization and Metallurgy in Southern Canaan. In *L'Urbanisation de la Palestine a l'age du bronze ancien*, edited by P. de Miroschedji, pp. 163–168. BAR International Series 527, Oxford.

Key, C. A. 1980. Trace Element Composition of the Copper and Copper Alloy Artifacts of the Nahal Mishmar Hoard. In *The Cave of the Treasure*, by P. Bar Adon, pp. 238–243. Israel Exploration Society, Jerusalem.

Lee, J. R. 1973. *Chalcolithic Ghassul: New Aspects and Master Typology.* Ph.D. dissertation, Hebrew University, Jerusalem.

Levy, T. E., and S. A. Rosen. 1987. The Chipped Stone Industry at Shiqmim: Typological Considerations. In *Shiqmim I*, edited by T. E. Levy, pp. 564–610. BAR International Series S356, Oxford.

Levy, T. E., and S. Shalev. 1989. Prehistoric Metalworking in the Southern Levant: Archaeometallurgical and Social Perspectives. *World Archaeology* 20:352–372.

Macalister, R. M. A. 1912. *The Excavation of Gezer II.* John Murray, London.

McConaughy, M. 1979. *Formal and Functional Analysis of Chipped Stone Tools from Bab edh Dhra, Jordan.* University Microfilms, Ann Arbor.

Macdonald, E. 1932. *Beth Peleth II.* British School of Archaeology in Egypt, London.

Moorey, P. R. S. 1988. The Chalcolithic Hoard from Nahal Mishmar, Israel, in Context. *World Archaeology* 20:171–189.

Muhly, J. D. 1980. The Bronze Age Setting. In *The Coming of the Age of Iron*, edited by T. A. Wertime and J. D. Muhly, pp. 25–68. Yale University Press, New Haven.

Neuville, R. 1930. Notes de prehistoire palestinienne. *Journal of the Palestine Oriental Society* 10:193-221.

Neuville, R. 1934/5. Les debuts de l'agriculture et la faucille préhistorique en Palestine. *Journal of the Jewish Palestine Exploration Society* 3:xvii-xlii.

Oakley, K. P. 1975. *Man the Tool-Maker.* British Museum of Natural History, London.

Parry, W. J., and R. L. Kelly. 1987. Expedient Core Technology and Sedentism. In *The Organization of Core Technology*, edited by J. K. Johnson and C. A. Morrow, pp. 285–304. Westview Press, Boulder.

Petrie, W. M. F. 1891. *Tell el Hesy.* Committee for the Palestine Exploration Fund, London.

Potaszkin, R., and K. Bar-Avi. 1980. A Material Investigation of the Metal Objects from the Nahal Mishmar Treasure. In *The Cave of the Treasure*, by P. Bar-Adon, pp. 235–237. Israel Exploration Society, Jerusalem.

Redman, C. L. 1978. *The Rise of Civilization.* Freeman, San Francisco.

Renfrew, C. 1984. The Anatomy of Innovation. In *Approaches to Social Archaeology*, edited by C. Renfrew, pp. 390–418. Edinburgh University Press, Edinburgh.

Rosen, S. A. 1984. The Adoption of Metallurgy in the Levant: A Lithic Perspective. *Current Anthropology* 25:504–505.

Rosen, S. A. 1986a. The Analysis of Trade and Craft Specialization in the Chalcolithic Period: Comparisons from Different Realms of Material Culture. *Michmanim* 3:21–32.

Rosen, S. A. 1986b. The Gezer Flint Caches, 1970–71. In *Gezer IV*, edited by W. G. Dever. Hebrew Union College, Jerusalem.

Rosen, S. A. 1987. The Potentials of Lithic Analysis in the Chalcolithic in the Northern Negev. In *Shiqmim I*, edited by T. E. Levy, pp. 295–312. BAR International Series 356, Oxford.

Rosen, S. A. 1988. Notes on the Flint Implements from Tell Yarmuth, 1980–1982. In *Yarmouth I*, by P. Miroschedji, pp. 135–142. Editions Recherche sur les Civilisations, Paris.

Rosen, S. A. 1989a. The Analysis of Early Bronze Age Chipped Stone Industries: A Summary Statement. In *L'urbanisation de la Palestine a l'age du bronze ancien*, edited by P. de Miroschedji, pp. 199–222. BAR International Series 527, Oxford.

Rosen, S. A. 1989b. The Origins of Craft Specialization: Lithic Perspectives. In *People and Culture in Change*, edited by I. Hershkovitz, pp. 107–114. BAR International Series 508, Oxford.

Rosen, S. A. n.d. Metals, Rocks, Specialization, and the Beginning of Urbanism in the Northern Negev. In *Biblical Archaeology Today II*. Israel Exploration Society, Jerusalem. In press.

Roshwalb, A. F. 1981. *Protohistory in the Wadi Ghazzeh: A Typological and Technological Study Based on the Macdonald Excavations.* Ph.D. dissertation, University of London, London.

Rothenberg, B., B. Tylecote, and P. J. Boydell, editors. 1978. *Chalcolithic Copper Smelting.* International Annual of Metallurgical Studies, London.

Rothenberg, B., and J. Glass. 1992. The Beginnings and Development of Early Metallurgy and the Settlement and Chronology of the Western Arabah from the Chalcolithic Period to Early Bronze Age IV. *Levant* 24:141–157.

Sahlins, M. 1972. *Stone Age Economics.* Aldine, Chicago.

Schick, T. 1978. Flint Implements. In *Early Arad*, edited by R. Amiran et al., pp. 58–63. Israel Exploration Society, Jerusalem.

Semenov, S. A. 1964. *Prehistoric Technology.* Translated by M. W. Thompson. Adams and Dart, Bath.

Schiffer, M. B., and J. M. Skibo. 1987. Theory and Experiment in the Study of Technological Change. *Current Anthropology* 28:595–622.

Shalev, S., Y. Goren, T. E. Levy, and J. P. Northover. 1992. A Chalcolithic Macehead from the Negev, Israel: Technological Aspects and Cultural Implications. *Archaeometry* 34:63–72.

Shalev, S., and P. Northover. 1987. Chalcolithic Metalworking from Shiqmim. In *Shiqmim I*, edited by T. E. Levy, p. 356- 71. BAR International Series 356, Oxford.

Shalev, S., and P. Northover. 1991. The Chalcolithic Metal Industry at Shiqmim and the Nahal Mishmar Hoard. *Institute for Archaeometallurgical Studies* 17 (June):4–5.

Sharp, L. 1952. Steel Axes for Stone-Age Australians. *Human Organization* 11:17–22.

Spratt, D. A. 1982. The Analysis of Innovation Processes. *Journal of Archaeological Science* 9:79–94.

Steensberg, A. 1943. *Ancient Harvesting Implements.* Copenhagen.

Waldbaum, J. 1978. *From Bronze to Iron.* Studies in Mediterranean Archaeology LIV. Paul Astroms, Goteborg.

Whallon, R. J. 1978. Threshing Sledge Flints: A Distinctive Pattern of Wear. *Paleorient* 4:319–324.

Yeivin, E. 1959. Flint Tools from Horvat Beter. *Atiqot* 2:40–44. (In Hebrew).

Zaccagnini, C. 1983. Patterns of Mobility among Ancient Near Eastern Craftsmen. *Journal of Near Eastern Studies* 42:245–264.

## Chapter 6

# Lithic Analysis and Questions of Cultural Complexity

## The Maya

JAY K. JOHNSON

### ABSTRACT

The ways that stone tool production and use were organized among the Maya and in the southeastern United States were not much different from one another, despite obvious differences in cultural complexity. In both regions, the bulk of the stone tools produced during the periods of sedentary, centralized control were flakes derived from ad hoc cores. One of the most vexing problems has been the role of craft specialization and the degree to which it was controlled by the elite among the Maya. An analysis of obsidian artifacts from the Classic site of Nohmul and a review of other data on stone tool manufacture lead to the conclusion that specialized production of elite paraphernalia may have been centrally controlled, while the manufacture of subsistence items was based on direct, consumer-producer relationships. A general lesson derived from looking at the Maya from the Southeast is the need to be careful about the way that the presence of hieroglyphs influences our reading of the lithic record.

### INTRODUCTION

There have been several overviews of the development of lithic analysis in Maya archaeology (Sheets 1977; Johnson 1985; Fowler 1991), and three collec-

**JAY K. JOHNSON** • Department of Sociology and Anthropology, University of Mississippi, University, Mississippi 38677.

*Stone Tools: Theoretical Insights into Human Prehistory*, edited by George H. Odell. Plenum Press, New York, 1996.

tions of essays dealing entirely or primarily with Maya lithic artifacts (Hester and Hammond 1976; McAnany and Issac 1989; Hester and Shafer 1991). What follows is not a general review of lithic analysis in southern Mesoamerica. Rather, it is a look at how stone tools have been used to examine aspects of complex social organization in this region. Since my own work with Maya stone tools has been intermittent, this will be an outside perspective, looking at the literature from the southeastern United States. In doing so, the stage will be set for the analysis of the obsidian artifacts recovered during several seasons of work at Nohmul, a major Maya site in northern Belize. Although the Maya were demonstrably more complex than the tribes and chiefdoms of the Southeast, it is difficult to measure that difference on the basis of stone tool production and use.

Remarkable parallels in the development of lithic analysis exist between southeastern (Johnson 1993) and Mayan archaeology. Stone tools were essentially ignored until the major site reports of the Carnegie Institute's work at Uaxactun (Ricketson and Ricketson 1937; Kidder 1947). Although chronology was clearly a major goal of the Carnegie program, little in the way of chronological control was achieved in the analysis of the lithic artifacts. In fact, Coe (1965:594), in his summary article on nonceramic artifacts in the *Handbook of Middle American Indians*, concluded that it is not possible to distinguish the Preclassic from the Late Classic period on the basis of stone tools.

Part of the problem was pointed out by Taylor (1948:124) when he noted that the initial Uaxactun typology (Ricketson and Ricketson 1937) was based on the classification developed for the stone tools from Pecos, New Mexico (Kidder 1932), and concluded that this could only be "for descriptive convenience and not for cultural elucidation." All of the major lithic analyses to follow well into the 1970s (Kidder 1947; Coe 1959; Willey et al. 1965; Willey 1972, 1978) derived their typological structure from the Uaxactun analysis (Johnson 1985). Part of the problem is that, although it can be done (Rovner 1975; Hester 1985; Gibson 1986), chronological assessment on the basis of the stone tools from a Maya site is often difficult because formal tools which are chronologically sensitive are remarkably rare.

Puleston (1969), in her Masters thesis on chert blade tools from Tikal, provided the first glimpse of what was to come in Maya lithic analysis. The primary emphasis in her analysis was function, as determined by microscopic examination using techniques introduced by Semenov (1964). John Witthoft was her thesis director, thereby having peripheral influence on lithic analysis in this region as well as the Southeast (Johnson 1993:41). Wear pattern studies have been relatively slow in coming to the Maya area, but have made major contributions in recent years (Hay 1978; Wilk 1978; Shafer 1983; Mallory 1984; Gibson 1986; Lewenstein 1987; Aldenderfer et al. 1989; Sievert 1992).

The second major trend in Maya lithic analysis came from another direction. Sheets (1972) and Rovner (1974) developed detailed production trajectory models for obsidian blade industries from the Maya area. These reflect a general trend in American lithic analysis (Bradley 1975; Collins 1975), which combined an increased interest in prehistoric behavior with a sophisticated knowledge of lithic technology. In the Maya region, the understanding of stone tool production techniques came primarily from Don Crabtree, as both Sheets and Rovner were summer flintknapping school alumni. The need to reconstruct behavior was one of the major concerns of the new archaeology, which was taking control of Maya archaeology at about this time (Sabloff 1990).

## AD HOC TOOLS

There remained an obstacle to putting this new-found ability to measure production activity on the level of prehistoric product distribution and tool manufacture. Once blade production has been described, there is little else that is technologically interesting in most lithic assemblages from the Southern Maya Lowlands. In stark contrast to the sophistication expressed in sculpture and architecture, the lithic artifacts from sites like Copan, Tikal and Palenque are rather rudimentary. And generally, there are not many of them. Kidder (1947:2) observed:

> In spite of the great amount of work that was done at Uaxactun, the total "take" of nonceramic utilitarian artifacts was very small. . . . One can only conclude that the Maya actually used relatively few chipped and ground stone tools and tools of bone, their place presumably having often been taken by implements of the extremely hard woods so abundant in the lowlands.

Mississippian period lithic assemblages from the Southeast, in contrast to Paleoindian and Archaic material, are similarly uninteresting. Part of the problem is the failure of archaeologists to recognize the nature of the technology. Parry and Kelly (1987) used data primarily from Mexico and the Southwest to argue that sedentism produced a fundamental shift in stone tool producing strategies. While bifaces were the focus of the lithic technology of hunters and gatherers, serving as core, knife and projectile point (Kelly 1988), sedentism allows the stockpiling of raw material and an emphasis on casual flake core technologies to produce the majority of the cutting edges used in day-to-day activities.

Although many of the participants in the second Maya Lithics Conference recognized the presence of what are generally called ad hoc or flake tools (Aldenderfer 1991; Clark and Bryant 1991; Hester et al. 1991; Mitchum 1991; Potter 1991), often as a sort of background noise against which the analysis of the more formal tools takes place, it is wear pattern studies that have demonstrated the importance of flake tools. Ad hoc tools were used to perform a broad range of activities in a sample of primarily house mound assemblages from the central Peten (Aldenderfer et al. 1989), where they appear to have made up the majority of the tool assemblage (Aldenderfer 1991:132). They constitute 65% of the artifacts showing wear from Cerros (Lewenstein 1987:132). If the mounds of chert nodules which Thompson (1991) describes for Becan are seen as stockpiles of raw material for flake cores as well as celt blanks, some of the analytical difficulties he discusses can be resolved.

Mostly low-to-medium quality chert and chalcedony is available throughout most of the Southern Maya Lowlands, and it seems likely that ad hoc tools were used in performing the majority of domestic and many of the craft activities from Preclassic to recent times. The need to recognize the importance of flake tools is underscored by the fact that two analysts (Fedick 1991; Aldenderfer 1991), working with similar data from the same region, reached opposite conclusions about the lithic assemblages from central Peten housemounds. On the basis of the distribution of a series of attributes which were selected to relate flakes to different stages of tool manufacture and maintenance, Fedick argued that tool production was more common away from major centers than at them. This led him to conclusions about elite control and the structure of Maya economy. He did consider an alternative that the early stage flakes are the result of ad hoc tool production, citing Boksenbaum's (1980) description of Preclassic obsidian flake tools from the Valley of Mexico, but he decided that this was unlikely, since none of the special flakes described by Boksenbaum were found. Aldenderfer's (1991) reconstruction of the production sequence makes it clear that nodule flaking, rather than nodule smashing, was a characteristic of ad hoc technology in the Peten. His analysis shows flake tools to be more common than tool production by-products in the housemounds away from the major centers.

## CRAFT SPECIALIZATION

Formal tools, primarily bifaces, are found in small numbers at Maya sites throughout the Lowlands. The most common form is a large, crude biface which is usually made from local chert and appears to have been used for a number of different tasks. The more finely produced and smaller bifaces are usually made from imported chert, which in the central Peten is often similar to material from northern Belize. In fact, the chert-bearing deposits of northern Belize provided the material for some of the most intensive stone tool production in Mesoamerica. The focus of this activity was the site of Colha, the discovery of which (Wilk 1975) was one of the factors that led to the first Maya Lithics Conference, held in Belize in 1976 (Hester and Hammond 1976). The fieldwork and analysis carried out for the next several years at Colha transformed our view of the Maya in several important ways (summarized in Shafer and Hester 1983, 1986; Shafer 1985, 1991; Hester 1985; and in several articles in Hester and Shafer 1991).

It became clear that lithic analysis could be used to answer questions about the origin and nature of Maya society. It had been assumed that the complexity expressed in Maya architecture and inscriptions would be reflected in social and economic organization. Craft specialization had been assumed (Adams 1970), and a meager amount of data had been brought to bear on the problem (Becker 1974). However, in no case was the evidence for craft specialization and workshops as clear as it was at Colha.

The sheer volume of the debitage, forming more than 100 mounds measuring up to 30 m in diameter and more than 1 m in depth, whose primary constituents are the flakes and production rejects from several specific technologies, would seem to make it clear that craft specialization is involved. However, as Shafer and Hester are aware, similar concentrations of material are possible in quarry areas as the consequence of repeated visits by non-specialists. For example, late Middle Archaic, Benton period tool production in the source area for high quality chert in northeast Mississippi and northwest Alabama resulted in extensive deposits of debitage (Johnson 1981). Although not as large as those at Colha, they were of sufficient size to prompt WPA archaeologists to designate a "workshop" level in the shell mounds they excavated along the banks of the Tennessee River in this region (Webb and DeJarnette 1942).

In addition to production volume, Shafer (1991:31) cited standardization of production and standardization of form as criteria for specialization. Again, the Benton data fulfill the expectations. The primary product was a distinctive biface without haft modification which was exported south onto the coastal plain or corner-removed at separate locations in the source area to form Benton Points. There is remarkable consistency in raw material and form, which allows Benton Points to be identified as such well beyond their source area. Moreover, the production sequence is one of the most formalized bifacial technologies to have been modeled (Johnson 1979, 1981, 1984).

One of the hallmarks of craft specialization is the production of more goods than are needed for local consumption (Clark 1989). One approach is to estimate the number of tools produced on the basis of the volume of debitage, number of cores, or number of failures. This total is divided by the number of years of occupation and the estimated population of the site. Shafer and Hester (1986:162) estimated that Late Preclassic tool production at Colha was 4.5 million, or 150 tools per male per year.

Another and perhaps more concrete approach to the problem is to document where the tools went. Colha appears to have been the primary supplier of several very specific tool forms for most of northern Belize outside the chert-bearing zone for most of the Preclassic through early Post-Classic occupation of the area (Shafer 1983; McAnany 1989b). There is, however, some evidence that the chert was also distributed in less than finished form (McSwain 1991).

Returning to Mississippi, there is clear evidence for production in excess of local needs. From 50% to 80% of the Benton points from sites for which we have data up to 160 km south of the source area are made of source area chert. In contrast, the range for the exchange of tools from Colha is estimated to extend 50 km from the site (McAnany 1989a). In addition, throughout northeast Mississippi caches of the unnotched preforms produced at the source area sites are fairly common, along with a broad range of remarkably well made, "eccentric" bifaces (Johnson and Brookes 1988, 1989). Their distribution is roughly the same as that of the utilitarian form. Ritual forms made from Colha material using techniques evident in the workshop debris extend between 75 and 100 km from the source area (McAnany 1989a; Gibson 1986).

This is not to make an argument for Middle Archaic craft specialization in the Southeast. There are several differences between Benton production and that carried out at Colha. Among these is scale. Although we will never know how many biface production locations there were in the Benton source area because most of them were covered by reservoirs during the 1930s, clearly the volume of production was not equal to that carried out at Colha.

There are also differences in the diversity of the forms produced. Discounting the ceremonial types, Benton period production focused on a single tool, the stemmed biface. Wear pattern analysis (Ahler 1983) indicates that this was a multipurpose tool. Several different tools were produced at Colha, with individual workshops specializing in specific forms. These forms have been related to specific and specialized agricultural tasks in northern Belize (Shafer 1985; McAnany 1989b). The spatial separation of production activities geared toward different end products would be difficult to understand if the Colha deposits were the result of individual tool production for personal consumption.

Finally, there is the nature of the deposits themselves to consider. While Benton period early stage production debris is sometimes located in the immediate vicinity of the chert outcrop and devoid of evidence for anything other than biface reduction, final stage production was carried out at what appear to have been hunting camps, containing evidence of tool use as well as tool production. During the Late Preclassic at Colha, very little other than debitage occurs in the workshop deposits. And although there has been some controversy over whether the Colha deposits are in primary or secondary context (Moholy-Nagy 1990; Hester and Shafer 1992), it seems obvious that they are the result of workshop production (cf. Mallory 1986).

So why bring up Middle Archaic biface production? It seems likely that if the Benton production and distribution network was discovered in Belize and dated to the Classic period, it would be presented as evidence for craft specialization. This technology is more structured and extensive than any chert-based industry other than that at Colha in the Southern Lowlands. We find what we expect. For example, Ford and Olson (1989) used ternary plots of size-graded debitage samples from several sites which were tested during a survey centering on the Belize River Valley near the boundary between Belize and Guatemala. They documented both biface and flake tool production activities and concluded that a broad segment of the formal tool manufacturing trajectory is evident in the river valley sites. Most upland sites, other than quarry locations, show low lithic densities, mostly smaller flakes. Many of these upland sites are located in the vicinity of El Pilar, one of the major centers in the region. Although they cited Ericson's (1984) discussion of quarry site staging, they concluded that there was centralized control of tool production in the uplands and opportunistic production in the valley. However, this pattern may have nothing to do with differences in centralized control. Valley chert resources were broadly spread, generally available river terrace gravels, while upland resources were nodule-bearing outcrops (Ford and Olson 1989:194) that were exploited at specific quarry locations. Early-stage reduction at point sources occurs regularly among hunters and gathers (Ericson 1984) where there is little likelihood of hierarchical organization.

Perhaps the most significant result of the Colha project is the documentation of intense and specialized craft production at a relatively minor site in terms of architecture and other markers of elite status. This was one of the factors that led Mallory (1986) to question the evidence for craft specialization at Colha. If the analysis of obsidian tools from Copan (Mallory 1984), one of the premier Lowland Maya sites, failed to produce much in the way of evidence for craft specialization, how could there be craft specialization at a small site like Colha? Building on recent reevaluations of Maya economics by Fry (1980), Freidel (1981), and Rice (1987), McAnany (1989a,b) outlined specific producer-consumer relationships between Colha and the ridge field-based agricultural communities around Pulltrouser Swamp and elsewhere in northern Belize. She characterized this as grass roots specialization using horizontal, trade partner mechanisms. A similar lack of centralized control is evident in the distribution of ceramics which have been assigned to different production groups. There is not much evidence that the complex social organization that is evident in the pyramids, tombs, and monuments controlled the production and distribution of subsistence goods, even those which resulted from craft specialization.

On the other hand, Aldenderfer's (1991a, 1991b; Aldenderfer et al. 1989) wear pattern and design theory analyses of chert tools from housemounds in the central Peten indicate that tool specialization directed toward stone and wood working was fairly common during the Late Classic. These specialized tools are found in a limited number of exclusively elite contexts (Aldenderfer 1991b:211), suggesting to Aldenderfer (1991a:137) the existence of attached craft specialization, albeit indirect in some cases (Aldenderfer et al. 1989:58), since specialized woodworking tools are found in elite contexts which are removed from major centers.

It is interesting to note that tool specialization data on similar stone tools from the premier Mississippian site of Cahokia in the midwestern United States (Yerkes 1983) have been interpreted as evidence for part-time specialization by some (Muller 1984; Prentice 1985), and full-time specialization by others (Yerkes 1983). In both places, it is likely that the product of the specialization was not basic subsistence items, but the paraphernalia used in differentiating status.

## OBSIDIAN IN THE LOWLANDS

This horizontal exchange model is far different from the expectations of the role of economics in the rise and maintenance of Maya civilization which were prevalent in the early 1970s. Perhaps the best expression of this viewpoint is Rathje's (1972) core-buffer zone model. According to this model, centralized social and political organization arose in the Southern Lowlands in response to a need to import basic subsistence commodities: obsidian, igneous rock, and salt. Obsidian became the easiest material to use in testing this hypothesis. The implications were clear: highland centers such as Kaminaljuyu arose because they controlled the source areas, and lowland centers such as Tikal grew because they controlled the trade and redistribution of the material.

The perceived importance of obsidian, in combination with the relative ease by which it can be assigned to specific highland sources using trace element analysis, has led to a remarkable number of obsidian sourcing studies (reviewed in Driess and Brown 1989). Although many of the early attempts to use these data in reconstructing trade routes were admittedly speculative (Hammond 1972), clear diachronic patterning does exist in the distribution of obsidian from various highland sources, and these patterns can be interpreted in terms of shifting routes and alliances (Sidrys and Kimberlin 1979; Moholy-Nagy et al. 1984; Fowler et al. 1989).

Although it is obvious that the Lowland Maya went to considerable trouble to obtain obsidian, it has become increasingly evident that this was not the critical resource it was thought to be. Most domestic tasks at Maya sites were performed using local chert flake tools. Studies of obsidian density per unit of excavation, obsidian-to-chert ratios, or cutting edge-to-mass have been used to evaluate the relative intensity of obsidian use at Lowland sites (Johnson 1976; Moholy-Nagy 1976; Sidrys 1979). These conclude that, by any measure, it is a scarce resource. This has prompted some workers (e.g., Sidrys 1976) to suggest that obsidian is a high status commodity. However, the general domestic context in which it is found in the Lowlands, in combination with wear pattern studies (Wilk 1978; Shafer 1983; Lewenstein 1987; Aldenderfer et al. 1989), indicate that however obsidian entered the system, it was used for a wide variety of everyday tasks once it got there. In functional terms, it was a non-essential resource (Mitchum 1989).

That is not to say that all segments of Maya society had equal access to obsidian. There are some data (Sidrys 1976; Johnson 1976; Rice 1987; Ford and Olson 1989) that suggest that sites which are ranked higher in terms of architecture and other status markers had greater access to obsidian in the Southern Lowland. On the other hand, intrasite data from Tikal (Moholy-Nagy 1989) seem to suggest otherwise. The practice of redepositing obsidian debitage in votive deposits within the site center at Tikal (Moholy-Nagy 1989, 1990) makes arguments for differential density difficult to evaluate. Massive ceremonial deposits uncovered during the excavation of the Great Plaza are estimated to account for 99% of all the obsidian recovered from Tikal (Moholy-Nagy 1994:72). Surely the elite structures in the central zone must have been drained of obsidian in order to gather so much material.

Copan appears to have taken advantage of its position on a trade route between the nearby upland obsidian sources and the rest of the Southern Lowlands. Although obsidian artifacts outnumber chert at the site (Valdez and Potter 1991), there is no correlation between obsidian density and site ranking (Mallory 1984). Blade production by-products occur widely, suggesting to Mallory (1984) that each household produced its own blades.

On the other hand, within the Highland source area at Kaminaljuyu, there appears from the primary deposits to have been centralized control of obsidian on a regional scale, since the smaller sites surrounding the center relied primarily on flake tools derived from water-borne cobbles of obsidian that were locally available (Hay 1978). There are conflicting interpretations of the obsidian industry within Kaminaljuyu. Michels (1976, 1979) argued that differences in the number of blade fragments per volume excavated can be related to craft specialization and that a small number of workshops were making blades for local use and export. Hay (1978) pointed out that blade production workshops should not have large numbers of blades, but rather large numbers of by-products. Moreover, his analysis of the distribution of cores and flakes suggests a situation similar to that at Copan, where each household manufactured its own blades. His computations of per annum production indicate that blade making would have been a part-time activity.

It is precisely these low numbers that caused Clark (1987) to question the conclusion that blade production was not a full-time activity at Kaminaljuyu. That is, according to Clark, technology makes demands on material, equipment and skill which could not be maintained on a part-time basis. He regarded the shift from flake to blade technology, which is particularly obvious on the lithic-poor coastal plains of Mesoamerica, as dependent on a centralization of authority that occurred during the Middle Formative in most places. In this way he accounted for an apparent enigma, the specialized production of a generalized tool.

The small amount of data on obsidian blade production from the Southern Lowlands seems to corroborate Clark's model for elite control of blade production. In a relatively small collection of lithic artifacts from the Palenque region (Johnson 1976), not only was obsidian relatively more common at first and second order sites, but obsidian collections from third order sites, those without ceremonial architecture, consisted entirely of blades. Cores and platform rejuvenation flakes were confined to Palenque and the second order sites.

Although small numbers of obsidian blades are a characteristic of most Lowland collections, Classic period obsidian workshop debris is relatively rare. Neivens and Libbey (1976) provide the most complete description available in their account of more than 12,000 artifacts from a single deposit at El Pozito, a medium-sized ceremonial center in northern Belize. The majority of these artifacts were blade fragments, although a large sample of platform rejuvenation flakes and core fragments were also found. Moholy-Nagy (1976) discussed secondary deposits of what appears to be obsidian workshop debris from Tikal. Dreiss (1988) summarized data on two relatively small samples from what are interpreted to be obsidian workshop deposits at Colha. Finally, Ford and Fedick (1992) reported, but did not describe, an obsidian workshop from the upper Belize River drainage.

## OBSIDIAN AT NOHMUL

Given the rarity of reported obsidian workshops in the Lowlands, the discovery at the site of Nohmul of obsidian blade production debris, although in secondary deposition, is of some importance. Nohmul, one of the largest ceremonial centers in northern Belize, was the focus of seven seasons of research under the direction of Norman Hammond (Hammond 1983, 1985; Hammond et al. 1985, 1987, 1988). Mapping and excavation included the ceremonial precinct, outlying residential areas, and raised fields.

As is typical of Lowland sites, almost all of the obsidian from Nohmul can be related to blade technology. Including surface collections, 4290 pieces of obsidian were recovered. All but four biface fragments were derived from blades or blade cores. The total amount of obsidian recovered was small until work began on the ball court during the third season (Hammond et al. 1987, 1988). By the end of the final season, the ball court had produced 3995 pieces of obsidian, 93.1% of the obsidian recovered at Nohmul, although the volume of the ball court excavations is a small fraction of the total amount dug at the site. This points to one of the problems in dealing with Maya ceremonial centers, i.e., they are large, complex and heterogeneous. Sample bias is always a difficulty.

In fact, the obsidian bearing deposits are unrelated to the use of the area as a ball court. Several plaster floors resulting from an earlier domestic use of the area were found during the excavation in the alleyway of the Terminal Classic/Early Post Classic ball court. Obsidian was relatively common in these deposits. One context in particular, designated C2072, consisted of a thin layer of fill between two of the floors relatively deep in the excavation, and showed sufficient concentration of obsidian to be designated as a workshop on the basis of field observations (Hammond et al. 1988:7).

In fact, C2072 produced 2733 pieces of obsidian, 63.7% of the total obsidian from Nohmul. Ceramics from this provenience indicate a Late Classic time of construction, dating to about A.D. 800. The question is, are these artifacts the result of workshop activity? Table 1 shows the breakdown of this assemblage into several different blade and flake categories. Blades that terminated by removing the core base constitute nearly half of the distal blade fragments. Although not a

Table 1. Obsidian Artifacts from C2072 at Nohmul

| Type | n | Mean width (mm) | Standard deviation |
|---|---|---|---|
| Blades | | | |
| Proximal | 367 | 8.12 | 1.94 |
| Medial | 1002 | 6.73 | 2.06 |
| Distal | | | |
| Regular Termination | 97 | 5.54 | 2.01 |
| Outrepassé | 94 | 6.20 | 1.55 |
| Flakes | | | |
| Failed Blade Initiation | 115 | 8.05 | 1.84 |
| Platform Rejuvenation | 28 | 8.75 | 1.84 |
| Core Base Removal | 28 | 8.04 | 2.91 |
| Other Blade Core Flakes | 251 | 8.46 | 2.17 |
| Non Blade Core Flakes | 751 | 7.49 | 2.49 |

serious mistake since the core can still be used, it is a waste of obsidian. These outrepassé flakes are particularly common in this collection.

More than a third of the flakes were clearly derived from a blade core on the basis of blade scars or core platform remnants on their dorsal surfaces. Of special interest are the flakes which resulted from unsuccessful attempts to remove a blade. These flakes have blade core platforms and blade scars on their dorsal surfaces that originated from the same platform. They are classified as flakes because they terminated a short distance from the platform; most are not much longer than they are wide.

Two special flake categories most likely relate to core rejuvenation. Platform rejuvenation flakes were struck perpendicular or near perpendicular to the long axis of the core and removed all or most of the original platform. This was done in order to establish a new platform, presumably for blade removal. These flakes suggest that rejuvenation at Nohmul was not a sure thing. Many of the resulting platforms would have been irregular, and some platform angles would have been obtuse. One such rejuvenated and used platform is preserved on a rejuvenation flake, documenting a second attempt at rejuvenation. On the basis of these flakes, it can be determined that the assemblage contained at least 28 cores in addition to the three core fragments to be discussed below. The width data show these cores to have been rather small.

There were also 28 flakes that removed the other end of the core. These may also have been attempts to establish another platform in order to continue blade removal. One core from elsewhere in the ball court excavation shows an unsuccessful attempt at core reversal using this technique. Because of the angle to the long axis formed by these flakes, blades could only be removed from one side of the core. A similar adjustment is seen on two other cores that were reversed when the facet left by an outrepassé blade removal was used as a platform. It may be that some outrepassé blades were intentional attempts to set up a platform on the other end of the core, particularly with the small, nearly exhausted cores that were being exploited at this location.

It seems clear that blade production and core rejuvenation were performed at this location. Some of the remainder of the flakes may be related to attempts to set up another platform for blade removal. However, none of the cores from the ball court are intact. All show considerable damage which can not be related to rejuvenation. The distribution of flake scars on the cores and the large number of flakes suggests that blade cores were converted to flake cores after every attempt to remove the final blade had failed. Although these flakes are small, Lewenstein (1987:167) found traces of wear on obsidian flakes 10 to 15 mm long. There is, incidently, no evidence for bipolar reduction in the Nohmul obsidian.

Before turning to intrasite comparisons, there is one other assemblage from the site center that deserves special attention. A Late Classic tomb discovered during the excavation of Structure 8 in the east group contained, among other things, jade, two chert eccentrics, four obsidian cores, and 14 whole blades. The cores are considerably larger than those found elsewhere at Nohmul and, although there is some evidence for rejuvenation, they are far from exhausted. In fact, one of them is a macrocore. Not only is it remarkably large, but the blade scars indicate that all blades were removed using percussion. This is the first reported percussion core from the Southern Lowlands. These were clearly valuable offerings.

When the obsidian artifacts are grouped into broad technological categories which reflect production and rejuvenation stages (Table 2), obvious patterning is revealed. Blade failures include outrepassé blades and failed blade initiation flakes. Rejuvenation artifacts are platform rejuvenation flakes and core base removal flakes. Surface collections are not included in this table.

The assemblage from C2072 is similar to the rest of the ball court material. Blades and, perhaps, flakes were produced in this locale and distributed to the rest of the site center, as well as to the periphery. In fact, the proportional breakdown of the obsidian from the site center when the ball court material is not included is nearly identical to the housemound material from the periphery. The tomb with its relatively large number of cores represents an assemblage pattern of its own.

The Nohmul data run counter to what was found at Kaminaljuyu and Copan near the obisdian source areas, but fit what little we know about obsidian blade production in the rest of the Southern Lowlands. That is, blade production was not a household activity. The location of the ball court material in the site center suggests the possibility of elite control. Is there evidence for elite control of access to obsidian?

Because of Pyburn's (1988) testing program in the site periphery, Nohmul offers a good opportunity to test for differential access to obsidian. Most of Pyburn's data come from small housemounds and habitation sites without platforms away from the site center. In order to measure differences in the relative abundance of obsidian at Nohmul, the ratio of sherds to obsidian was computed

**Table 2. Obsidian Broken Down by Provenience and Production Class**

| Count | Cores | Blade blades | Rejuv. failures | Core flakes | Blade flakes | Row flakes | Total |
|---|---|---|---|---|---|---|---|
| Tomb | 4 | 14 | 0 | 0 | 0 | 0 | 18 |
| C2072 | 3 | 1466 | 209 | 56 | 251 | 751 | 2736 |
| Ball Court | 7 | 916 | 79 | 25 | 127 | 115 | 1269 |
| Center | 6 | 71 | 0 | 0 | 3 | 3 | 83 |
| Periphery | 1 | 137 | 0 | 0 | 4 | 11 | 153 |
| Total | 21 | 2604 | 288 | 81 | 385 | 880 | 4259 |

for those proveniences where sherd count data were available. This approach was used in an unsuccessful search for chronological trends in a small collection of obsidian from Cuello (Johnson 1991).

There is some patterning in the distribution of this ratio. For example, unit C2072 yielded 2736 pieces of obsidian and 4023 sherds, giving a sherd-to-obsidian ratio of 1.47. The remainder of the ball court proveniences for which we have sherd counts had 12.40 sherds per piece of obsidian. Site center locations excluding the ball court had a ratio of 85.18, while the ratio at the site periphery was 123.21. The obsidian density measure parallels the obsidian assemblage data (Table 2). That is, C2072 and the rest of the ball court are similar by both measures, thereby distinguishing this workshop area from the rest of the site. And, as would be expected if there was elite control of access, there is relatively less obsidian in the periphery test pits. Unfortunately, sherd counts from operation 61, an extensive habitation site excavation in the periphery (Pyburn 1988; Hammond et al. 1988), are not available at this time.

## CONCLUSIONS

Perhaps the first lesson to be drawn from this review is that even when dealing with the artifacts of complex societies, lithic analysis must begin by modeling production activity by means of detailed study of flakes, rejects and tools. A clear understanding of what actually is being made is necessary before hypotheses about state level economics can be tested.

Secondly, Maya lithic technology needs to be viewed from a more general perspective. Similar system-controlling variables were at work among the Maya as were operative elsewhere, *regardless of level of complexity.* Without a broad view, patterns which can be easily explained without reference to chiefdom or state level organization are interpreted as evidence for centralized control of production and distribution.

Finally, the importance of lithic analysis, once a stepchild in Maya archaeology, is clear. Classic period data on Maya stone tool production can best be understood by distinguishing subsistence-based activities from ritual production and use. The former relies primarily on local resources and there is little evidence for elite control of this aspect, even when craft production arises. On the other hand, exotic raw materials, acquired through extensive, long distance trade networks, were the primary focus of ritual activity. The acquisition, production, and distribution of ritual tools appears to have been strongly centralized.

Obsidian would seem to fall in between. That is, it is an exotic raw material from a remote resource area. Classic period blade production in the Southern Lowlands has been documented only at a few locations, most of which are situated near the site center as defined on the basis of ceremonial architecture. The distribution of obsidian within sites and regions in the Lowlands seems to be hierarchical. All of this points to elite control. However, once the blades were redistributed, the majority were used and discarded as though they were chert flakes derived from an ad hoc core technology. That is, they became everyday, subsistence tools.

There is an interesting parallel in the midwestern United States where, during the Middle Woodland period, blades made from exotic, high quality chert were broadly distributed and used for domestic chores (Yerkes 1990). Morrow (1987) has argued that these blades served as markers of social ties, and that certainly may have been a factor. However, Odell (1994) has shown that blades from mortuary contexts in the Lower Illinois Valley were used for a rather narrow range of activities, in contrast with blades from domestic deposits in the same region. It seems likely that the raw material and the technology were imported for ritual purposes. However, once the cores and the expertise are available in a given situation, it is possible to produce many blades, perhaps more than would be needed in the rituals. The surplus was probably distributed among the villagers, and the blades were put to a wide variety of uses.

Obsidian blade production, use, and distribution among the Classic Maya may have been the result of similar activity. Hayden (1987:182–183) reviewed the ethnohistoric data on ritual use of obsidian and documented a parallel use of glass among the contemporary highland Maya (Deal and Hayden 1987). The importance of obsidian in burial ritual in the Southern Lowlands is expressed in the massive debitage deposits associated with tombs at Tikal (Moholy-Nagy 1976, 1989). In fact, it would be surprising if obsidian blades were not used in the blood ritual of the Maya, given their suitability for the task. The relatively small size of the C2072 blades (Table 3) makes ritual use reasonable.

It seems likely that obsidian cores were imported into the Southern Lowlands to be reduced to blades by specialists for the primary purpose of producing blades for ritual. The surplus that resulted was redistributed to the general populace along kin lines and according to status. The blades were used for subsistence activities and, perhaps, household ritual. Their numbers were never sufficient to replace chert tools but, given the obvious importance of ritual among the Maya, there was a constant supply.

**Table 3. Obsidian Blade Widths Broken Down by Provenience**

| Provenience | n | Mean width (mm) | Standard deviation |
|---|---|---|---|
| Ball court | 2524 | 7.61 | 2.42 |
| Center | 85 | 10.18 | 2.38 |
| Periphery | 137 | 10.14 | 2.77 |
| Surface | 21 | 12.95 | 2.64 |

If obsidian is viewed in this light, it is clear that the primary specialization among the classic Maya, and the only specialization controlled by the elite, was ritual. The specialists were attached to and supported by elites (Brumfiel 1987; Earle 1987; McAnany 1989b) in order to produce the elaborate paraphernalia, including the pyramids and inscriptions, which were used to document and maintain their relative status. The interpretation of Maya economics which is emerging is political rather than functional (McAnany 1989b; Hayden and Gargett 1990). In this scenario, specialization and long distance trade were the result, rather than the cause, of status differentiation (Freidel 1981; Lewenstein 1987), and solidarity was mechanical rather than organic (de Montmollin 1989). This perspective is much different from that of even a few years ago and requires an entirely different evolutionary framework. Lithic analysis has been essential in rejecting earlier models of Maya civilization and in establishing the current one.

## ACKNOWLEDGMENTS

First, I thank Norman Hammond for making the obsidian from Nohmul available to me for study. Laura Kosakowsky was remarkably patient and generous in explaining the Nohmul provenience system and providing sherd count data. Rich Stallings helped in the study of this material. The paper has profited from the discussions which went on during the sessions and beyond at Tulsa and I thank the participants, particularly George Odell and Michael Shott, for making useful comments on earlier drafts of this paper. Hattula Moholy-Nagy, Beth Misner, and an anonymous reviewer also made helpful suggestions, many of which I was wise enough to heed.

## REFERENCES

Adams, R. E. W. 1970. Suggested Classic Period Occupational Specialization in the Southern Maya Lowlands. In *Monographs and Papers in Maya Archaeology*, edited by W. R. Bullard, Jr., pp.

487–502. Papers of the Peabody Museum of Archaeology and Ethnology, vol. 61. Harvard University, Cambridge.

Ahler, S. A. 1983. Microwear Analysis and Evaluation of the Chipped Stone Tool Classification System for the University of West Florida Midden Mound Project. In *Archaeological Investigations in the Upper Tombigbee Valley, Mississippi, Phase I,* edited by J. A. Bense, Vol. 4:IIIE1-IIIE36. University of West Florida, Office of Cultural and Archaeological Research, Report of Investigations, Number 3.

Aldenderfer, M. 1991a. The Structure of Late Classic Lithic Assemblages in the Central Peten Lakes Region, Guatemala. In *Maya Stone Tools: Selected Papers from the Second Maya Lithic Conference,* edited by T. R. Hester and H. J. Shafer, pp. 119–142. Prehistory Press, Madison.

Aldenderfer, M. 1991b. Functional Evidence for Lapidary and Carpentry Craft Specialties in the Late Classic of the Central Peten Lakes Region. *Ancient Mesoamerica* 2:205–214.

Aldenderfer, M. S., L. R. Kimball, and A. Sievert. 1989. Microwear Analysis in the Maya Lowlands: The Use of Functional Data in a Complex-Society Setting. *Journal of Field Archaeology* 16:47–60.

Becker, M. J. 1973. Archaeological Evidence for Occupational Specialization among the Classic-Period Maya at Tikal, Guatemala. *American Antiquity* 38:396–406.

Boksenbaum, M. W. 1980. Basic Mesoamerican Stone-Working: Nodule Smashing? *Lithic Technology* 9:12–26.

Bradley, B. A. 1975. Lithic Reduction Sequences: A Glossary and Discussion. In *Lithic Technology: Making and Using Stone Tools,* edited by E. Swanson, pp. 5–14. Mouton, The Hague.

Brumfiel, E. M. 1987. Elite and Utilitarian Crafts in the Aztec State. In *Specialization, Exchange, and Complex Societies,* edited by E. M. Brumfiel and T. K. Earle, pp. 102–118. Cambridge University Press, Cambridge.

Clark, J. 1987. Politics, Prismatic Blades, and Mesoamerican Civilization. In *The Organization of Core Technology,* edited by J. K. Johnson and C. A. Morrow, pp. 259–284. Westview Press, Boulder.

Clark, J. 1989. Hacia una Definicion de Tallares. In *La Obsidiana en Mesoamerica,* edited by M. Gaxiola G. and J. Clark, pp. 213–218. Instituto Nacional de Anthropologia e Historia, Mexico City.

Clark, J., and D. Bryant. 1991. The Production of Chert Projectile Points at Herba Buena, Chiapas, Mexico. In *Maya Stone Tools: Selected Papers from the Second Maya Lithic Conference,* edited by T. R. Hester and H. J. Shafer, pp. 85–102. Prehistory Press, Madison.

Coe, W. R. 1959. *Piedras Negras Archaeology: Artifacts, Caches, and Burials.* Museum Monographs, University Museum, University of Pennsylvania, Philadelphia.

Coe, W. R. 1965. Artifacts of the Maya Lowlands. In *Archaeology of Southern Mesoamerica,* edited by G. R. Willey and R. Wauchope, pp. 594–602. Handbook of Middle American Indians, vol. 3. University of Texas Press, Austin.

Collins, M. B. 1975. Lithic Technology as a Means of Processual Inference. In *Lithic Technology: Making and Using Stone Tools,* edited by E. Swanson, pp. 15–34. Mouton, The Hague.

Deal, M., and B. Hayden. 1987. The Persistence of Pre-Columbian Lithic Technology in the Form of Glassworking. In *Lithic Studies among the Contemporary Highland Maya,* edited by B. Hayden, pp. 235–331. University of Arizona Press, Tuscon.

de Montmollin, O. 1989. *The Archaeology of Political Structure: Settlement Analysis in a Classic Maya Polity.* Cambridge University Press, Cambridge.

Dreiss, M. L. 1988. *Obsidian at Colha, Belize: A Technological Analysis and Distributional Study Based on Trace Element Data. Papers of the Colha Project 1:4,* San Antonio.

Dreiss, M. L., and D. O. Brown. 1989. Obsidian Exchange Patterns in Belize. In *Prehistoric Maya Economies,* edited by P. A. McAnany and B. L. Isaac, pp. 57–90. Research in Economic Anthropology, Supplement 4, JAI Press, Inc., Greenwich.

Earle, T. K. 1987. Specialization and the Production of Wealth: Hawaiian Chiefdoms and the Inka Empire. In *Specialization, Exchange, and Complex Societies*, edited by E. M. Brumfiel and T. K. Earle, pp. 64–75. Cambridge University Press, Cambridge.

Ericson, J. E. 1984. Towards the Analysis of Lithic Production Systems. In *Prehistoric Quarries and Lithic Production*, edited by J. E. Ericson and B. A. Purdy, pp. 1–10. Cambridge University Press, Cambridge.

Fedick, S. 1991. Chert Tool Production and Consumption among Classic Period Maya Households. In *Maya Stone Tools: Selected Papers from the Second Maya Lithic Conference*, edited by T. R. Hester and H. J. Shafer, pp. 103–118. Prehistory Press, Madison.

Ford, A., and S. Fedick. 1992. Prehistoric Maya Settlement Patterns in the Upper Belize River Area: Initial Results of the Belize River Archaeological Settlement Survey. *Journal of Field Archaeology* 19:35–49.

Ford, A., and K. Olson. 1989. Aspects of Ancient Maya Household Economy: Variation in Chipped Stone Production and Consumption. In *Prehistoric Maya Economies*, edited by P. A. McAnany and B. L. Isaac, pp. 185–214. Research in Economic Anthropology, Supplement 4, JAI Press, Inc., Greenwich.

Fowler, W. R., Jr. 1991. Lithic Analysis as a Means of Processual Inference in Southern Mesoamerica: A Review of Recent Research. In *Maya Stone Tools: Selected Papers from the Second Maya Lithic Conference*, edited by T. R. Hester and H. J. Shafer, pp. 1–19. Prehistory Press, Madison.

Fowler, W. R., Jr., A. A. Demarest, H. V. Michel, and F. H. Stross. 1989. Sources of Obsidian from El Mirador, Guatemala: New Evidence on Preclassic Maya Interaction. *American Anthropologist* 91:158–168.

Freidel, D. A. 1981. The Political Economies of Residential Dispersion among the Lowland Maya. In *Lowland Maya Settlement Patterns*, edited by W. Ashmore, pp. 371–384. University of New Mexico Press, Albuquerque.

Fry, R. E. 1980. Models of Exchange for Major Shape Classes of Lowland Maya Pottery. In *Models and Methods in Regional Exchange*, edited by R. E. Fry, pp. 3–18. Society for American Archaeology Papers 1. Washington D.C.

Gibson, E. 1986. *Diachronic Patterns of Lithic Production, Use and Exchange in the Southern Maya Lowlands*. Ph.D. dissertation, Department of Anthropology, Harvard University.

Hammond, N. 1972. Obsidian Trade Routes in the Mayan Area. *Science* 178:1092–1093.

Hammond, N. 1983. Nohmul, Belize: 1982 Investigations. *Journal of Field Archaeology* 10:245–254.

Hammond, N. 1985. *Nohmul: A Prehistoric Maya Community in Belize. Excavations 1973–1983*. BAR International Series 250, Oxford.

Hammond, N., C. Clark, M. Horton, M. Hodges, L. McNatt, L. J. Kosakowsky, and K. A. Pyburn. 1985. Excavation and Survey at Nohmul, Belize, 1983. *Journal of Field Archaeology* 12:177–200.

Hammond, N., S. Donaghey, C. Gleason, J. C. Staneko, D. Van Tuerenhout, and L. J. Kosakowsky. 1987. Excavations at Nohmul, Belize, 1985. *Journal of Field Archaeology* 14:257–281.

Hammond, N., K. A. Pyburn, J. Rose, J. C. Staneko, and D. Muyskens. 1988. Excavation and Survey at Nohmul, Belize, 1986. *Journal of Field Archaeology* 15:1–15.

Hay, C. A. 1978. *Kaminaljuyu Obsidian: Lithic Analysis and the Economic Organization of a Prehistoric Maya Chiefdom*. Ph.D. Dissertation, Department of Anthropology, Pennsylvania State University.

Hayden, B. 1987. Past to Present Uses of Stone Tools and Their Effects on Assemblage Characteristics. In *Lithic Studies among the Contemporary Highland Maya*, edited by B. Hayden, pp. 160–234. University of Arizona Press, Tucson.

Hayden, B., and R. Gargett. 1990. Big Man, Big Heart?: A Mesoamerican View of the Emergence of Complex Society. *Ancient Mesoamerica* 1:3–20.

Hester, T. R. 1985. The Maya Lithic Sequence in Northern Belize. In *Stone Tool Analysis: Essays in Honor of Don E. Crabtree*, edited by M. G. Plew, J. C. Woods and M. G. Pavasic, pp. 187–210. University of New Mexico Press, Albuquerque.

Hester, T. R., and N. Hammond (editors). 1976. *Maya Lithic Studies: Papers from the 1976 Belize Field Symposium*. Center for Archaeological Research, University of Texas, San Antonio.

Hester, T. R., and H. J. Shafer. 1992. Lithic Workshops Revisited: Comments on Moholy-Nagy. *Latin American Antiquity* 3:243–248.

Hester, T. R., and H. J. Shafer (editors). 1991. *Maya Stone Tools: Selected Papers from the Second Maya Lithic Conference*. Prehistory Press, Madison.

Hester, T. R., H. J. Shafer, and T. Berry. 1991. Technological and Comparative Analyses of the Chipped Stone Artifacts from El Pozito, Belize. In *Maya Stone Tools: Selected Papers from the Second Maya Lithic Conference*, edited by T. R. Hester and H. J. Shafer, pp. 67–84. Prehistory Press, Madison.

Johnson, J. K. 1976. *Chipped Stone Artifacts from the Western Maya Periphery*. Ph.D. Dissertation, Department of Anthropology, Southern Illinois University, Carbondale.

Johnson, J. K. 1979. Archaic Biface Manufacture: Production Failures, a Chronicle of the Misbegotten. *Lithic Technology* 8:25–35.

Johnson, J. K. 1981. *Lithic Procurement and Utilization Trajectories: Analysis, Yellow Creek Nuclear Power Plant Site, Tishomingo County, Mississippi, Vol. II*. Publications in Anthropology 28. Tennessee Valley Authority, Knoxville.

Johnson, J. K. 1984. Measuring Prehistoric Quarry Site Activity in Northeastern Mississippi. In *Prehistoric Chert Exploitation: Studies from the Midcontinent*, edited by B. M. Butler and E. E. May, pp. 225–235. Center for Archaeological Investigations, Southern Illinois University, Carbondale. Occasional Paper 2.

Johnson, J. K. 1985. Typological Structure in Maya Lithic Analysis: A Historic Perspective. In *Contributions to the Archaeology and Ethnology of Greater Mesoamerica*, edited by W. J. Folan, pp. 188–204. Southern Illinois Press, Carbondale.

Johnson, J. K. 1991. Obsidian: A Technological Analysis. In *Cuello: An Early Classic Maya Community in Belize*, edited by N. Hammond, pp. 169–172. Cambridge University Press, Cambridge.

Johnson, J. K. 1993. Lithic Analysis. In *The Development of Southeastern Archaeology*, edited by J. K. Johnson, pp. 36–52. University of Alabama Press, Tuscaloosa.

Johnson, J. K., and S. O. Brookes. 1988. Rocks from the Northeast: Archaic Exchange in North Mississippi. *Mississippi Archaeology* 23:53–63.

Johnson, J. K., and S. O. Brookes. 1989. Benton Points, Turkey Tails and Cache Blades: Middle Archaic Exchange in the Midsouth. *Southeastern Archaeology* 8:134–145.

Kelly, R. L. 1988. The Three Sides of a Biface. *American Antiquity* 53:717–734.

Kidder, A. V. 1932. *The Artifacts of Pecos*. Southwestern Expedition, Phillips Academy, Papers 6. Yale University Press, New Haven.

Kidder, A. V. 1947. *The Artifacts of Uaxactun, Guatemala*. Carnegie Institution of Washington, Publication 576.

Lewenstein, S. 1987. *Stone Tool Use at Cerros*. University of Texas Press, Austin.

McAnany, P. A. 1989a. Stone-Tool Production and Exchange in the Eastern Maya Lowlands: The Consumer Perspective from Pulltrouser Swamp, Belize. *American Antiquity* 54:332–346.

McAnany, P. A. 1989b. Economic Foundations of Prehistoric Maya Society: Paradigms and Concepts. In *Prehistoric Maya Economies*, edited by P. A. McAnany and B. L. Isaac, pp. 347–372. Research in Economic Anthropology, Supplement 4, JAI Press, Inc., Greenwich.

McAnany, P. A., and B. L. Isaac (editors). 1989. *Prehistoric Maya Economies*. Research in Economic Anthropology, Supplement 4, JAI Press, Inc., Greenwich.

McSwain, R. 1991. A Comparative Evaluation of the Producer-Consumer Model for Lithic Exchange in Northern Belize, Central America. *Latin American Antiquity* 2:337–354.

Mallory, J. K. 1984. *Late Classic Maya Economic Specialization: Evidence from the Copan Obsidian Assemblage*. Ph.D. Dissertation, Department of Anthropology, Pennsylvania State University.

Mallory, J. K. 1986. "Workshops" and "Specialized Production" in the Production of Maya Chert Tools: A Response to Shafer and Hester. *American Antiquity* 51:152–157.

Michels, J. W. 1976. Some Sociological Observations on Obsidian Production at Kaminaljuyu, Guatemala. In *Maya Lithic Studies: Papers from the 1976 Belize Field Symposium*, edited by T. R. Hester and N. Hammond, pp. 109–118. Center for Archaeological Research, University of Texas, San Antonio.

Michels, J. W. 1979. *The Kaminaljuyu Chiefdom*. Pennsylvania State University Monograph Series on Kaminaljuyu. Pennsylvania State University Press, University Park.

Mitchum, B. 1989. Obsidian as a Non-Essential Resource. In *La Obsidiana en Mesoamerica*, edited by M. Gaxiola G. and J. Clark, pp. 375–378. Instituto Nacional de Anthropologia e Historia, Mexico City.

Mitchum, B. 1991. Lithic Artifacts from Cerros, Belize: Production, Consumption, and Trade. In *Maya Stone Tools: Selected Papers from the Second Maya Lithic Conference*, edited by T. R. Hester and H. J. Shafer, pp. 45–54. Prehistory Press, Madison.

Moholy-Nagy, H. 1976. Spatial Distribution of Flint and Obsidian Artifacts at Tikal, Guatemala. In *Maya Lithic Studies: Papers from the 1976 Belize Field Symposium*, edited by T. R. Hester and N. Hammond, pp. 91–108. Center for Archaeological Research, University of Texas, San Antonio.

Moholy-Nagy, H. 1989. Who Used Obsidian at Tikal? In *La Obsidiana en Mesoamerica*, edited by M. Gaxiola G. and J. Clark, pp. 379–390. Instituto Nacional de Anthropologia e Historia, Mexico City.

Moholy-Nagy, H. 1990. The Misidentification of Mesoamerican Lithic Workshops. *Latin American Antiquity* 1:268 279.

Moholy-Nagy, H. 1994. *Tikal Material Culture: Artifacts and Social Structure at a Classic Lowland Maya City*. Ph.D. Dissertation, Department of Anthropology, University of Michigan.

Moholy-Nagy, H., F. Asaro, and F. H. Stross. 1984. Tikal Obsidian: Sources and Typology. *American Antiquity* 49:104–117.

Morrow, C. A. 1987. Blades and Cobden Chert: A Technological Argument for their Role as Markers of Regional Identification during the Hopewell Period in Illinois. In *The Organization of Core Technology*, edited by J. K. Johnson and C. A. Morrow, pp. 119–150. Westview Press, Boulder.

Muller, J. 1984. Mississippian Specialization and Salt. *American Antiquity* 49:489–507.

Neivens, M., and D. Libbey. 1976. An Obsidian Workshop at El Pozito, Northern Belize. In *Maya Lithic studies: Papers from the 1976 Belize Field Symposium*, edited by T. R. Hester and N. Hammond, pp. 137–149. Center for Archaeological Research, University of Texas, San Antonio.

Odell, G. H. 1994. The Role of Stone Bladelets in Middle Woodland Society. *American Antiquity* 59:102–120.

Parry, W. J., and R. L. Kelly. 1987. Expedient Core Technology. In *The Organization of Core Technology*, edited by J. K. Johnson and C. A. Morrow, pp. 285–304. Westview Press, Boulder.

Potter, D. R. 1991. Lithic Analysis as a Means of Processual Inference in Southern Mesoamerica: A Review of Recent Research. In *Maya Stone Tools: Selected Papers from the Second Maya Lithic Conference*, edited by T. R. Hester and H. J. Shafer, pp. 21–29. Prehistory Press, Madison.

Prentice, G. 1985. Economic Differentiation among Mississippian Farmsteads. *Midcontinental Journal of Archaeology* 10:77–122.

Puleston, O. S. 1969. *Functional Analysis of a Workshop Tool Kit from Tikal.* M.A. Thesis, Department of Anthropology, University of Pennsylvania.

Pyburn, K. A. 1988. *The Settlement of Nohmul: Development of a Prehispanic Maya Community in Northern Belize.* Ph.D. Dissertation, Department of Anthropology, University of Arizona.

Rathje, W. L. 1972. Praise the Gods and Pass the Metates: A Hypothesis of the Development of Lowland Rainforest Civilization in Middle America. In *Contemporary Archeology,* edited by M. Leone, pp. 365–392. Southern Illinois University Press, Carbondale.

Rice, P. M. 1987. Economic Change in the Lowland Maya Late Classic Period. In *Specialization, Exchange, and Complex Societies,* edited by E. M. Brumfiel and T. K. Earle, pp. 76–85. Cambridge University Press, Cambridge.

Ricketson, O. G., Jr., and E. B. Ricketson. 1937. *Uaxactun, Guatemala, Group E, 1926–1931.* Carnegie Institution of Washington, Publication 477.

Rovner, I. 1974. Evidence for a Secondary Obsidian Workshop at Mayapan, Yucatan. *Newsletter of Lithic Technology* 3:19–27.

Rovner, I. 1975. *Lithic Sequences from the Maya Lowlands.* Ph.D. Dissertation, Department of Anthropology, University of Wisconsin-Madison.

Sabloff, J. A. 1990. *The New Archaeology and the Ancient Maya.* W. H. Freeman, New York.

Semenov, S. A. 1964. *Prehistoric Technology.* Translated by M. W. Thompson. Barnes and Noble Books, New York.

Shafer, H. J. 1983. The Lithic Artifacts of the Pulltrouser Area: Settlement and Fields. In *Pulltrouser Swamp: Ancient Maya Habitat, Agriculture, and Settlement in Northern Belize,* edited by B. L. Turner II and P. D. Harrison, pp. 212–245. University of Texas Press, Austin.

Shafer, H. J. 1985. A Technological Study of Two Maya Lithic Workshops at Colha, Belize. In *Stone Tool Analysis: Essays in Honor of Don E. Crabtree,* edited by M. G. Plew, J. C. Woods and M. G. Pavasic, pp. 277–315. University of New Mexico Press, Albuquerque.

Shafer, H. J. 1991. Late Preclassic Formal Stone Tool Production at Colha, Belize. In *Maya Stone Tools: Selected Papers from the Second Maya Lithic Conference,* edited by T. R. Hester and H. J. Shafer, pp. 31–44. Prehistory Press, Madison.

Shafer, H. J., and T. R. Hester. 1983. Ancient Maya Chert Workshops in Northern Belize. *American Antiquity* 48:519–543.

Shafer, H. J., and T. R. Hester. 1986. Maya Stone-Tool Craft Specialization and Production at Colha, Belize: Reply to Mallory. *American Antiquity* 51:158–166.

Sheets, P. D. 1972. A Model of Mesoamerican Obsidian Technology Based on Preclassic Workshop Debris in El Salvador. *Ceramica de Cultura Maya* 8:17--33.

Sheets, P. D. 1977. The Analysis of Chipped Stone Artifacts in Southern Mesoamerica. *Latin American Research Review* 12:139–158.

Sidrys, R. V. 1976. Classic Maya Obsidian Trade. *American Antiquity* 41:449–464.

Sidrys, R. V., and J. Kimberlin. 1979. Use of Mayan Obsidian Sources through Time: Trace Element Data from El Balsamo, Guatemala. *Journal of Field Archaeology* 6:116–122.

Sievert, A. K. 1992. *Maya Ceremonial Specialization: Lithic Tools from the Sacred Cenote at Chichen Itza, Yucatan.* Prehistory Press, Madison.

Taylor, W. W. 1948. *A Study of Archaeology.* American Anthropological Association, Memoir 69.

Thompson, M. 1991. Flaked Celt Production at Becan, Campeche, Mexico. In *Maya Stone Tools: Selected Papers from the Second Maya Lithic Conference,* edited by T. R. Hester and H. J. Shafer, pp. 143–154. Prehistory Press, Madison.

Valdez, F., and D. R. Potter. 1991. Chert Debitage from the Harvard Copan Excavations: Descriptions and Comments. In *Maya Stone Tools: Selected Papers from the Second Maya Lithic Conference*, edited by T. R. Hester and H. J. Shafer, pp. 203–206. Prehistory Press, Madison.

Webb, W. S., and D. L. DeJarnette. 1942. *An Archaeological Survey of Pickwick Basin in the Adjacent Portion of the States of Alabama, Mississippi and Tennessee*. Bureau of American Ethnology, Bulletin 129, Washington, D.C.

Wilk, R. 1975. Superficial Examination of Structure 100, Colha. In *Archaeology in Northern Belize: British Museum-Cambridge University Corozal Project, 1974–75 Interim Report*, edited by N. Hammond, pp. 152–173. Centre of Latin American Studies, University of Cambridge.

Wilk, R. 1978. Microscopic Analysis of Chipped Flint and Obsidian. In *Excavations at Seibal: Artifacts*, edited by G. R. Willey, pp. 139–145. Memoirs of the Peabody Museum of Archaeology and Ethnology, vol. 14, no. 1, Harvard University.

Willey, G. R. 1972. *The Artifacts of Altar de Sacrificios*. Papers of the Peabody Museum of Archaeology and Ethnology, vol. 64, no. 1, Harvard University.

Willey, G. R. 1978. *Excavations at Seibal: Artifacts*. Memoirs of the Peabody Museum of Archaeology and Ethnology, vol. 14, no. 1, Harvard University.

Willey, G. R., W. R. Bullard, Jr., J. B. Glass, and J. C. Gifford. 1965. *Prehistoric Maya Settlements in the Belize Valley*. Papers of the Peabody Museum of Archaeology and Ethnology, vol. 54, Harvard University.

Yerkes, R. W. 1983. Microwear, Microdrills, and Mississippian Craft Specialization. *American Antiquity* 48:499–518.

Yerkes, R. W. 1990. Using Microwear Analysis to Investigate Domestic Activities and Craft Specialization at the Murphy Site, a Small Hopewell Settlement in Licking County, Ohio. In *The Interpretive Possibilities of Microwear Studies*, edited by B. Graslund, H. Knutsson, K. Knutsson and J. Taffinder, pp. 167–176. Societas Archaeologica Upsaliensis, AUN 14, Uppsala, Sweden.

Chapter 7

# The Role of Chipped Stone in the Political Economy of Social Ranking

MICHAEL S. NASSANEY

## ABSTRACT

Chipped stone tools can figure prominently in the political economy of social ranking and the transformation of social relationships. For instance, incipient elites may try to aggrandize themselves by controlling access to lithic resources or the organization of tool production. In this paper I examine three aspects of stone tool technology—raw material acquisition, labor allocation, and productive intensification—to explore how lithic artifacts were implicated in the integration and disintegration of Plum Bayou culture in central Arkansas (ca. A.D. 700–950). The analyses expose longitudinal changes in the organization of technology which suggest rudimentary attempts at control and/or intensification. Despite these efforts, socially ranked individuals apparently failed to monopolize raw materials or the production process.

## INTRODUCTION

Social scientists have long been interested in understanding the development of social forms from egalitarian hunter-gatherers to large scale, multi-ethnic

**MICHAEL S. NASSANEY** • Department of Anthropology, Western Michigan University, Kalamazoo, Michigan 49008–5032.

*Stone Tools: Theoretical Insights into Human Prehistory*, edited by George H. Odell. Plenum Press, New York, 1996.

states. Anthropological archaeologists, in particular, have examined the processes that contributed to social ranking from various theoretical perspectives employing numerous case studies (see Bender 1985; Lee 1990; contributions to Upham 1990). The archaeological and ethnohistorical records suggest that social ranking emerged repeatedly throughout the world during the past 10,000 years, and perhaps earlier. Ranked societies differ from egalitarian social forms by having fewer positions of valued status than there are individuals capable of filling them (Fried 1967:109). Since high-ranking individuals in nonstratified societies cannot sustain their positions through the monopolization of force, they must attempt to extend their authority into various domains of social life, stone-tool technologies notwithstanding. The purpose of this paper is to explore the role of chipped stone tools and their production in the political economy of social ranking. Stone may be implicated because authority is reinforced materially and symbolically by controlling access to, and legitimizing control over, some combination of resources, people, and/or places essential for social reproduction. Because individuals frequently contested attempts to control the material world, however, social ranking was often resisted, creating cyclicity in human histories (see Nassaney 1992a).

## THE POLITICAL ECONOMY OF SOCIAL RANKING

Elsewhere I have reviewed a broad range of theoretical approaches to social ranking (Nassaney 1992b:14–47). Service (1971) and Fried (1967) can be considered representative of two oppositional viewpoints, integration and conflict (Haas 1982), although other dichotomies could easily be drawn. I find a political-economic framework, which develops ideas from historical materialism within an ecological context, analytically useful for investigating the emergence of social ranking. A political-economic approach examines the production and distribution of wealth and power within and between social groups (see Wolf 1982:7–8; Roseberry 1989:44; Cobb 1993) and attempts to identify what basic resources were being contested and how that struggle is manifest archaeologically.

I am interested in how positions of differential prestige are created and transformed in the context of the material world. An important premise is the relative concentration of power in society, by which I mean the ability of agents to alter the conditions of their existence in specific social and material contexts (Miller and Tilley 1984:5). A corollary of this premise is the degree to which power can be employed to mobilize social surplus toward specific ends. Although the definition of surplus has attracted considerable attention (cf. Harris 1959; Dalton 1960, 1963; Wolf 1966:5–10; Paynter 1982:23), I use the concept of "surplus" to designate the results of human production over and above that needed to reproduce the labor force. Thus the critical issues revolve around how individual agents exercise the power to extract and accumulate surplus and how their strategies are resisted.

Individual and social differences that form the potential basis for inequality can be traced back to the origins of humanity (Bender 1989). In so-called egalitarian societies, however, individuals are unable to act upon these differences or manipulate them toward their own benefit. Tensions and potential social inequalities along the lines of age and gender, for instance, are dampened primarily through reciprocity. Constraints upon mobility, as among sedentary horticulturists, provide conditions that may allow accumulation which can threaten reciprocity even when individuals of rank are still bound by kinship arrangements (Wolf 1982:97).

Accumulation and generosity stand in opposition, however, creating a dynamic tension between those who seek to accumulate surplus and those who are forced to relinquish it. "The well developed chiefdom creates for itself the dampening paradox of stoking rebellion by funding its authority" (Sahlins 1963:298). This contradiction is often resolved in favor of the producers who are able to divorce themselves from niggardly chiefs and realign themselves with other (initially) less costly kinship relations (Sahlins 1963:298). External factors can also undermine leadership in nonstratified societies. For example, emerging interests on the periphery of society may challenge local lineage heads, threatening their abilities to satisfy demands for surplus and labor.

## MATERIAL DIMENSIONS OF SOCIAL RANKING

Material objects, such as stone tools and debitage, are implicated in social relations. Agents insinuate their messages through material symbols at many different spatial scales from arrangements of ceramic designs (e.g., Wobst 1977; Hegmon 1989) to macro-regional settlement patterns (e.g., Paynter 1982; Kus 1983). Theoretically, the content of each of these scales underwrites different social behaviors with regard to the loci of accumulation and the process of labor mobilization.

Archaeologists have postulated a number of ways by which individuals can transform egalitarianism to establish positions of rank. Change in the organization of production is one means. In egalitarian societies, all members enjoy usufruct rights to the land and whatever resources they can extract from it; access to basic resources is constrained predominantly by the friction of distance. Under conditions of increased territoriality and decreased mobility, however, strategic groups can exert control over raw materials needed by others for social reproduction. When access costs are borne disproportionately among members of society, differential access constitutes a potential lever of social inequality.

Consumption, or use, is another consideration. In egalitarian societies, artifact usage differs among individuals only according to age, gender, and achieved rank. Unequal access to and display of particular objects by segments of a society may heighten tensions and oppositions internal to kin groups, as when elders and juniors contest objects of personal adornment. In reality, the objects themselves may only symbolize what is really being contested—namely, positions of leadership and authority. There is also potential opposition between men and women.

Surplus can also be monopolized in the production process from raw material acquisition to the finished product. While recent empirical studies have attempted to link chipped stone tool form and assemblage composition to changes in mobility patterns (e.g., Binford 1979; Parry and Kelly 1987), risk avoidance (Torrence 1989), political organization (Clark 1987), and the sexual division of labor (Sassaman 1992), variation may also be linked to a spatial or technical division of labor. Indeed, embryonic forms of specialization often co-occurred with the emergence of ranking (Fried 1967:129–130; see also Clark 1991). Some forms of specialization can force producers to relinquish control over their labor product so that control is maintained only by those who command the process of transforming a raw material into a useable object.

In sum, material objects are essential, active elements in the creation, maintenance, and signification of social identities and relationships (Wobst 1977; Hodder 1982). To the degree that the material world substantiates and replicates the social order, analyses of material objects provide an entry point for investigating how material culture is implicated in changing social relations. The manufacture, adoption, accumulation, distribution, use, display, and discard of material objects serve to define social distinctions and boundaries, as well as to reproduce and maintain social divisions (e.g., see Tilley 1984:116; Gero 1989).

## THE POLITICAL ECONOMY OF STONE

There is a considerable range of variation in the kinds of material objects that can come under the control of elite interests. In fact, strategic resources are contextually defined and groups continually negotiate which objects constitute preciosities and sumptuary goods.[1] Furthermore, evidence of some classes of goods is seldom preserved archaeologically (e.g., bear grease, feathered headdresses, fur cloaks, salt). While there are many possible strategic resources that one could analyze, I will examine the archaeological evidence for the production and distribution of chipped stone tools. Incipient elites may have attempted to manipulate the production or flow of these objects. I will provide a brief background for my archaeological case in the American Southeast before discussing the theoretical and methodological underpinnings that link each of the analyses to the political economy of social ranking. The literature contains additional details regarding the historical and archaeological context of the Plum Bayou culture of central Arkansas (e.g., Rolingson 1982a, 1990; Nassaney 1994).

## PLUM BAYOU CULTURE: A BRIEF DESCRIPTION

Plum Bayou culture was first defined from archaeological investigations at the Toltec Mounds site (3LN42) in the Arkansas River Lowland (Rolingson 1982a). Toltec is one of the largest multiple mound and plaza complexes in the Lower Mississippi Valley; it consists of 18 mounds, 10 of which are carefully arranged around a large plaza (Rolingson 1990b:Figure 7). A semi-circular earthen embankment 1.6 km long encloses the site features in an area of approximately 42 ha. The Bureau of American Ethnology originally examined Toltec in the 1880s (Thomas 1894:243–245); modern investigations began nearly a century later. The major mound-building occupations at Toltec date unequivocally to the Coles Creek period (ca. A.D. 700–1000), proving that the site's final form was not a Mississippian achievement, as investigators once believed (e.g., Phillips 1970:916–917).

Plum Bayou culture comprises numerous sites throughout the region that display Baytown and Coles Creek-period artifact assemblages (see Nassaney 1994 for a discussion of diagnostic artifacts). My research has focused on studying outlying sites that were occupied from the late Marksville through late Mississippi periods (ca. A.D. 200–1700) to understand how changes in the size, content, and distribution of settlements contributed to the consolidation and dissolution of Plum Bayou culture (Nassaney 1992b).

An increase in site size, site density, and scale of integration in the region corresponds with the period of mound construction at Toltec, the establishment of a loosely-integrated settlement hierarchy, and the expansion of Plum Bayou culture over most of east-central Arkansas (Nassaney 1991). At the beginning of the Mississippi period (ca. A.D. 1000–1100), another significant change occurred in regional demography and settlement integration—mound construction ceased and site densities decreased dramatically, suggesting regional abandonment. Later Mississippi-period (ca. A.D. 1400–1700) groups occupied the Lowland, though they cannot be linked historically with the Coles Creek-period antecedents who built the mounds at Toltec.

Preliminary studies have been conducted on some Plum Bayou culture artifacts. Ceramic and lithic artifacts dominate most assemblages, although faunal and floral remains also occur. Diagnostic Coles Creek-period ceramic types are popular at Toltec and related sites (Stewart-Abernathy 1982; Nassaney 1994). Chert gravels, novaculite preforms, and quartz crystals were used as raw materials for chipped stone tools, including dart points, arrow points, knives, drills, adzes, and celts. Pecked, ground, and polished tools such as hammerstones, celts, plummets, and gorgets were made from quartzite, sandstone, siltstone, hematite or syenite, and various igneous stones. Many of these raw materials originated beyond the Arkansas River Lowland in the West Gulf Coastal Plain, Ozark Plateau, and Ouachita Mountains (Figure 1). Finally, Plum Bayou subsistence was based

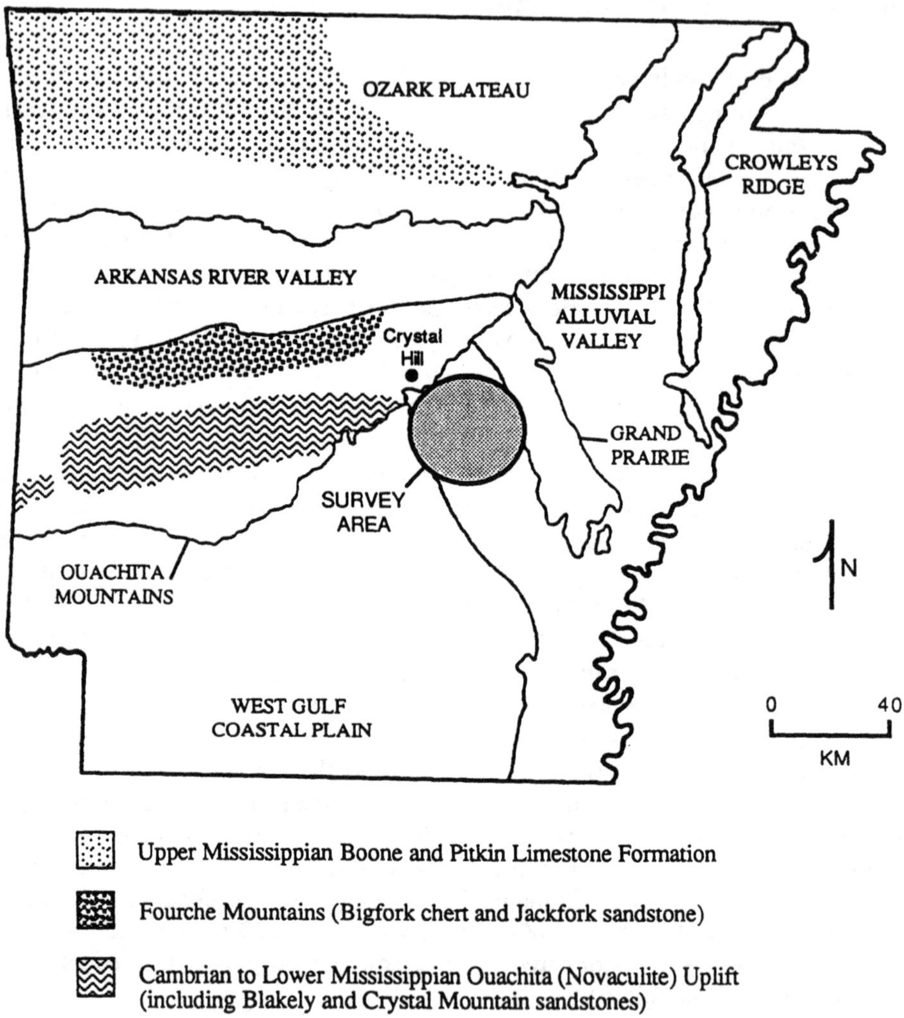


**Figure 1.** Geological sources of lithic raw materials used by the aboriginal inhabitants of central Arkansas.

on hunting and gathering wild plant and animal foods, with the addition of some native cultigens (Fritz 1988; Nassaney 1991:190–192).

Toltec is interpreted as a regional civic and ceremonial center for Plum Bayou culture during the Coles Creek period. The hierarchical settlement pattern and evidence for labor mobilization to build the mounds from a large, dispersed population in the Lowland suggests a ranked social formation. How was authority constituted and compliance enforced? Can these processes be interpreted from the archaeological record? How were the production, distribution, and consumption of chipped stone tools implicated?

## CHIPPED STONE TOOLS: RAW MATERIAL ACQUISITION, LABOR ALLOCATION, AND THE INTENSIFICATION OF PRODUCTION

In a very general sense, economy is a system for the acquisition, production, distribution, and consumption of resources and finished goods. The relationships of these subsystems and the role of individuals within the economy vary according to particular historical and political circumstances. In egalitarian societies, the division of labor is based on age and gender; a technical division of labor is underdeveloped. Individuals maintain equal access to resources and the means of production either by provisioning themselves or through reciprocal exchange. Everyone may perform essentially the same tasks, and the fruits of labor are shared equally in society. Production is generally for use rather than for exchange (Lee 1988:255). When exchange occurs, it takes the form of reciprocity (e.g., Sahlins 1972:191–196; Weissner 1977).

More complex divisions of labor may arise by restructuring labor allocation to various tasks (see Tosi 1984). In a technical division of labor, an economic task (e.g., biface production) is segmented into a series of steps or stages that may be performed by different individuals (Braverman 1974). This forces some individual producers to relinquish control, because they do not see the raw material transformed into a usable object—control is maintained only by those who command the production process. Technological organization of this type takes the form of a series of "specialists" who produce the various objects in the stages of production. For instance, different producers might participate in early and late stages of biface production. These production steps can co-occur at a specific location, or they can occur in different places such as specialized workshops, denoting a spatial division of labor.

Another form of a spatial division of labor occurs when completely different goods are made by different producers. For instance, some producers might make dart points or bifacial knives, whereas others may produce arrow points (cf. Seeman 1985; Cross 1990). In this sense the spatial division of labor results in differential "output per capita for a given product within the population sampled"—what Tosi (1984:23) defines as craft specialization.[2]

A spatial or technical division of labor may emerge in society when some persons are better than others at performing certain tasks. Local or regional specializations in exchange goods are reinforced by the friction of distance, which becomes more pronounced when mobility decreases. To some extent, incipient specialization merely represents an elaboration of the division of labor based on age and gender so long as less mobile individuals or groups can maintain relationships of interdependence and equivalence. Strategic individuals, however, can manipulate the flow of goods by controlling access to spatially-restricted resources or by transforming negative reciprocity to their benefit to create and realize varying exchange rates. These attempts allow them to monopolize surplus and accrue commodities, which further reinforces their status. Under these conditions, specialization ceases to be a complementary division of labor, and social equivalence can become ideologically transformed into socially meaningful differences. Furthermore, these conditions promote the opportunity for high ranking individuals to withdraw from the primary productive labor force (Fried 1967:131). Monopolization and resource control have distinctive spatial patterns in the archaeological record that differ from those associated with unrestricted access. The challenge is to identify how control is expressed materially and how it can be inferred archaeologically.

I approach these issues by investigating three aspects of chipped stone tool production in central Arkansas: (1) raw material acquisition; (2) labor allocation; and (3) productive intensification. To identify patterns of acquisition, I examine the natural distribution of raw materials in the region to determine if the materials were accessible to all segments of society. I then analyze the distribution of lithic materials in archaeological assemblages to interpret how stone was acquired and if its acquisition was controlled by central institutions. The proportions of raw materials by count and weight prove to be particularly useful in this evaluation.

Labor allocation can be studied in the material world by examining archaeological indicators for a spatial or technical division of labor. Variation in the distribution of production, for instance, may suggest production for exchange that could be manipulated by segments of society. This analysis requires an understanding of the reduction trajectories used in the production of chipped stone tools (e.g., see Johnson 1993a). What kinds of stone tools are being produced, and where does production take place? Are all stages in production spatially auto-correlated, and what do deviations from these patterns suggest?

Finally, some lithic acquisition and production strategies are more labor intensive (i.e., require more labor inputs) than others. Intentional thermal alteration is one such practice. Is thermal alteration spatially or temporally correlated with other strategies that support social ranking? I develop more detailed expectations for each of these lines of analysis below.

## Resource Acquisition and Spatial Accumulation

In nonstratified societies, resource acquisition and exchange relationships are embedded within a wider arena of social relations. Archaeologists can monitor the role of objects in exchange systems by comparing the form, content, and diversity of archaeological assemblages to study resource acquisition and distribution in the material world as they pertain to human interaction and social integration (e.g., Brumfiel and Earle 1987; Brown et al. 1990). Investigations of the political economy of exchange relationships are oriented toward understanding whether or not segments of society monopolize and control access to particular resources (e.g., Cobb 1989, 1993:60–65).

In egalitarian societies, direct acquisition or reciprocal down-the-line exchange are the most prevalent tactics for getting resources to consumers. When resources are obtained by direct acquisition, raw material frequencies are inversely proportional to the distance from their sources; they exhibit clinal patterns due to the friction of distance (Renfrew 1975; Paynter 1982:32–44). In this model, exchange networks need not be invoked to explain the frequency distribution of raw materials. When a decrease in mobility restricts direct access to raw materials, exchange is needed to overcome the friction of distance. Clinal distributional patterns are also expected in this model for some material dimensions, as in the proportion of a particular raw material type to another, for instance. Qualitative differences might be expected along other dimensions, however. For example, under conditions of reciprocal exchange most early stage reduction may occur at the source area, with progressively later stage debitage occurring further from the source (see Johnson 1989).

The material manifestations of exchange monopolies can take the form of discontinuous distributions of objects or other patterns that deviate from clines (e.g., Renfrew 1975). For instance, high ranking individuals who controlled material objects in order to reinforce positions of status often resided at central places. Thus ideologically charged or otherwise desirable objects may occur at central places in greater frequencies than at smaller, less differentiated sites in the settlement hierarchy (e.g., Frankenstein and Rowlands 1978; Welch 1986).

## Raw Material Acquisition

The aboriginal inhabitants of the Arkansas River Lowland and surrounding areas used predominantly chert, novaculite, and quartz crystal to make chipped stone tools. Tables 1 and 2 show the percentages of raw material types for 20 assemblages in central Arkansas, grouped chronologically into the Marksville, Baytown-Coles Creek, and Mississippi periods, based on associated ceramics. The samples derive from sites investigated during the Plum Bayou Survey project, published site reports, and several large surface collections that were sorted by Survey personnel at the Toltec Mounds Research Station (see Figures 2 and 3). I distinguish those samples that were recovered using systematic procedures and also take into account the biasing effects of using counts or weights in the interpretations.

### *Chert*

Chert is the predominant raw material employed in lithic manufacturing activities. It typically occurs in gravel forms. Gravels were transported and

**Table 1. Distribution of chipped stone raw materials by count for select central Arkansas sites. Systematic samples are shown in bold. Key to cultural affiliation by period: MK=Marksville, BY=Baytown, CC=Coles Creek, WD= Woodland, and MS=Mississippi**

| Site | Sample size (n) | Cultural affiliation | Chert (%) | Novaculite (%) | Quartz crystal (%) |
|---|---|---|---|---|---|
| **3PU115** | 1059 | MK | 67.9 | 30.1 | 2.0 |
| **3LN342** | 729 | MK | 78.6 | 20.3 | 1.1 |
| 3PU106 | 338 | MK-MS | 68.9 | 24.6 | 6.5 |
| **3LN212** | 126 | BY | 61.1 | 31.7 | 7.1 |
| **3LN342** | 5410 | BY,CC | 82.1 | 14.8 | 3.1 |
| **3LN182** | 98 | CC | 88.8 | 8.2 | 3.0 |
| **3PU3** | 3699 | CC | 77.5 | 17.9 | 4.6 |
| **3PU332** | 333 | CC | 76.3 | 19.2 | 4.5 |
| **3PU330** | 1215 | CC | 82.9 | 7.7 | 9.4 |
| **3LN42**[a] | 3009 | BY,CC | 79.6 | 4.3 | 16.1 |
| **3LN42**[b] | 26439 | BY,CC | 85.8 | 6.5 | 7.8 |
| **3PU252**[c] | 11298 | BY,CC,MS | 79.9 | 16.9 | 3.2 |
| 3LN182[d] | 252 | CC | 52.8 | 7.1 | 40.1 |
| 3LN200 | 58 | CC | 51.7 | 39.7 | 8.6 |
| 3PU298 | 171 | CC | 83.0 | 8.8 | 8.2 |
| 3PU301 | 174 | CC | 80.5 | 12.6 | 6.9 |
| 3LN20 | 3057 | MK-MS | 64.8 | 22.7 | 12.5 |
| 3PU163 | 186 | CC,MS | 60.8 | 18.3 | 21.0 |
| 3LN27 | 516 | CC,MS | 69.6 | 23.3 | 7.2 |
| **3PU17** | 405 | CC,MS | 92.6 | 4.9 | 2.4 |
| **3PU167** | 64 | CC,MS | 93.8 | 6.3 | 0.0 |

[a]This assemblage includes only debitage in surface collections from the eastern half of the site (Stanfill 1980:Table 3).
[b]Mound D lithics as reported by Hoffman (1982b:Table 1).
[c]Waddell et al. (1987).
[d]Riggs collection.

deposited naturally far from their parent sources, which are probably the Upper Mississippian Boone and Pitkin limestone formations of the northern Arkansas Ozarks (Figure 1; see Sabo et al. 1982:10–11; Manger 1986:210). They are highly rounded, with a tan-to-light brown and occasionally blue-gray, relatively smooth, hard cortex. Chert was probably available locally along the Arkansas River and its former channels, though its precise source locations have not been identified. Gravels almost identical to those in the Mound D artifact collections have been noted during dredging operations in the vicinity of Toltec (Hoffman 1982a:40). Gravel bars may be visually obscured today by siltation associated with modern land-clearing and agricultural activities.

**Table 2. Distribution of chipped stone raw materials by weight for select central Arkansas sites. Systematic samples are shown in bold. Key to cultural affiliation by period: MK=Marksville, BY=Baytown, CC=Coles Creek, WD= Woodland, and MS=Mississippi**

| Site | Sample size (g) | Cultural affiliation | Chert (%) | Novaculite (%) | Quartz crystal (%) |
|---|---|---|---|---|---|
| **3PU115** | 2285.8 | MK | 91.4 | 7.0 | 1.6 |
| **3LN342** | 1670.3 | MK | 95.2 | 3.9 | 0.9 |
| 3PU106 | 4188.4 | MK-MS | 65.9 | 21.4 | 12.7 |
| **3LN212** | 296.5 | BY | 52.2 | 35.3 | 12.5 |
| **3LN342** | 10477.3 | BY,CC | 91.0 | 4.5 | 4.5 |
| **3LN182** | 45.3 | CC | 88.5 | 5.1 | 6.4 |
| **3PU3** | 10204.8 | CC | 86.3 | 8.4 | 5.4 |
| **3PU332** | 1449.6 | CC | 87.6 | 8.6 | 3.8 |
| **3LN42**[a] | 12507.4 | BY,CC | 88.3 | 3.4 | 8.3 |
| **3PU330** | 3484.6 | CC | 89.3 | 4.9 | 5.8 |
| **3PU252**[b] | 7480.0 | BY-MS | 68.7 | 17.7 | 13.6 |
| **3CN117**[c] | 10676.0 | MK,CC,MS | 95.4 | 1.5 | 3.1 |
| 3LN182[d] | 45.3 | CC | 54.0 | 7.8 | 38.1 |
| 3LN200 | 605.0 | CC | 60.2 | 24.0 | 15.8 |
| 3PU298 | 4728.7 | CC | 94.6 | 3.6 | 1.8 |
| 3PU301 | 1903.5 | CC | 87.0 | 10.1 | 2.9 |
| 3LN20 | 19255.7 | MK-MS | 70.4 | 14.1 | 15.6 |
| 3PU163 | 1034.1 | CC,MS | 64.3 | 13.8 | 21.9 |
| 3LN27 | 5069.3 | CC,MS | 85.6 | 10.8 | 3.5 |
| **3PU17** | 1580.4 | CC,MS | 98.0 | 1.6 | 0.4 |
| **3PU167** | 357.1 | CC,MS | 95.1 | 4.9 | 0.0 |

[a]This assemblage includes only debitage in surface collections from the eastern half of the site (Stanfill 1980:Table 3).
[b]Waddell et al. (1987).
[c]This assemblage includes only debitage (Hemmings 1985:33).
[d]Riggs collection.

### *Novaculite*

The most common aboriginally exploited lithic raw material in the Ouachitas is novaculite (Etchieson 1989). It outcrops in bedded form throughout the Ouachita Mountains west of the survey area (Figure 1). Extensive novaculite ridges have been documented from Oklahoma into central Arkansas (Holmes 1891; Jenny 1891; Branner 1927). The large extent of this major formation and its occurrence in outcrops, talus slopes, and stream terraces as redeposited cobbles make it difficult to identify any one source for novaculite in the absence of geochemical sourcing analyses. Most of the material in archaeological assemblages is probably from a bedded source, because cortical surfaces are very rare (Hemmings 1985:32, and Waddell et al. 1987).

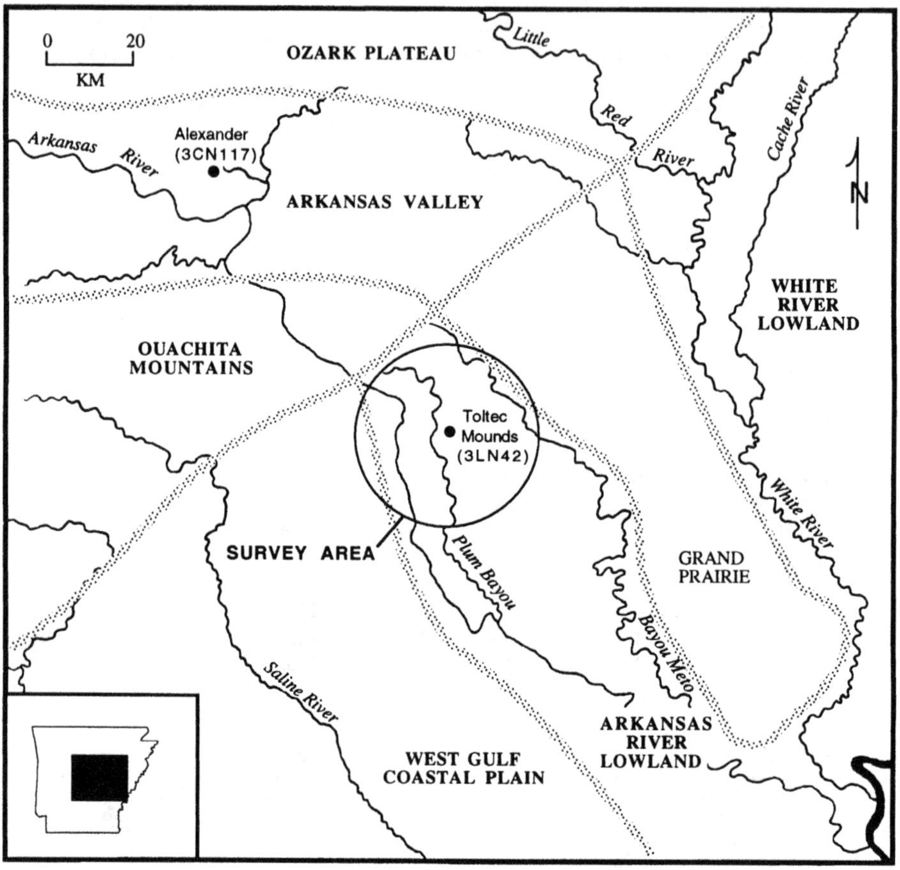


**Figure 2.** Physiographic diversity of central Arkansas, showing the locations of the survey area and the Toltec Mounds (3LN42) and Alexander (3CN117) sites.

## *Quartz Crystal*

Quartz crystal, the official state mineral, was used extensively by aboriginal inhabitants in the Midsouth. The principal deposits of high-grade quartz are found in the Blakely and Crystal Mountain sandstones of the Ouachita Mountains, although other deposits are found throughout exposures of Paleozoic shales, sandstones, and cherts along the central Ouachita Mountains (Williams 1959:74; Hoffman 1982a:41; Howard 1986). Major crystal-bearing veins (though possibly of lesser quality) include a concentration in North Little Rock at Crystal Hill, about 30 km northwest of Toltec (Figure 1). Quartz crystal arrow points, dart points, cores, flakes, and hammerstones have been recovered archaeologically from Plum Bayou sites. Bifacially chipped quartz crystal flakes and the tips of quartz crystals

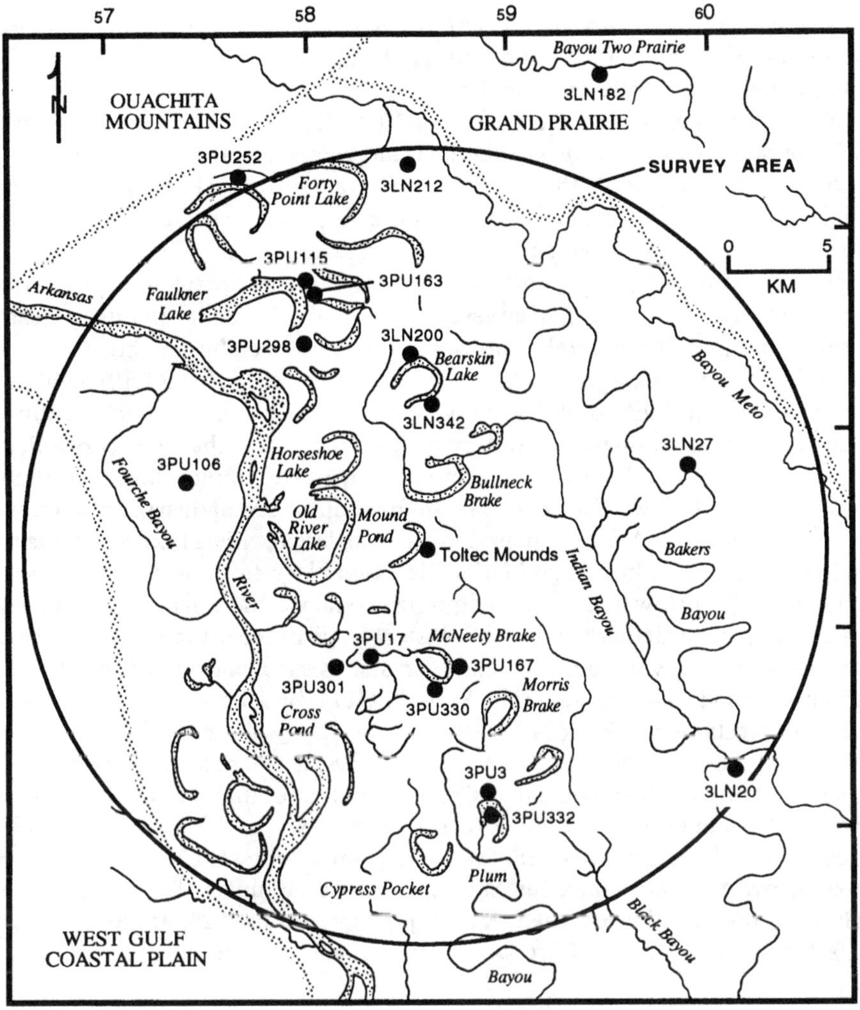


**Figure 3.** Physical geography of the Arkansas River Lowland, showing the locations of sites discussed in the text.

served as cutting and engraving tools respectively (Waddell et al. 1987; Etchieson 1989; Rolingson 1990b: 35).

## Analysis of Raw Material Frequencies in Archaeological Assemblages

Chert is the predominant chipped stone raw material in the 20 assemblages available for this analysis (Tables 1 and 2). It ranges in frequency from 51.7% to

93.8% by count and 54.0% to 94.6% by weight. Although chert dominates all of these assemblages, its *relative* contribution (by count) to chipped stone assemblages increased from the Marksville through the late Mississippi periods, particularly in systematically collected samples.

Novaculite ranges in frequency from 4.3% to 31.7% by count and from 1.6% to 35.3% by weight. On average, novaculite is more prevalent in unsystematic than in systematic samples, with the exception of Fitzhugh (3LN212). This pattern probably reflects collector biases.[3] Novaculite is poorly represented (4.9% to 6.3%) at sites that have Mississippi-period components (i.e., 3PU167, 3PU17). This pattern stands in sharp contrast to Marksville sites, in which novaculite comprises from 20.3% to 30.1% of an assemblage. Thus novaculite counts decrease over time, though this temporal trend is not as pronounced by weight.

All but one assemblage contains quartz crystal; it ranges from 1.1% to 40.1% by count and from 0.9% to 38.1% by weight. Unsystematic collections exhibit greater proportions of quartz crystal by weight and count than the controlled, systematic samples (Tables 1 and 2). Among controlled samples, however, there is a clear increase, followed by a decrease, in the proportions of quartz crystal from the Marksville through Mississippi periods. For instance, only 21 pieces of quartz crystal were identified in a large Marksville assemblage from St. Marks Church (3PU115). Similarly low frequencies of quartz crystal were found at Bearskin Lake South (3LN342), particularly along the western edge of the artifact scatter where Marksville ceramic types occur. Quartz crystal is also poorly represented in Mississippi-period assemblages from Keo (3PU17), and no quartz crystal was found at McNeely East (3PU167).[4]

The frequency data suggest that quartz crystal reached the height of its popularity during the Baytown-Coles Creek period. Sites that are contemporaneous with Toltec contain from 3.0% to 40.1% quartz crystal, with a mean of 11.7%. The lowest and the highest frequencies derive from Winfrey (3LN182), reflecting the use of two different collection strategies. If we eliminate the higher value, which we know to be biased, the mean of quartz crystal is 8.2% and the median is 7.1% for the remaining 15 Baytown-Coles Creek assemblages.[5]

## Interpretation of Variation in the Distribution of Raw Materials

The proportions of raw material types in the analyzed assemblages exhibit temporal and spatial variation. What factors can explain the variance, and is the variance related to attempts by elites to accumulate surplus or control the organization of production?

### *Chert*

If we assume that local chert gravels were widespread along tributaries of the Arkansas River throughout the survey area prior to channel modification, then the archaeological distribution of chert in Arkansas River Lowland sites closely corresponds with its natural distribution. Access to this resource seems to have been unrestricted during the Marksville through Mississippi periods—anyone could acquire chert and use it in culturally appropriate ways. The relative proportions of chert artifacts (by count and weight) suggest that chert was readily accessible to all members of society (Sassaman 1992).

There is a tendency for the proportions of lithic raw materials to vary through time. A relative increase in the proportion of chert must be examined in the context of the strategies used to reduce chert cobbles into workable tools (see below). With the exception of the Fitzhugh site (3LN212), deviations from this temporal pattern are probably due to collection biases. Chert debitage, for instance, is often ignored in unsystematic collecting procedures. Thus smaller pieces of chert are underrepresented, even though they are ubiquitous in most archaeological assemblages in the survey area. In contrast, I suspect that novaculite is overrepresented in unsystematic collections, because it was often disproportionately collected in the form of finished core tools such as large bifaces. From this perspective, the low frequency of chert at Fitzhugh is noteworthy. Fitzhugh is located in a backswamp area more than 1 km from any permanent water course and associated gravel bars. Chert cobbles were probably absent or, at best, rare in proximity to the site. The paucity of locally available cherts would have required the site inhabitants to equip themselves with tool kits manufactured from high-quality raw material (i.e., novaculite), and these tools were prepared in advance (Nassaney and Hoffman 1992). Chert was apparently readily available within the catchments of the Toltec Mounds and Coy sites, the two large centers in the survey area. There is no indication, however, that chert was being controlled or monopolized at either of these sites. In fact,

based on its archaeological distribution, it does not appear that chert was controlled in central Arkansas during the late Marksville through late Mississippi periods.

### *Novaculite*

Novaculite was used throughout the prehistoric sequence; however, its relative contribution to chipped stone assemblages decreased through time.[6] The smaller natural size of novaculite in comparison to chert suggests differences in the use of these raw materials (Nassaney 1992b:Table 7.5). Novaculite in archaeological contexts typically consists of relatively small, late stage products; its initial reduction occurred outside the survey area at the site of procurement (Hemmings 1985:32).

In addition to these temporal trends, there are some indications that mechanisms other than direct access or down-the-line exchange were established to acquire novaculite among contemporaneous Plum Bayou culture sites. This observation is based on the variable proportions of novaculite in systematically obtained assemblages. Although some of the variation may be clinal (i.e., explained by distance from source), this proposition cannot be evaluated until the precise source areas of novaculite can be identified.

Several sites have relatively high proportions of novaculite. The Fitzhugh site (3LN212) inhabitants used novaculite in high proportions because of the paucity of locally available raw materials. The Clear Lake site (3PU3) is a small center that may represent a location where novaculite was differentially used or controlled. Similar proportions of raw materials in sites 3PU3 and 3PU332 suggest that these localities, separated only by Clear Lake, may have formed a single community. The remaining two moderately sized sites (3LN342 and 3PU252) may have had disproportionate access to non-local raw materials. The Ink Bayou site (3PU252) is located in the northwest portion of the survey area in close proximity to the Ouachita source area (see Figure 3).

The archaeological distribution of novaculite is not isomorphic with its natural distribution; it is slightly clustered (rather than uniformly or clinally distributed) over the landscape during the Baytown-Coles Creek period. This variation may reflect differential access or use. If clustering reflects political-economic exchange relationships, the pattern does not coincide with the locations associated with mound construction and labor mobilization. Thus it may constitute evidence for a sequential hierarchy outside the control of social relations at Toltec. The interpretations of the observed patterns are provocative, but must remain tentative until further spatial and temporal studies are conducted.

### *Quartz Crystal*

Much of the variation in the distribution of quartz crystal in central Arkansas can be accounted for by collection strategy, proximity to Crystal Hill, cultural utility, and acquisition strategies. For example, quartz crystal is probably overrepresented in unsystematic samples because it was clearly visible or aesthetically pleasing to some collectors. The high proportions of quartz crystal at Winfrey (3LN182) in the Riggs collection is a case in point, particularly in light of the controlled, excavated sample from the site.[7] Most of the other unsystematic collections are comparable to the controlled samples, although the mean weight of quartz crystal is consistently greater in unsystematic collections (Nassaney 1992b:Table 7.5). Sample size does not explain the variation in the proportion of quartz crystal; relatively small and large samples contain both high and low frequencies (cf. 3LN42, 3LN182, 3LN342, and 3PU163).

Unlike the situation with chert, the distance of quartz from its source may be an important variable in the composition of lithic assemblages. If Crystal Hill is assumed to be the source area, then the frequency of quartz crystal in an archaeological assemblage can be related to distance from the source to determine the importance of proximity.

**Table 3. Distribution of quartz crystal by count, weight, and mean weight from select central Arkansas sites, and their distances (km) from Crystal Hill and the Toltec Mounds site. Systematic samples are shown in bold. Key to cultural affiliation by period: MK=Marksville, BY=Baytown, CC=Coles Creek, WD=Woodland, and MS=Mississippi**

| Site | Cultural affiliation | Distance from Crystal Hill | Distance from Toltec mounds | Quartz crystal ct (%) | Quartz crystal wt (%) | Quartz crystal mean wt (g) |
|---|---|---|---|---|---|---|
| **3PU115** | MK | 17.5 | 14.4 | 2.0 | 1.6 | 1.7 |
| **3LN342** | MK | 26.0 | 7.5 | 1.1 | 0.9 | 2.0 |
| 3PU106 | MK,MS | 19.0 | 11.9 | 6.5 | 12.7 | 24.1 |
| **3LN212** | BY | 22.5 | 18.8 | 7.1 | 12.5 | 4.1 |
| **3LN342** | BY,CC | 26.0 | 7.5 | 3.1 | 4.5 | 2.8 |
| **3LN182** | CC | 33.3 | 24.3 | 3.0 | 6.4 | 1.0 |
| **3PU252**[a] | BY,CC,MS | 14.3 | 20.2 | 3.2 | 13.6 | — |
| **3PU3** | CC | 40.0 | 12.5 | 4.6 | 5.4 | 3.2 |
| **3PU332** | CC | 40.5 | 13.1 | 4.5 | 3.8 | 3.7 |
| **3LN42**[b] | BY,CC | 29.3 | 0.0 | 16.1 | 8.3 | 2.1 |
| **3LN42**[c] | BY,CC | 29.3 | 0.0 | 7.8 | — | — |
| **3PU330** | CC,MS | 34.5 | 6.6 | 9.4 | 5.8 | 1.8 |
| **3CN117**[d] | MK,CC,MS | 55.3 | 80.0 | — | 3.1 | — |
| 3LN182[e] | CC | 33.3 | 24.3 | 40.1 | 38.1 | 2.7 |
| 3LN200 | CC | 24.8 | 9.4 | 8.6 | 15.8 | 19.1 |
| 3PU298 | CC | 15.0 | 15.0 | 8.2 | 1.8 | 6.0 |
| 3PU301 | CC | 30.5 | 6.8 | 6.9 | 2.9 | 4.7 |
| 3LN20 | MK-MS | 48.3 | 18.7 | 12.5 | 15.6 | 7.8 |
| 3PU163 | CC,MS | 18.3 | 14.3 | 21.0 | 21.9 | 5.8 |
| 3LN27 | CC,MS | 40.0 | 14.4 | 7.2 | 3.5 | 4.9 |
| **3PU17** | CC,MS | 31.3 | 5.3 | 2.4 | 0.4 | 0.7 |
| **3PU167** | CC,MS | 34.5 | 5.6 | 0.0 | 0.0 | 0.0 |

[a]Waddell et al. (1987).
[b]This assemblage includes only debitage in surface collection from the eastern half of the site (Stanfill 1980:Table 3).
[c]Mound D (Hoffman 1982b:Table 1).
[d]Debitage only (Hemmings 1985:33).
[e]Riggs collection.

Table 3 shows the straight line distances between Crystal Hill and 20 archaeological sites in central Arkansas that were occupied from the Marksville through Mississippi periods (A.D. 200–1700). Most of these sites have Plum Bayou components, although earlier and later sites are also included to evaluate diachronic changes in quartz crystal use. The relationship of quartz crystal (by count, mean weight, and weight) to distance from Crystal Hill is shown graphically in Figures 4–6. These graphs show that distance from Crystal Hill is a poor predictor of the percentage of quartz crystal in an assemblage by count ($r^2$=0.001; Figure 4) and mean weight ($r^2$=0.048; Figure 5). A weak negative correlation, however,

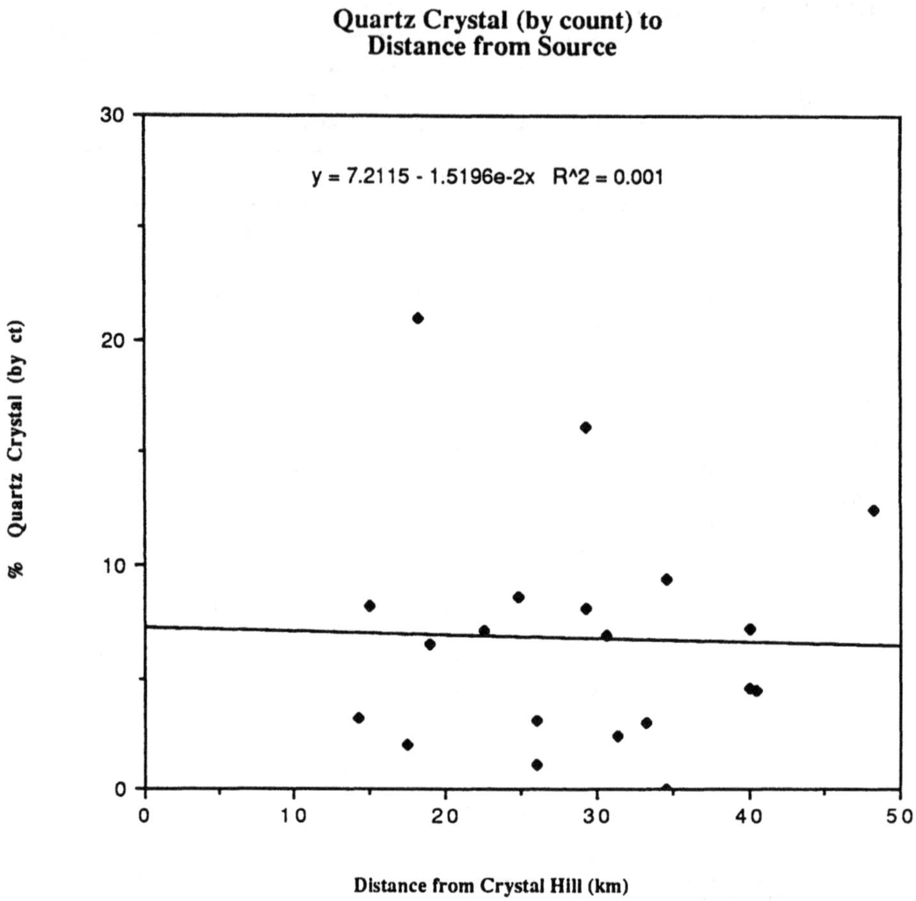


**Figure 4.** Relationship of quartz crystal to distance from source by count.

obtains between distance from source and weight ($r^2$=0.060; Figure 6). The relationship can be strengthened if we eliminate a few of the outliers. Four of the sites that have less quartz than predicted (3PU115, 3LN342, 3PU17, and 3PU167) were occupied during the Marksville and Mississippi periods; thus they can be eliminated on chronological grounds. The Chownings collected a small sample (n=14) of quartz pieces from site 3PU298, whereas their collection, combined with Miller's, from Coy (3LN20) has a disproportionately high number of large pieces of quartz. If we also exclude these aberrant sites from the analysis, distance from source explains more than 56% of the variation ($r^2$=0.565; Figure 7) in the proportional weight of quartz crystal among the 14 remaining sites—a reasonably strong negative correlation. This relationship suggests that quartz crystal was

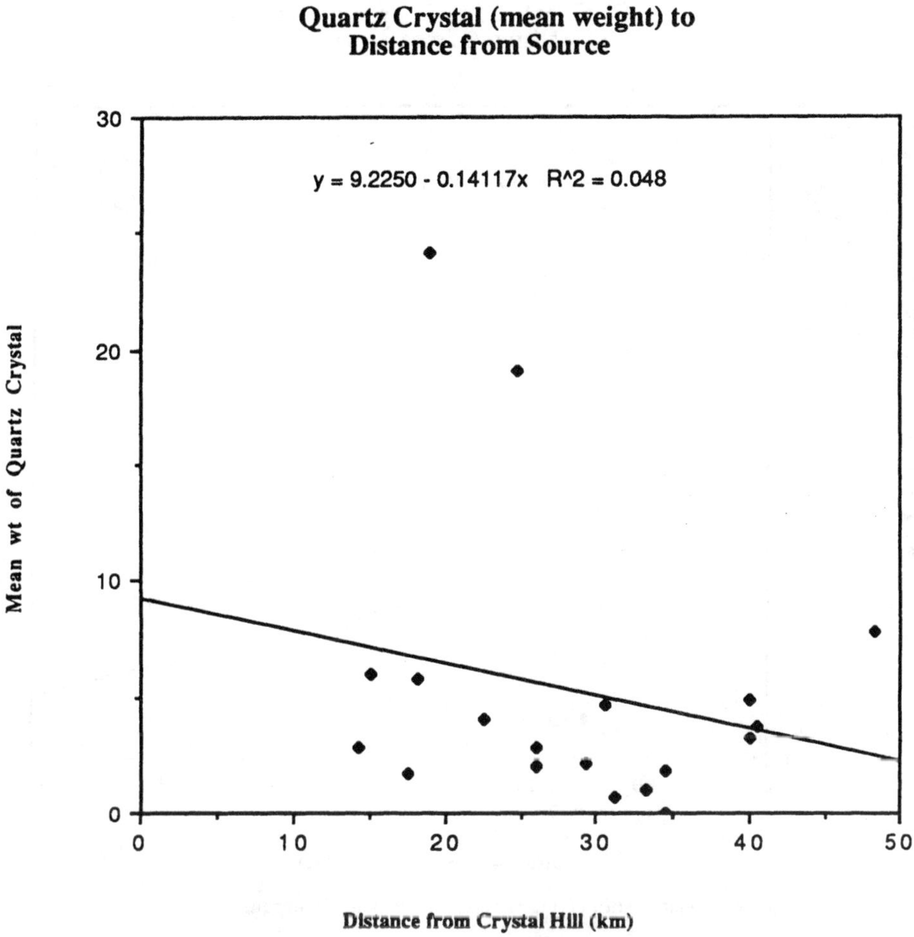


Figure 5. Relationship of quartz crystal to distance from source by mean weight.

acquired either directly from its source or through down-the-line reciprocal exchange *in most cases* during the Baytown-Coles Creek period.

Several other sites appear to have more quartz than predicted by this linear regression. For instance, the County Dairy Farm site (3PU163) contains an inordinate amount of quartz by weight (21.9%) and count (21.0%), despite its close proximity to Crystal Hill (19 km). Bearskin North (3LN200) also has higher proportions of quartz than expected by weight (15.8%). Interestingly, both of these sites were collected by the Chownings and Miller, raising serious questions about sample bias.

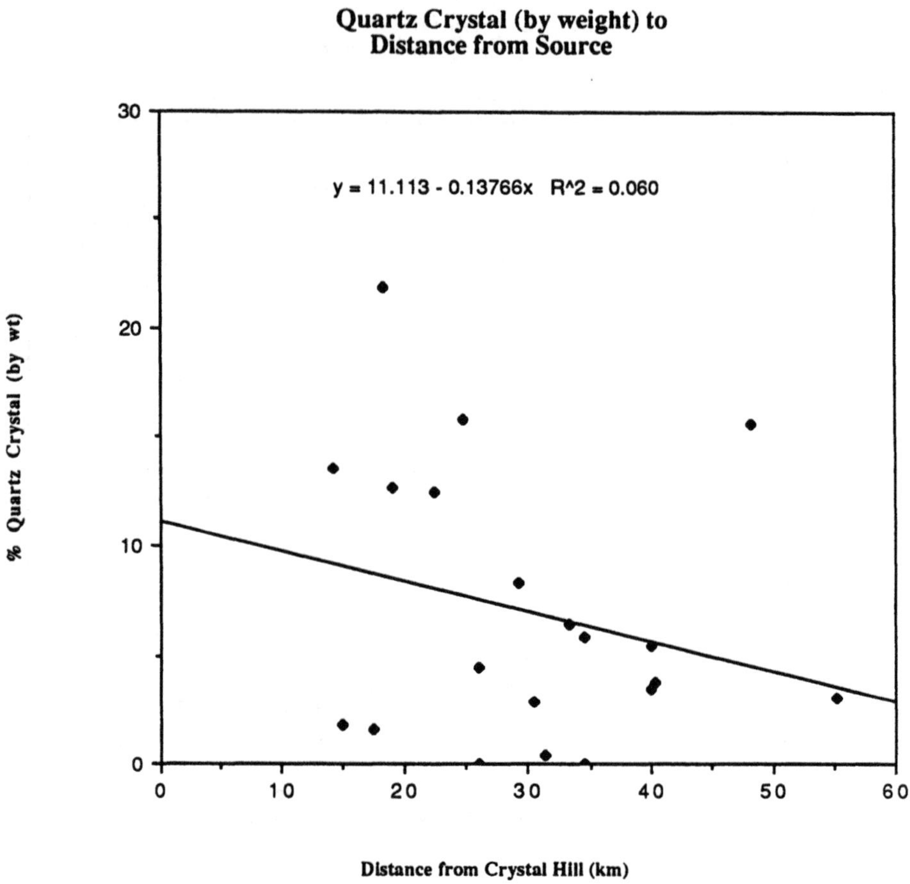


**Figure 6.** Relationship of quartz crystal to distance from source by weight.

The distribution of quartz by count is more ambiguous (Figure 4). Apart from County Dairy Farm (3PU163), the off-mound surface collections east of the plaza at Toltec have the highest percentage of quartz among assemblages. Quartz occurs with nearly twice the frequency in this assemblage as in the Toltec Mound D collection, implying intra-site functional differentiation. If the mounds adjacent to the plaza (e.g., D, E, and S) supported specialized activities such as ceremonial feasting (Rolingson 1990a), then quartz crystal reduction may have been a specialized production activity localized *away* from Toltec's ceremonial plaza. The segregation of quartz crystal reduction activities at Toltec raises the possibility that some sites in the survey area may have been provisioned with quartz crystal from Toltec. There is no relationship, however, between distance from Toltec and

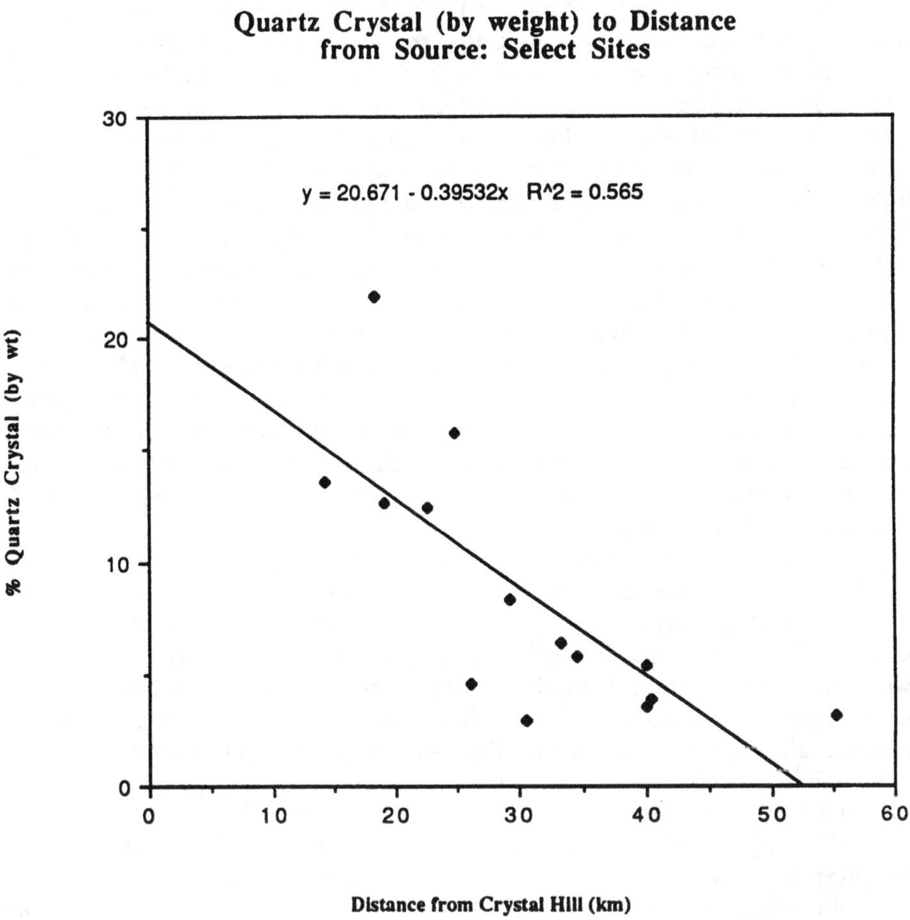


Figure 7. Relationship of quartz crystal to distance from source by weight for select sites in central Arkansas.

quantity of quartz crystal by count, weight, or mean weight (Nassaney 1992a: Figure 7.6).

There is potentially significant variation in the mean weight of quartz crystal which cannot be explained by distance from Crystal Hill. For instance, inhabitants of the Coy Mound site (3LN20), which is contemporaneous with Toltec and served as the major ceremonial center in the eastern half of the survey area, had access to quartz crystal, either via Toltec or directly from the source, despite its distance from Crystal Hill (48 km). Collection strategy also contributes to the variation in the mean weight of quartz crystal in some assemblages, however. For example, the mean size of quartz recovered in a small, excavated sample screened through

1/4-inch mesh is 1.0 g at the Winfrey site (3LN182), whereas the average weight of a piece of quartz in the Riggs collection from the same site is 2.7 g.

The quantities of quartz crystal at the Coy Mound and the Toltec Mounds sites, measured along multiple dimensions (e.g., count, weight, mean weight), underscore the importance of quartz for social reproduction at large multiple mound centers in Plum Bayou culture. Likewise, the frequency distributions indicate that quartz crystal reaches the height of its popularity during the Baytown-Coles Creek period. Marksville and Mississippi-period assemblages generally exhibit low proportions of quartz crystal, suggesting that the material was either inaccessible or of little importance in technological and social reproduction in the periods before and after the Plum Bayou cultural integration. Quartz crystal occurs so infrequently at Marksville and Mississippi-period sites that it may be associated with Baytown or Coles Creek-period use of these same sites. Mechanisms had become well established for acquiring quartz and other non-local raw materials during the Baytown-Coles Creek period, but they withered after the decline of Toltec (ca. post-A.D. 950) (Rolingson 1982a:89,93).[8]

Quartz crystal constitutes 3.1% (by weight) of the Alexander assemblage (3CN117), despite its distance from Crystal Hill (55 km); this implies that quartz crystal was an important resource for the inhabitants of Alexander. It seems unlikely that quartz was obtained through down-the-line exchange; individuals were either provisioning themselves or obtaining the raw material by exchange from nearer the source. Quartz crystal also occurs at sites with Coles Creek Incised ceramics along the northern and northeastern peripheries of Plum Bayou culture. Although quantitative measures are not available, quartz crystal occurs in archaeological assemblages located >100 km from its source areas.[9]

Crystals were clearly desirable raw materials in Plum Bayou culture. It is tempting to suggest that they represented prestige items that were imbued with symbolic significance by individuals of rank. Indeed, the ethnographic record provides some interesting clues regarding the role of quartz crystals in the social and ideological domains of native life throughout the Americas, including the southeastern United States (Hall 1979; Reichel-Dolmatoff 1979, 1981). Objects as powerful as quartz crystals would have been ideal commodities for socially ranked individuals to monopolize for status reinforcement.

Does the evidence for the archaeological distribution of quartz crystals support this interpretation? The distribution of quartz crystal over the regional landscape suggests that its acquisition or reduction may have been intensified or partially controlled by segments of society. For instance, individuals at sites in close proximity to Crystal Hill (e.g., 3PU163, 3PU252) may have been chipping quartz crystal and passing on finished products to centralized locations in the settlement hierarchy, such as Toltec and Coy (Waddell et al. 1987). This interpretation is based on the intra-site spatial clustering of quartz crystal debitage at Ink Bayou and the quantities of finished tools at County Dairy Farm, which imply specialized workshop activities. This interpretation does not preclude the possibility that attached specialists (Earle 1981) were chipping quartz crystal at Toltec in areas away from the plaza to produce objects that symbolized offices of rank for a patron elite.

Another potential workshop area is site 3PU106, situated in close proximity to Granite Mountain, where extensive outcrops of various dike rocks, including syenite and pulaskite, can be found today. Miller's collection from 3PU106 contains a wide variety of chipped and ground stone materials, including chert, novaculite, quartz crystal, sandstone, quartzite, hematite, syenite, siltstone, bauxite, lamprophyre, and other unidentified igneous and sedimentary stones in all stages of production (Nassaney 1992b:Table 4.2). The diversity of material suggests that this site was an important workshop for the production of chipped and ground stone tools (Nassaney 1992b:162–163). The few diagnostic ceramics from the site range from the Marksville to the Mississippi periods, suggesting that this location was utilized for stone tool production over several centuries. Site 3PU106 may have been a lithic workshop for the production of the many finished stone tools that were valued farther to the east, much as the Hale site functioned in the production of Mill Creek chert hoes during the Mississippi period in southwestern Illinois (Cobb 1989:83).

## SUMMARY

Aboriginal inhabitants used chert, novaculite, and quartz crystal to manufacture chipped stone tools during the late Marksville through late Mississippi periods in central Arkansas. The change in relative percentages of raw materials from systematic collections by period are shown in Figure 8. Locally available gravel chert was increasingly selected over other raw materials through time. Throughout the periods under study, chert was obtained directly from secondary gravel deposits along the Arkansas River, its former channels, and tributaries.

Novaculite decreased in proportion to chert over time. The low mean weight of novaculite suggests that it was used differently from chert, and it may represent predominantly late stage tool finishing and maintenance activities. Inordinate amounts of novaculite that would indicate monopolization remain unidentified, although there is some potentially significant variation in the spatial distribution of this resource which may be related to productive control and a spatial division of labor.

Quartz crystal increased and then decreased in frequency from the Marksville to the Mississippi periods, reaching the height of its popularity during the Baytown-Coles Creek period. This pattern suggests that quartz was implicated in the production and reproduction of social ranking in central Arkansas. There is a moderately strong relationship between the proportion of quartz crystal in an assemblage (by weight) and distance from its parent source. Neither the mechan-

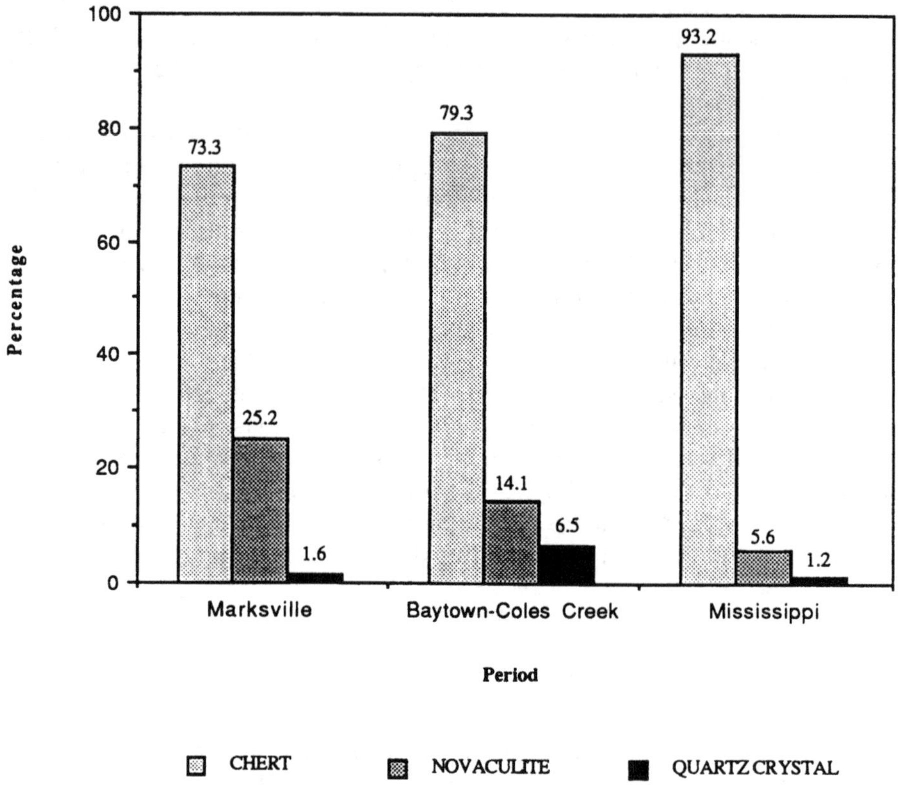


**Figure 8.** Average percentages of raw material by period for select archaeological assemblages in central Arkansas.

isms of direct procurement, down-the-line exchange, nor centralized control from Toltec can account for all of the variation in the distribution of quartz crystal in central Arkansas. If individual agents were attempting to monopolize access to quartz crystal in central Arkansas, their control was tentative at best. More likely, individuals were pursuing different strategies, including direct acquisition, reciprocal exchange, and specialized production and accumulation, to varying degrees of success, to procure and control the distribution of quartz crystals. We now need to explore whether or not strategies were used to control technological organization in the production of chipped stone tools.

## Technological Organization: Intensification and the Division of Labor

Resource monopolization can be interpreted from the archaeological record by examining archaeological indicators for intensification and spatial or technical division of labor. Variation in the distribution of production, for instance, may suggest production for exchange that can be manipulated by segments of society. Alternatively, the labor process can be segmented so that individual producers must relinquish control because they cannot monitor production from raw material to finished product. Control ultimately falls into the hands of those who command the production process. Production can be intensified by increasing output per capita. In order to evaluate if intensification and a spatial or technical division of labor were being used by incipient elites to control surplus, we must first consider the process by which stone tools were being made and where these activities occurred on the landscape.

## Lithic Reduction Strategies in Central Arkansas

In her studies of the Toltec Mound D stone artifacts, Teresa Hoffman (1982a,b) proposed that the shift from dart points to arrow points could be monitored stratigraphically, and that each of these tool forms was associated with different reduction strategies. Empirical evidence from the Marksville through Mississippi periods supports a gradual change in reduction techniques from an emphasis on a core tool to a flake tool technology[10] (Nassaney 1992a). This shift correlates with the adoption of the bow and arrow, decreased mobility inferred from the settlement data, and changes in subsistence strategies (Nassaney 1992b:216–221).

The core tool manufacturing process was oriented toward the production of bifacially chipped, multi-purpose hafted tools (Figure 9). The initial form of the object was a relatively large, flat pebble, which was bifacially reduced using freehand percussion with a hard hammer. Decortication flakes constitute the major form of debitage in the initial stage of reduction. In the next stage, preforms were thinned and shaped into large bifaces, mainly through soft-hammer percussion. The principal waste products from these activities were bifacial thinning flakes, tertiary flakes, and minor amounts of decortication flakes. Large bifaces were given their final shaping through a combination of soft-hammer percussion and pressure flaking. Debitage was comprised of bifacial thinning and pressure flakes.

Unlike the core tool sequence, flake (or expedient) tool manufacturing exhibits a more diverse orientation towards the production of tools, with a greater variety of tools (principally arrow points, drills, and flake tools) represented in most stages of manufacture (Figure 10). Whereas flakes of various shapes constitute the by-products of the core tool technique, the flake tool technique was used to produce flakes—cores were exhausted and discarded. Decortication and interior flakes were detached from small, rounded pebbles using bipolar percussion. Some of these flakes were used without further modification, while others were shaped into small bifaces through soft-hammer percussion and marginal pressure

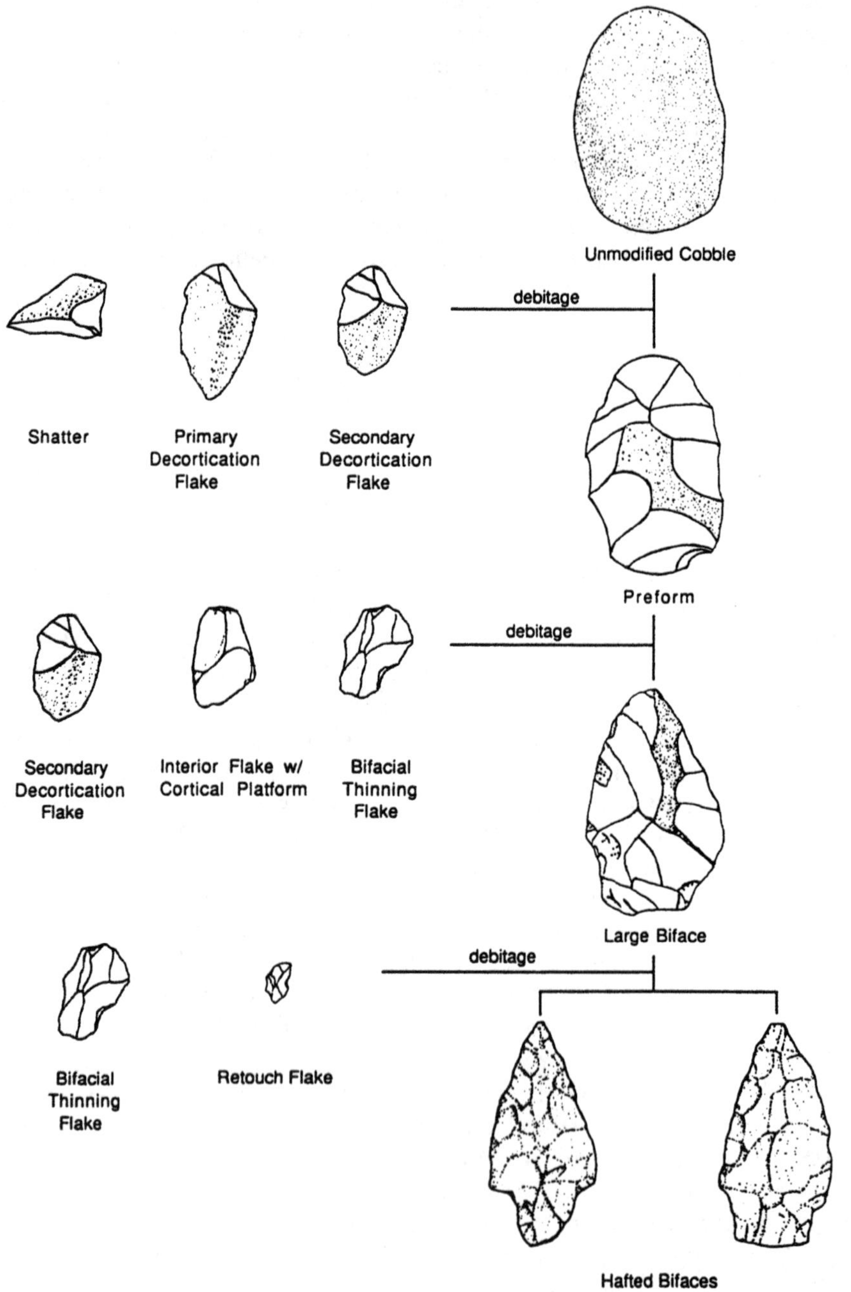


**Figure 9.** Chert core tool manufacturing sequence.

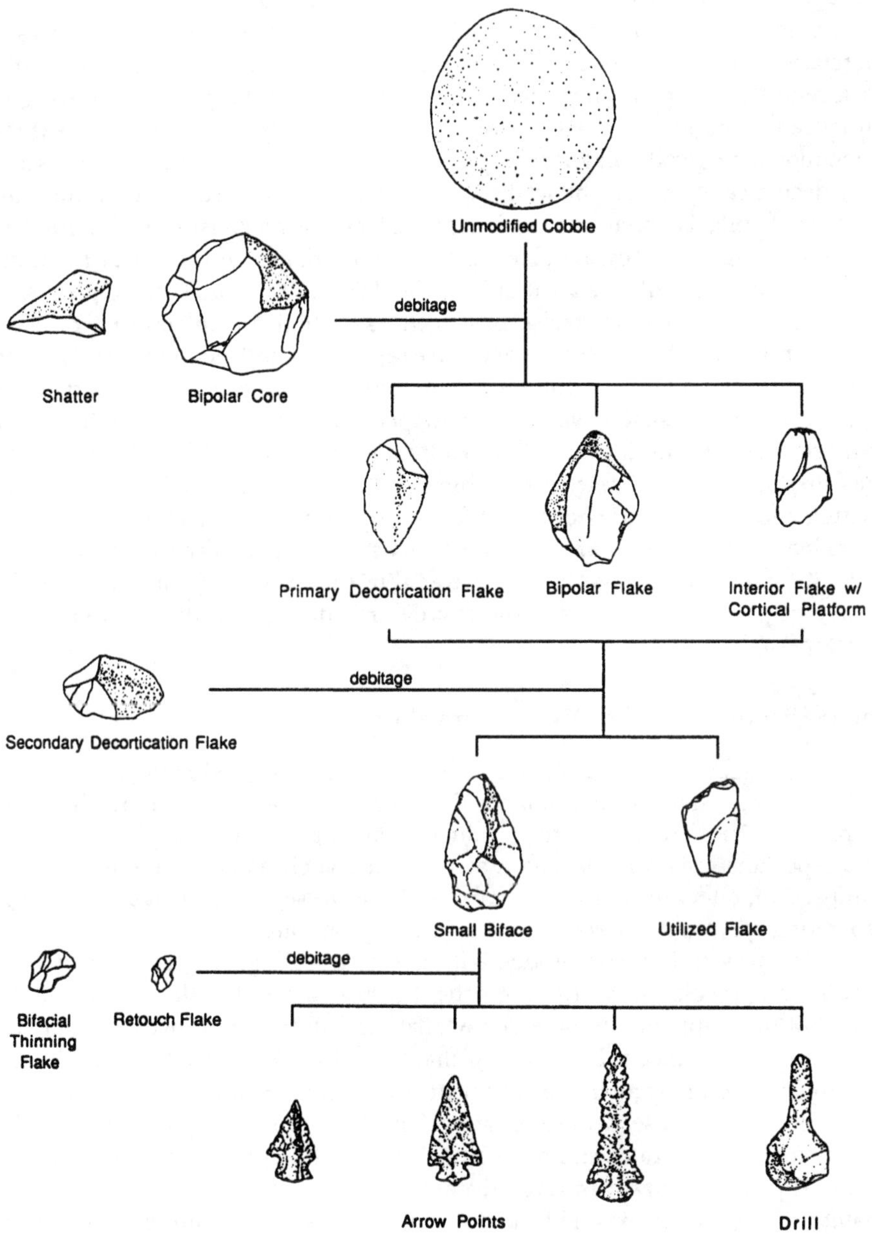


Figure 10. Chert flake tool manufacturing sequence.

flaking (Hoffman 1982b:48–50). The small bifaces were given final shaping through pressure flaking to produce arrow points and drills.

Since both core and flake tool reduction strategies are subtractive processes, there are similarities between the two which mask some of the variation within an assemblage. For instance, decortication flakes may be produced during the initial reduction stages of either sequence. Fortunately, each trajectory exhibits some morphologically diagnostic by-products. Tabular or freehand cores, secondary decortication flakes, bifacial thinning flakes (with cortical and noncortical platforms), bifacial preforms, and large bifaces are diagnostic of the core tool technique. Bipolar cores, bipolar flakes, retouch flakes with cortical platforms, and arrow points/drills are attributed to the flake tool manufacturing process.

Novaculite tools were also produced using both manufacturing processes (Hoffman 1982a:50–53). Researchers have repeatedly noticed, however, that early stages in the production of novaculite core tools seldom occur in the survey area, suggesting that novaculite was initially shaped elsewhere and obtained in partially reduced form (Hemmings 1985:32; Hoffman 1982a:50). Although quantitative measures are not available, novaculite and non-local cherts (including cores, shatter, and large flakes) occur most frequently in preceramic chipped stone tool assemblages; novaculite cores (of almost any size) are exceedingly rare at ceramic sites. It appears that large bifacial novaculite preforms were produced at their source of procurement outside the survey area during the Baytown and Coles Creek periods.

## Intensification and the Division of Labor

A strategy for increasing surplus is to increase productivity or create a technical or spatial division of labor that forces producers to relinquish control. A spatial division of labor in the production of chipped stone tools would lead to local specialists who made different products, such as core or flake tools. In comparison, a technical division of labor serves to segment the labor process by separating early and late reduction stages, for example.

One way to increase productivity is to ensure higher success rates in the production process, by controlling either the raw material or the worker (Braverman 1974). Knappers can increase raw material control by heat-treating chert or novaculite to enhance their chipping characteristics (Rick 1978). Since energy is required to thermally alter chert, heat treatment can be considered a labor-intensive process. In this sense, heat treatment is a way to intensify production, since higher tool success rates can be expected. It is difficult to determine if intensification is justified, however, given the amount of energy required to heat the raw materials (e.g., time expenditures for collecting wood, building fires) and the failure rates involved with the process (e.g., application of excessive heat).[11]

Thermal alteration can also be seen as a conservation measure to increase the probability of obtaining the intended product. This interpretation would apply if appropriate raw materials were in short supply. If the cost of acquiring non-local resources exceeded the amount of energy needed to heat local cherts, then thermal alteration would be a rational economic decision.

## Analysis of the Technological Organization of Chert and Novaculite Reduction

I classified the chipped stone assemblages from eight sites in the survey area according to the manufacturing processes illustrated in Figures 9 and 10 to determine if reduction activities were oriented toward the production of core tools or flake tools (Table 4). The sites are ordered chronologically in the tables that follow, and are separated into Marksville, Baytown-Coles Creek, and Mississippi-period groups. The eight sites selected for this analysis are the only ones from the survey locality with comparable data. One of these sites (3LN342) represents four clusters of material that can be spatially segregated (Nassaney 1992b:163–166). Cluster 1 corresponds with collection units 1 and 2 (CU 1 and 2), whereas Clusters 2, 3, and 4 correspond with CU 4, 6, and 9 respectively. Because these clusters

**Table 4. Frequency of chipped stone by technology and raw material type. Major components are shown in bold. Key to abbreviations: CA=Cultural affiliation, MK=Marksville, BY=Baytown, CC=Coles Creek, MS=Mississippi, CG=Chert gravel, N/NLC=Novaculite/Non-local chert, C:F=Core:Flake Tool Products**

| | | Core tool products | | | Flake tool products | | | Ratio |
|---|---|---|---|---|---|---|---|---|
| Provenience | CA | CG | N/NLC | Subtotal | CG | N/NLC | Subtotal | C:F |
| 3PU115 | **MK** | 95 | 57 | 152 | 143 | 5 | 148 | 1.03 |
| 3LN342 | | | | | | | | |
| Cluster 1 | **MK** | 118 | 14 | 132 | 73 | 0 | 73 | 1.81 |
| Subtotal | | 213 | 71 | 284 | 216 | 5 | 221 | 1.29 |
| 3LN212 | **BY** | 18 | 20 | 38 | 3 | 1 | 4 | 9.50 |
| 3LN342 | | | | | | | | |
| Cluster 2 | **BY-CC** | 76 | 13 | 89 | 51 | 3 | 54 | 1.65 |
| Cluster 3 | **BY-CC** | 114 | 19 | 133 | 79 | 5 | 84 | 1.58 |
| Cluster 4 | **BY-CC** | 80 | 9 | 89 | 100 | 1 | 101 | 0.88 |
| 3PU3 | **CC** | 403 | 156 | 559 | 542 | 28 | 570 | 0.98 |
| 3PU332 | **CC** | 44 | 16 | 60 | 84 | 3 | 87 | 0.69 |
| 3PU330 | **CC,MS** | 135 | 23 | 158 | 221 | 4 | 225 | 0.70 |
| Subtotal | | 870 | 256 | 1126 | 1080 | 45 | 1125 | 1.00 |
| 3PU167 | CC,**MS** | 13 | 2 | 15 | 21 | 1 | 22 | 0.68 |
| 3PU17 | CC,**MS** | 28 | 6 | 34 | 133 | 2 | 135 | 0.25 |
| Subtotal | | 41 | 8 | 49 | 154 | 3 | 157 | 0.31 |
| Total | | 1124 | 335 | 1459 | 1450 | 53 | 1503 | |

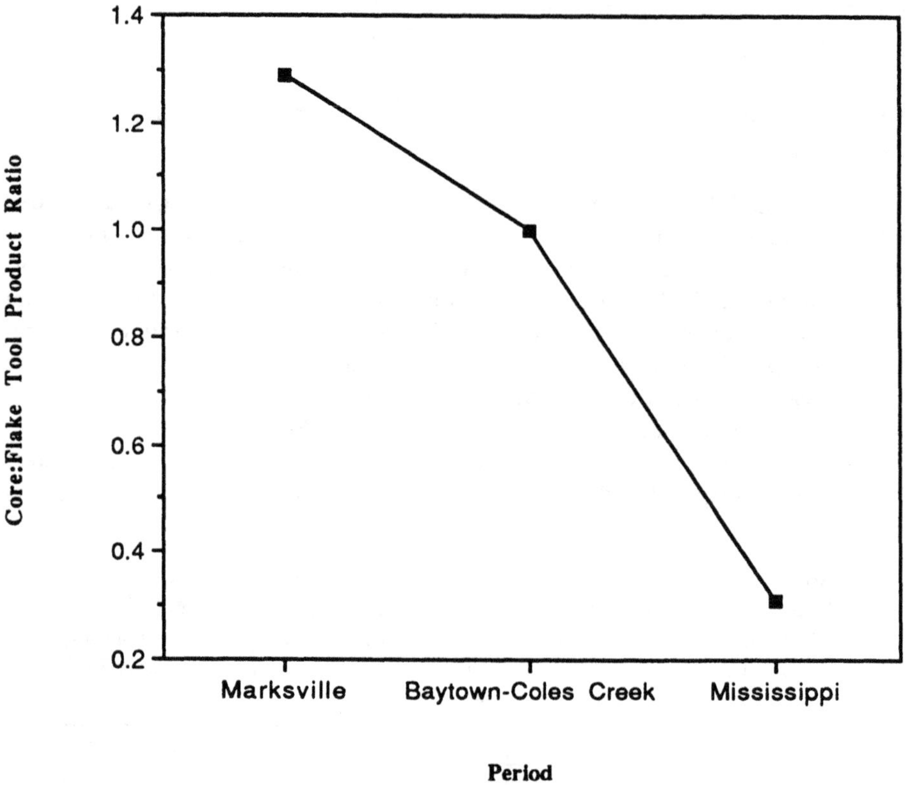


**Figure 11.** Mean core tool:flake tool ratios for select archaeological assemblages in central Arkansas by cultural period.

appear to have distinct functional-chronological characteristics, I chose to examine them separately in this analysis to explore the nature of intra-site variation in the organization of production. Thus I use a total of 11 provenience units for this analysis.

The purpose of the analysis is to compare the spatial and temporal dimensions of the core and flake technologies to determine if they reflect a spatial division of labor. Table 5 shows the frequency of chipped lithic artifacts by reduction stages (early, late) for the same components to determine if a technical division of labor was established to monopolize social surplus. Lastly, Table 6 shows the frequency of heat-treated chert gravel among these 11 components.[12]

There is a decrease through time in the ratio of core tool:flake tool products from these sites (Table 4). Taking the proportion of core tool to flake tool products for each of the periods yields ratios of 1.29 (Marksville), 0.98 (Baytown-Coles Creek), and 0.31 (Mississippi) (Figure 11). Although the high proportion of core

**Table 5. Distribution of chipped stone by reduction stage and raw material type. Major components are shown in bold. Key to abbreviations: CA=Cultural affiliation, MK=Marksville, BY=Baytown, CC=Coles Creek, MS=Mississippi, CG=Chert gravel, N/NLC=Novaculite/Non-local chert, E:L=Early:Late Stage Products**

| | | Early stage products | | | Late stage products | | | Ratio |
|---|---|---|---|---|---|---|---|---|
| Provenience | CA | CG | N/NLC | Subtotal | CG | N/NLC | Subtotal | E:L |
| 3PU115 | **MK** | 277 | 17 | 294 | 88 | 115 | 203 | 1.45 |
| 3LN342 | | | | | | | | |
| Cluster 1 | **MK** | 219 | 5 | 224 | 94 | 48 | 142 | 1.58 |
| Subtotal | | 496 | 22 | 518 | 182 | 163 | 345 | 1.50 |
| 3LN212 | **BY** | 23 | 4 | 27 | 13 | 23 | 36 | 0.75 |
| 3LN342 | | | | | | | | |
| Cluster 2 | **BY-CC** | 174 | 9 | 183 | 55 | 33 | 88 | 2.08 |
| Cluster 3 | **BY-CC** | 217 | 11 | 228 | 75 | 41 | 116 | 1.97 |
| Cluster 4 | **BY-CC** | 268 | 7 | 275 | 95 | 13 | 108 | 2.55 |
| 3PU3 | **CC** | 1040 | 97 | 1137 | 469 | 267 | 736 | 1.54 |
| 3PU332 | **CC** | 136 | 13 | 149 | 47 | 22 | 69 | 2.16 |
| 3PU330 | **CC,MS** | 571 | 30 | 601 | 118 | 28 | 146 | 4.12 |
| Subtotal | | 2429 | 171 | 2600 | 872 | 427 | 1299 | 2.00 |
| 3PU167 | CC,**MS** | 39 | 6 | 45 | 1 | 4 | 5 | 9.00 |
| 3PU17 | CC,**MS** | 289 | 4 | 293 | 13 | 9 | 22 | 13.32 |
| Subtotal | | 328 | 10 | 338 | 14 | 13 | 27 | 12.52 |
| Total | | 3253 | 203 | 3456 | 1068 | 603 | 1671 | |

to flake tools at the Fitzhugh site (3LN212) is anomalous for the Baytown period, the technological characteristics of its lithic assemblage are consistent with an occupation of limited duration at a site located away from a suitable local raw material source (Nassaney and Hoffman 1992). The Bearskin Lake South site (3LN342) exhibits both temporal and synchronic variation in lithic technology which may be potentially significant. The ratio of core tool:flake tool products decreases at the site from west to east, beginning with Cluster 1.

Raw materials were not used indiscriminately in the production of core and flake tools. Novaculite and non-local cherts comprise 23.0% of all core tool products, but only 3.5% of the flake tool products. This pattern suggests that lithic technology and raw material choice were interdependent.

Table 5 shows the frequency of chipped lithic artifacts by reduction stage and raw material type for 11 components. I distinguish between early and late production stages (see Figures 9 and 10). Early stage products consist of cores, decortication flakes, bipolar flakes, secondary decortication flakes, interior flakes with cortical platforms, and shatter. I identify bifacial thinning flakes, retouch or pressure flakes, and finished tools (e.g., hafted bifaces, arrow points) as late stage products.

There is an increase in the ratio of early:late stage products over time. This pattern becomes most pronounced in the Mississippi period (Figure 12). Taking the proportion of early to late stage products for each of the three periods yields ratios of 1.50 (Marksville), 2.00 (Baytown-Coles Creek), and 12.50 (Mississippi). However, novaculite and non-local cherts exhibit significantly different patterns, which are masked when the raw material types are aggregated. For instance, chert occurs in *early* stages of production in more than three out of four cases (75.3%), whereas the reverse pattern holds for novaculite and non-local cherts; they occur in *late* stages of production nearly as frequently (74.8%). These patterns suggest that finishing, retouch, and maintenance of novaculite and non-local cherts took place on-site throughout the periods under investigation, but initial reduction occurred predominantly at the source, which is outside the survey area (Hem-

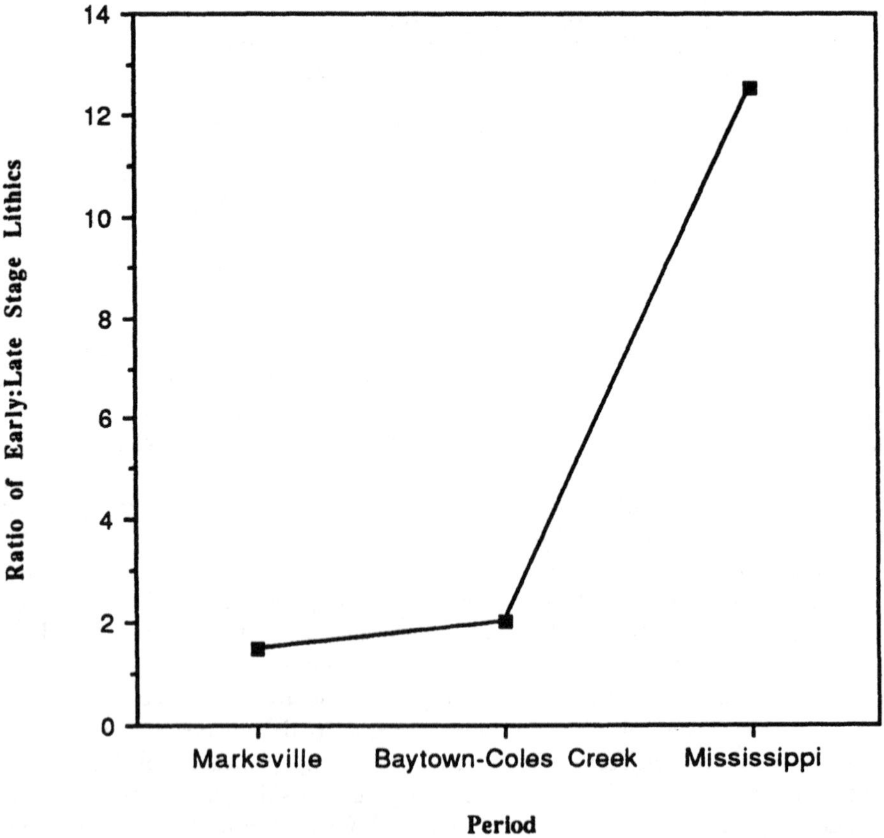


Figure 12. Mean early stage:late stage product ratios for select archaeological assemblages in central Arkansas by cultural period.

mings 1985:32 and Hoffman 1982b:59). In contrast, chert gravel was initially reduced on site after it was obtained locally. Furthermore, chert reduction consisted of the application of direct force, often using a bipolar technique to produce flakes that could be used as expedient tools. This technology seldom required the retouch or further modification that produces late stage products. As indicated earlier, the tendency to make expedient tools was most prevalent during the Mississippi period. Novaculite and non-local cherts comprise 36.1% of all late stage products, underscoring their role in the production and maintenance of bifacial core tools.

Table 6 shows the frequency of thermally altered chert by technology for 11 components in the survey area. Taking the proportion of thermally altered to non-thermally altered chert gravel for each of the three periods yields ratios of 0.21 (Marksville), 0.78 (Baytown-Coles Creek), and 0.28 (Mississippi). There is a clear increase, followed by a decrease, in the ratio of heated:non-heated cherts through time (Figure 13). Approximately half of the thermally-altered cherts derive from a core tool technology (50.2%), but only 39.3% of all flake tool products were heat treated. Unmodified cobbles were apparently adequate to detach expedient flake tools from bipolar cores, particularly during the Marksville

**Table 6. Distribution of thermally altered gravel chert by technology. Major components are shown in bold. Key to abbreviations: CA=Cultural affiliation, MK=Marksville, BY=Baytown, CC=Coles Creek, MS=Mississippi, TA:NTA= Thermally altered:Non-thermally altered**

| | | Thermally altered | | | Non-thermally altered | | | |
|---|---|---|---|---|---|---|---|---|
| Provenience | CA | Core | Flake | Subtotal | Core | Flake | Subtotal | Ratio TA:NTA |
| 3PU115 | **MK** | 14 | 22 | 36 | 81 | 121 | 202 | 0.18 |
| 3LN342 | | | | | | | | |
| Cluster 1 | **MK** | 24 | 14 | 38 | 94 | 59 | 153 | 0.25 |
| Subtotal | | 38 | 36 | 74 | 175 | 180 | 355 | 0.21 |
| 3LN212 | **BY** | 5 | 2 | 7 | 13 | 1 | 14 | 0.50 |
| 3LN342 | | | | | | | | |
| Cluster 2 | **BY-CC** | 38 | 20 | 58 | 38 | 31 | 69 | 0.84 |
| Cluster 3 | **BY-CC** | 57 | 34 | 91 | 37 | 45 | 82 | 1.11 |
| Cluster 4 | **BY-CC** | 40 | 43 | 83 | 40 | 57 | 97 | 0.86 |
| 3PU3 | **CC** | 218 | 244 | 462 | 185 | 298 | 483 | 0.96 |
| 3PU332 | **CC** | 12 | 16 | 28 | 32 | 68 | 100 | 0.28 |
| 3PU330 | **CC**,MS | 61 | 55 | 116 | 74 | 166 | 240 | 0.48 |
| Subtotal | | 431 | 414 | 845 | 419 | 666 | 1085 | 0.78 |
| 3PU167 | CC,**MS** | 4 | 2 | 6 | 9 | 19 | 28 | 0.21 |
| 3PU17 | CC,**MS** | 9 | 26 | 35 | 19 | 107 | 126 | 0.28 |
| Subtotal | | 13 | 28 | 41 | 28 | 126 | 154 | 0.27 |
| Total | | 482 | 478 | 960 | 622 | 972 | 1594 | |

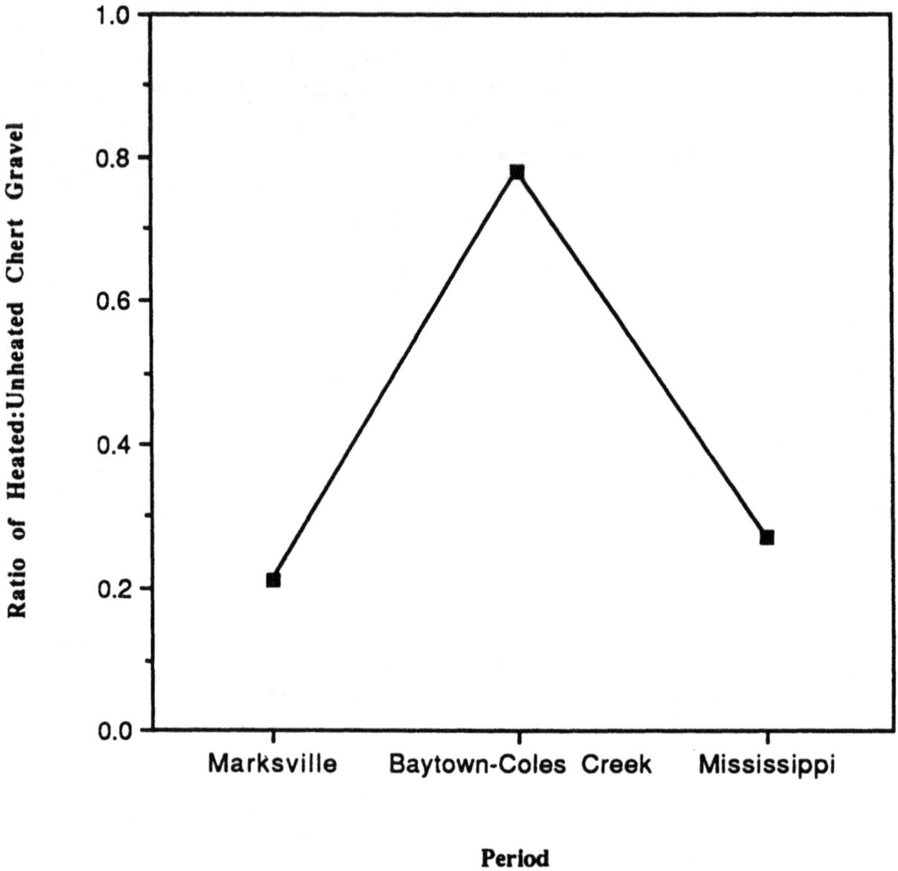


**Figure 13.** Mean thermally altered:non-thermally altered chert gravel ratios for select archaeological assemblages in central Arkansas by cultural period.

and Mississippi periods. Heat treatment was used with nearly equal frequency to produce flake and core tools during the Baytown-Coles Creek period. The McNeely South (3PU330) and Nipps (3PU332) sites, however, are notable exceptions.

## Interpretations of Chert and Novaculite Technological Organization

There is a clear diachronic pattern of decreasing proportions of core tools and tool products at the expense of artifacts associated with an expedient (flake) technology (Figure 11). Furthermore, the observed patterns are consistent with expectations for longitudinal changes associated with decreased mobility (e.g., Parry and Kelly 1987). While we do not entirely understand why mobility decreased, it may be related to other strategies to mobilize labor at the regional scale.

Marksville-period groups, in contrast to their Archaic-stage predecessors, began to use chert gravel for the production of chipped stone bifaces and some expedient tools (see House 1980). Novaculite was used in ways that were consistent with a core tool technology in the Marksville through Mississippi periods. Settlement locations and substantial mobility would have allowed Marksville-period groups to maintain direct access to the high quality, readily available lithic resources of the nearby Ouachita Mountains in the absence of institutionalized control or exchange relationships. The high proportions of core tool products in Marksville-period assemblages suggest that: (1) groups of this period were more mobile than later occupants of the region, and/or (2) bow-and-arrow technology had not yet been introduced. The Marksville-period data are consistent with relatively egalitarian social relations, whereby production was predominantly for use and everyone had access to the means of production. There is no clear evidence for individuals attempting to intensify production by increasing output.

There is some indication of a technical division of labor, however, since the early stages of novaculite reduction occurred outside the survey area, presumably at the source. This pattern was maintained through the late Mississippi period. As mobility decreased, some individuals could have maintained access to high quality raw materials and used these strategic resources to enhance their social status.

There is greater reliance on chert gravels to produce flake tools and, to a lesser extent, core tools among the more sedentary groups of the Baytown-Coles Creek period. Access to small, locally available chert gravels could not be easily controlled and may have undermined attempts to institutionalize exchange networks needed to supply non-source areas with non-local chipped stone. Some groups may have chosen not to participate in these reciprocal exchanges in order to avoid (increasing?) acquisition costs. This interpretation contradicts the idea that Plum Bayou culture sites contain abundant non-local igneous and sedimentary stone *obtained through long-distance exchange*. While lithic resources from *non-local geological source areas* are frequently encountered at Baytown-Coles Creek period sites, it is not clear if these materials could have been collected locally. While some objects may have been exchanged into the study area, some groups may have chosen to avoid these exchange relationships by "searching out localized sources of rare stone" that may have been distributed as redeposited gravels along the Arkansas River, its tributaries, and former channels (see Rolingson 1982b:93). A more detailed geological study needs to be conducted to determine the frequency and distribution of "non-local" raw materials among the riverine gravel formations of the survey locality.

Another strategy that Baytown-Coles Creek groups may have used to avoid the costs incurred by high quality raw material was the intentional thermal alteration of gravel cherts. The high incidence of heat-treated cherts during the Baytown-Coles Creek period suggests that local groups chose to improve the flaking characteristics of lower quality, and perhaps less costly, raw materials.

The variation in the ratios of core tool:flake tool products, early stage:late stage lithic products, and thermally-altered:non-thermally altered chert gravels suggests that there could have been some inter-site spatial division of labor to control the production process. For instance, core tool products occur more than twice as frequently in Cluster 2 at the Bearskin Lake South site (3LN342) than at Nipps (3PU332). In comparing early stage and late stage ratios, Clear Lake (3PU3) and McNeely South (3PU330) show interesting contrasts. Activities at McNeely South are oriented toward the initial reduction of gravel cherts, whereas by-products from the finishing and maintaining of novaculite tools are better represented at Clear Lake. Clear Lake exhibits a relatively high proportion of heat treated chert. Yet despite the similarities in raw materials previously discussed between Clear Lake and Nipps, thermal alteration was infrequent at the Nipps site.

There may also have been some attempts at intra-site household specialization. For instance, variation in the ratios of core tool and flake tool products at the Bearskin Lake South site suggests that contemporaneous households, represented by discrete clusters of artifacts, may have specialized in the production of either core or flake tools. The ratio of heated:non-heated chert in Cluster 4 is significantly greater than anywhere else on the site when cherts from core, flake, and ambiguous technologies are combined (Nassaney 1992a:Figure 7.6; see also Nassaney 1992b:Appendix C). Furthermore, only in this cluster do flake tools and their by-products exceed core tools and their by-products (Nassaney 1992a:Figure 7.7). Lastly, the changing proportions of lithic raw material types indicate that chert gravel was overwhelmingly preferred to novaculite and non-local cherts (Nassaney 1992a:Figure 7.8). Activities in Cluster 4 appear to have been oriented toward the production of thermally altered flake tools derived from chert gravel. This may represent evidence for a spatial division of labor—one tactic for monopolizing surplus that was available to individuals in Plum Bayou culture.

Heat treatment, like the use of quartz crystal, seems to have reached its greatest popularity during the Baytown-Coles Creek period. Heat treatment was apparently important for producing flake tools (e.g., dart and arrow points) from often nonhomogeneous chert gravel. This pattern stands in marked contrast to the Marksville and Mississippi periods, when heat was used less frequently to alter the chipping characteristics of chert in the production of both flake and core tools.

## CONCLUSIONS

In this article, I used a political-economic model of social ranking to illuminate the material dimensions associated with the Plum Bayou culture in central Arkansas. I compared the results from several analyses of chipped stone with the implications of the model to determine if control over, or intensification of, resource acquisition, production and/or exchange were being used to underwrite social ranking. Throughout the analyses, I emphasized the different strategies that agents use to create, support, and resist social ranking in the process of manipulating the material world.

Aboriginal inhabitants of central Arkansas used chert, novaculite, and quartz crystal to manufacture chipped stone tools during the late Marksville through late Mississippi periods. Processes of resource acquisition, production, and exchange, however, did not remain constant throughout this interval. Populations increasingly relied on chert gravel, obtained directly from its secondary sources in gravel bars along the Arkansas River and its tributaries. Novaculite, which entered the survey locality in roughed-out preforms or blanks, decreased in relative frequency from the late Archaic through Mississippi periods. Quartz crystal reached the height of its popularity during the Baytown-Coles Creek period. Access to chert was unrestricted and could be acquired directly. In contrast, novaculite was probably obtained through reciprocal exchange, whereas quartz crystal may have been accumulated at, and redistributed from, some sites in the region.

There are clear diachronic changes in chipped stone technology in the survey area, marked by increasing proportions of expedient (flake) tools and tool products at the expense of artifacts associated with a core tool technology. Marksville-period sites contain a significant number of core tool reduction by-products from the manufacture of large, bifacial, multi-purpose tools. Novaculite bifaces were predominantly finished and maintained at sites in the survey area; initial reduction was performed elsewhere, presumably at the source. Heat was seldom used to alter the chipping characteristics of local cherts in biface production.

By the late Baytown period, a flake tool technology was increasingly adopted. This technology utilized thermal alteration to produce small "core" tools (i.e., arrow points and drills) from local gravel cherts. Heat treatment appears to have been an attempt to conserve a resource made relatively scarce by the decline in mobility during this period. Moreover, the high incidence of thermal alteration at Baytown-Coles Creek sites may represent a strategy of resistance whereby local groups could undermine exchange networks and associated costs by transforming the suitability of readily available but marginal raw materials. By the Mississippi period, a flake tool technology was well established and expedient tools were common among these sedentary populations. Thermal alteration occurred infrequently; apparently, it was not needed to produce the expedient tools that predominate in Mississippi-period assemblages.

Deviations from these tendencies may suggest strategies to mobilize surplus or resist its extraction. For instance, lithic reduction activities at Bearskin Lake South (3LN342) exhibit significant intra-site diversity, suggesting a spatial division of labor whereby some households were working large bifacial cores while others were utilizing an expedient technology and thermal alteration to produce arrow points and drills. This pattern may be indicative of incipient specialization. Likewise, quartz crystal reduction may have been spatially restricted within areas at Toltec and Ink Bayou. The internal configuration of the Toltec Mounds site suggests significant differences in the spatial organization of intra-site activities. The high proportions of quartz crystal away from the plaza in the eastern half of the site may represent workshop areas for attached specialists who were working quartz for patron elites. At Ink Bayou, quartz crystal is spatially restricted to a discrete area behind the habitation structure, suggesting limited but intensive episodes of reduction (Waddell et al. 1987). Together, these data suggest that quartz crystal was implicated in the reproduction of Plum Bayou culture.

In sum, there appear to be rudimentary attempts to control and/or intensify the production of chipped stone tools during the Baytown-Coles Creek period by segments of Plum Bayou society. It appears unlikely, however, that socially ranked individuals successfully monopolized any of these lines of production, with the possible exception of quartz crystal. Individuals may have resisted these tactics with counter-strategies to maintain some degree of autonomy.

## ACKNOWLEDGMENTS

This paper is a revised version of a chapter in my doctoral dissertation completed at the University of Massachusetts-Amherst. I appreciate the insightful comments made on various drafts of the work by Bob Paynter, Dena Dincauze, Martha Rolingson, Ken Sassaman, John Clark, Brad Koldehoff, George Odell, Michael Shott, Brian Hayden, and an anonymous reviewer. The final typescript was prepared with equipment provided by a grant from the New Faculty Research Support Program, Western Michigan University. I also thank George Odell for organizing the conference in which this paper was discussed, and for supporting my participation. The critical comments of the conference participants are also gratefully acknowledged.

## NOTES

1. Clark (1991:3) argues that it is the "ability to influence or control the value and meaning of things [which] provides one means of gaining, marking, and retaining differential status and privilege."
2. Tosi (1984:23–24) uses six categories of increasingly complex forms of labor allocation to examine the development of craft specialization. The first three are applicable to this study. In his first category, societies lack craft specialization. Each household has direct access to the means of production and no labor is allocated to craft production outside the immediate control of the household. Craft specialization emerges in Tosi's second category at the intra-site scale. Differential output within sites is based on variation in the allocation of labor. Finally, inter-set-

tlement specialization occurs when "differentiation between settlements reflects larger volumes of local exchange and higher economic integration over a larger territory" (Tosi 1984:24). He goes on to discuss several pertinent archaeological indicators for craft production, including facilities (e.g., pottery kilns), tools for manufacture (e.g., hammerstones; anvils), residues (e.g., lithic debitage), and semifinished products (e.g., bifacial preforms). Indicators such as these can be used to investigate the spatial division of labor in Plum Bayou culture by examining their specific forms and distribution within and between sites in the region.

3. There has been a bias toward the collection of complete tools (and away from debitage) by amateurs in the region. Since novaculite was often the preferred raw material for large dart points, it seems to be overrepresented in amateur collections, particularly by proportional weight.
4. Admittedly, the density of surface material at 3PU167 was low; however, we intensively surveyed an area greater than 8 ha. The Keo site (3PU17) yielded only 10 pieces of quartz crystal, and they may be associated with the site's Coles Creek-period Plum Bayou culture component.
5. The mean weight of chert, novaculite, and quartz crystal was initially tabulated in this analysis; this measure is more difficult to summarize than the counts and weights. It is important to note that most of the amateur collections contain larger pieces of chipped stone on average than those in the controlled samples. It appears that using mean weight as a measure of artifact size to characterize an assemblage is highly sensitive to collection strategy and survey conditions.
6. Although preceramic assemblages are not quantified in this study, Archaic-period groups depended upon high quality raw materials such as novaculite and non-gravel cherts (e.g., Boone chert) that naturally occur in size ranges suitable for large, hafted bifacial forms.
7. Riggs (personal communication, 1990) indicates that he frequently observed and collected quartz crystal at the expense of other chipped stone debris during repeated visits to Winfrey. The Chownings and Miller may also have emphasized the collection of quartz crystal, given the high proportions at 3PU163, 3LN20, and 3LN200.
8. There is an emerging consensus that quartz crystal is a hallmark of Plum Bayou culture. The representation of quartz crystal by count and weight among Baytown-Coles Creek period sites in the survey area supports this interpretation and suggests that crystals were implicated in creating and reinforcing unequal social relations. Furthermore, quartz crystals appear to be geographically restricted to Plum Bayou sites; they are rare at Coles Creek culture sites in Louisiana, even though they commonly occur at Poverty Point sites in the same region (Marvin Jeter, personal communication, 1989; see also Johnson 1993b).
9. For instance, quartz crystals have been found in the lower Cache Basin associated with Late Woodland occupations, and "a concentration of quartz crystal fragments was observed near one of the mounds at the [contemporaneous] Dogtown complex" (House 1975:159). Several sites with grog-tempered ceramics along the Little Red River in north-central Arkansas (Cleburne and Van Buren counties) have also yielded quartz crystal chipping debris (Waddell 1987).
10. I equate core tools with formal, standardized, or curated tools, whereas flake tools are synonymous with informal, unstandardized, or expedient tool forms (Binford 1979). These definitions by no means imply that expedient tools or flakes are not tools; rather, they refer to differences in their reduction trajectories leading to the selection of different raw material forms, percussion tools, by-products, and end-products (cf. Gero 1991:164–166).
11. A certain threshold of production may be required before thermal alteration is routinely used (e.g., see Clark [1987] on the production of prismatic blades).
12. I do not analyze the prevalence of thermally altered novaculite for two reasons. First, modern knappers indicate that novaculite was invariably altered by heat, because it is extremely difficult to chip in its natural state (Don Dickson, personal communication, 1990). Second, unlike gravel cherts, which exhibit obvious reddening when heated, novaculite takes on a lustrous appearance

with very subtle color changes, making it difficult to identify thermally-altered novaculite macroscopically.

## REFERENCES


Bender, B. 1985. Emergent Tribal Formations in the American Midcontinent. *American Antiquity* 50:52–62.

Bender, B. 1989. The Roots of Inequality. In *Domination and Resistance,* edited by D. Miller, M. Rowlands, and C. Tilley, pp. 83–95. Unwin Hyman, London.

Binford, L. R. 1979. Organization and Formation Processes: Looking at Curated Technologies. *Journal of Anthropological Research* 35:255–273.

Branner, G. C. 1927. *Outlines of Arkansas' Mineral Resources.* Bureau of Mines, Manufactures, and Agriculture and State Geological Survey, Little Rock.

Braverman, H. 1974. *Labor and Monopoly Capital.* Monthly Review Press, New York.

Brown, J. A., R. A. Kerber, and H. D. Winters. 1990. Trade and the Evolution of Exchange Relations at the Beginning of the Mississippian Period. In *The Mississippian Emergence,* edited by B. Smith, pp. 251–280. Smithsonian Institution Press, Washington, D. C.

Brumfiel, E. M., and T. Earle (editors). 1987. *Specialisation, Exchange, and Complex Societies.* Cambridge University Press, Cambridge.

Clark, J. E. 1987. Politics, Prismatic Blades, and Mesoamerican Civilization. In *The Organization of Core Technology,* edited by J. K. Johnson and C. A. Morrow, pp. 259–284. Westview Press, Boulder.

Clark, John E. 1991. Craft Specialization and the Emergence of Rank Societies. Paper presented at the 90th annual meeting of the American Anthropological Association, Chicago.

Cobb, C. R. 1989. An Appraisal of the Role of Mill Creek Chert Hoes in Mississippian Exchange Systems. *Southeastern Archaeology* 8:79–92.

Cobb, C. R. 1993. Archaeological Approaches to the Political Economy of Nonstratified Societies. In *Archaeological Method and Theory,* Volume 5, edited by M. B. Schiffer, pp. 43–100. University of Arizona Press, Tucson.

Cross, J. R. 1990. *Specialized Production in Non-Stratified Society: An Example from the Late Archaic Northeast.* Ph.D. dissertation, Department of Anthropology, University of Massachusetts, Amherst. University Microfilms International, Ann Arbor.

Dalton, G. 1960. A Note of Clarification on Economic Surplus. *American Anthropologist* 62:483–490.

Dalton, G. 1963. Economic Surplus, Once Again. *American Anthropologist* 64:389–394.

Earle, T. 1981. Comment on "Evolution of Specialized Pottery" by P. M. Rice. *Current Anthropology* 22:230–231.

Etchieson, M. 1989. Prehistoric Use of Geological Resources in the Ouachita Mountains. Paper presented at the 54th annual meeting of the Society for American Archaeology, Atlanta.

Frankenstein, S., and M. J. Rowlands. 1978. The Internal Structure and Regional Context of Early Iron Age Society in Southwestern Germany. *University of London, Institute of Archaeology Bulletin* 15:73–112.

Fried, M. 1967. *The Evolution of Political Society.* Random House, New York.

Fritz, G. J. 1988. Adding the Plant Remains to Assessments of Late Woodland/Early Mississippi Period Plant Husbandry. Paper presented at the 53rd annual meeting of the Society for American Archaeology, Phoenix.

Gero, J. M. 1989. Assessing Social Information in Material Objects: How Well Do Lithics Measure Up? In *Time, Energy and Stone Tools*, edited by R. Torrence, pp. 92–105. Cambridge University Press, Cambridge.

Gero, J. M. 1991. Genderlithics: Women's Roles in Stone Tool Production. In *Engendering Archaeology*, edited by J. M. Gero and M. W. Conkey, pp. 163–193. Basil Blackwell, Oxford.

Haas, J. 1982. *The Evolution of the Prehistoric State*. Columbia University Press, New York.

Hall, R. L. 1979. In Search of the Ideology of the Adena-Hopewell Climax. In *Hopewell Archaeology*, edited by D. Brose and N. Greber, pp. 258-265. Kent State University Press, Kent, Ohio.

Harris, M. 1959. The Economy Has No Surplus? *American Anthropologist* 61:189–199.

Hegmon, M. 1989. The Styles of Integration: Ceramic Style and Pueblo I Integrative Architecture in Southwestern Colorado. In *The Architecture of Social Organization in Prehistoric Pueblos*, edited by W. D. Lipe and M. Hegmon, pp. 125–142. Crow Canyon Archaeological Center, Cortez, Colorado, Occasional Paper 1.

Hemmings, E. T. 1985. Analysis of Materials. In *The Alexander Site*, edited by E. T. Hemmings and J. H. House, pp. 27–48. Arkansas Archeological Survey, Research Series, No. 24. Fayetteville.

Hodder, I. 1982. *Symbols in Action*. Cambridge University Press, Cambridge.

Hoffman, T. L. 1982a. *Chipped Stone Tool Manufacturing Processes in Mound D at the Toltec Mounds Site (3LN42)*. M. A. thesis, Department of Anthropology, University of Arkansas.

Hoffman, T. L. 1982b. Lithic Technology at Toltec: Preliminary Results from Mound D. In *Emerging Patterns of Plum Bayou Culture*, edited by M. A. Rolingson, pp. 54–59. Arkansas Archeological Survey, Research Series 18. Fayetteville.

Holmes, W. H. 1891. Aboriginal Novaculite Quarries in Garland County, Arkansas. *American Anthropologist* 4(old series):313–316.

House, J. H. 1975. Summary of Archaeological Knowledge Updated with Newly Gathered Data. In *The Cache River Archeological Project*, assembled by M. B. Schiffer and J. H. House, pp. 153-162. Arkansas Archeological Survey, Research Series, No. 8. Fayetteville.

House, J. H. 1980. Archaic Occupation in the Arkansas River Lowland. *Arkansas Archeological Society, Field Notes* 171:5–10.

Howard, J. M. 1986. *Arkansas Quartz Crystals*. Pamphlet prepared by the Arkansas Geological Commission, Little Rock.

Jenny, W. P. 1891. Ancient Novaculite Mines Near Magnet Cove, Hot Springs County, Arkansas. *American Anthropologist* 4(old series):316–318.

Johnson, J. K. 1989. The Utility of Production Trajectory Modeling as a Framework for Regional Analysis. In *Alternative Approaches to Lithic Analysis*, edited by D. O. Henry and G. H. Odell, pp. 119–138. Archaeological Papers of the American Anthropological Association, No. 1.

Johnson, J. K. 1993a. North American Biface Production Trajectory Modeling in Historic Perspective. *Plains Anthropologist* 38:151–162.

Johnson, J. K. 1993b. Poverty Point Period Crystal Drill Bits, Microliths, and Social Organization in the Yazoo Basin, Mississippi. *Southeastern Archaeology* 12:59–64.

Kus, S. M. 1983. The Social Representation of Space: Dimensioning the Cosmological and the Quotidian. In *Archaeological Hammers and Theories*, edited by J. A. Moore and A. S. Keene, pp. 277-298. Academic Press, New York.

Lee, R. 1988. Reflections on Primitive Communism. In *Hunters and Gatherers 1: History, Evolution, and Social Change*, edited by T. Ingold, D. Riches, and J. Woodburn, pp. 252–268. St. Martin's Press, New York.

Lee, R. 1990. Primitive Communism and the Origin of Social Inequality. In *The Evolution of Political Systems: Sociopolitics in Small-Scale Sedentary Societies*, edited by S. Upham, pp. 225–246. Cambridge University Press, Cambridge.

Manger, W. L. 1986. Regional Geological Setting and the Availability of Various Raw Materials. In *Village Creek: An Explicitly Regional Approach to Cultural Resources*, by T. C. Klinger, pp. 204–213. Arkansas Archeological Survey, Research Report, No. 26. Fayetteville.

Miller, D., and C. Tilley (editors). 1984. *Ideology, Power, and Prehistory*. Cambridge University Press, Cambridge.

Nassaney, M. S. 1991. Spatial-Temporal Dimensions of Social Integration during the Coles Creek Period in Central Arkansas. In *Stability, Transformation, and Variation: The Late Woodland Southeast*, edited by M. S. Nassaney and C. R. Cobb, pp. 177-220. Plenum Press, New York.

Nassaney, M. S. 1992a. Communal Societies and the Emergence of Elites in the Prehistoric American Southeast. In *Lords of the Southeast: Social Inequality and the Native Elites of Southeastern North America*, edited by A. Barker and T. R. Pauketat, pp. 111–143. Archeological Papers of the American Anthropological Association, No. 3.

Nassaney, M. S. 1992b. *Experiments in Social Ranking in Central Arkansas*. Ph.D. dissertation, Department of Anthropology, University of Massachusetts, Amherst. University Microfilms International, Ann Arbor.

Nassaney, M. S. 1994. The Historical and Archaeological Context of Plum Bayou Culture in Central Arkansas. *Southeastern Archaeology* 13:36–55.

Nassaney, M. S., and R. Hoffman. 1992. Archaeological Investigations at the Fitzhugh Site (3LN212): A Plum Bayou Culture Household in Central Arkansas. *Midcontinental Journal of Archaeology* 17:139–165.

Parry, W. J., and R. L. Kelly. 1987. Expedient Core Technology and Sedentism. In *The Organization of Core Technology*, edited by J. K. Johnson and C. A. Morrow, pp. 285–304. Westview Press, Boulder.

Paynter, R. 1982. *Models of Spatial Inequality*. Academic Press, New York.

Phillips, P. 1970. *Archaeological Survey in the Lower Yazoo Basin, Mississippi, 1949–1955*. Papers of the Peabody Museum of Archaeology and Ethnology 60. Harvard University, Cambridge.

Reichel-Dolmatoff, G. 1979. Desana Shaman's Rock Crystals and the Hexagonal Universe. *Journal of Latin American Lore* 5:117-128.

Reichel-Dolmatoff, G. 1981. Things of Beauty Replete With Meaning—Metals and Crystals in Columbian Indian Cosmology. In *Sweat of the Sun, Tears of the Moon*, edited by D. H. Seligman, pp. 17–33. Natural History Museum Alliance of Los Angeles County, Los Angeles.

Renfrew, C. 1975. Trade as Action at a Distance. In *Ancient Civilization and Trade*, edited by J. A. Sabloff and C. C. Lamberg-Karlovsky, pp. 3–59. University of New Mexico Press, Albuquerque.

Rick, J. W. 1978. *Heat-Altered Cherts of the Lower Illinois Valley: An Experimental Study in Prehistoric Technology*. Northwestern University Archeological Program, Prehistoric Records, No. 2. Evanston.

Rolingson, M. A. 1982a. The Concept of Plum Bayou Culture. In *Emerging Patterns of Plum Bayou Culture*, edited by M. A. Rolingson, pp. 87–93. Arkansas Archeological Survey, Research Series 18. Fayetteville.

Rolingson, M. A. 1990a. The 1989 Society Dig at the Toltec Mounds State Park. *Arkansas Archeological Society, Field Notes* 233:6–12.

Rolingson, M. A. 1990b. The Toltec Mounds Site: A Ceremonial Center in the Arkansas River Lowland. In *The Mississippian Emergence*, edited by B. D. Smith, pp. 27–49. Smithsonian Institution Press, Washington, D.C.

Rolingson, M. A. (editor). 1982b. *Emerging Patterns of Plum Bayou Culture*. Arkansas Archeological Survey, Research Series 18. Fayetteville.

Roseberry, W. 1989. *Anthropologies and Histories: Essays in Culture, History, and Political Economy.* Rutgers University Press, New Brunswick, NJ.

Sabo, G., III, D. B. Waddell, and J. H. House. 1982. *A Cultural Resource Overview of the Ozark-St. Francis National Forests, Arkansas.* Arkansas Archeological Survey, Fayetteville. Submitted to the U.S.D.A. Forest Service, Ozark-St. Francis National Forests, Russellville, Arkansas.

Sahlins, M. 1963. Poor Man, Rich Man, Big-Man, Chief: Political Types in Melanesia and Polynesia. *Comparative Studies in Society and History* 5:285–303.

Sahlins, M. 1972. *Stone Age Economics.* Aldine, New York.

Sassaman, K. E. 1992. Lithic Technology and the Hunter-Gatherer Sexual Division of Labor. *North American Archaeologist* 13:249–262.

Seeman, M. F. 1985. Craft Specialization and Tool Kit Structure: A Systemic Perspective on the Midcontinental Flint Knapper. In *Lithic Resource Procurement: Proceedings for the Second Conference on Prehistoric Chert Exploitation,* edited by S. C. Vehik, pp. 7–36. Center for Archaeological Investigations, Occasional Paper No. 4. Southern Illinois University, Carbondale.

Service, E. 1971. *Primitive Social Organization: An Evolutionary Perspective.* Second Edition. Random House, New York.

Stanfill, A. L. 1980. *Lithic Artifact Analysis of the 1976 Surface Collection from the Toltec Site.* M.A. thesis, Department of Anthropology, University of Arkansas.

Stewart-Abernathy, J. 1982. Ceramic Studies at the Toltec Mounds Site: Basis for a Tentative Cultural Sequence. In *Emerging Patterns of Plum Bayou Culture,* edited by M. A. Rolingson, pp. 44–53. Arkansas Archeological Survey, Research Series 18. Fayetteville.

Tilley, C. 1984. Ideology and the Legitimation of Power in the Middle Neolithic of Southern Sweden. In *Ideology, Power, and Prehistory,* edited by D. Miller and C. Tilley, pp. 111–146. Cambridge University Press, Cambridge.

Thomas, C. 1894. Report on the Mound Explorations of the Bureau of Ethnology. *Annual Report of the Bureau of Ethnology,* 1890-1891, 12:17-742.

Torrence, R. 1989. Retooling: Towards a Behavioral Theory of Stone Tools. In *Time, Energy and Stone Tools,* edited by R. Torrence, pp. 57–66. Cambridge University Press, Cambridge.

Tosi, M. 1984. The Notion of Craft Specialization and Its Representation in the Archaeological Record of Early States in the Turanian Basin. In *Marxist Approaches in Archaeology,* edited by M. Spriggs, pp. 22–52. Cambridge University Press, Cambridge.

Upham, S. (editor). 1990. *The Evolution of Political Systems: Sociopolitics in Small-Scale Sedentary Societies.* Cambridge University Press, Cambridge.

Waddell, D. B. 1987. *A Cultural Resources Survey of the Proposed Wastewater Treatment Project Right-of-Way, Clinton, Van Buren County, Arkansas.* Arkansas Archeological Survey, Fayetteville. Submitted to the Clinton Water and Sewer Department, Clinton, Arkansas.

Waddell, D. B., J. H. House, M. Colburn, F. King, and M. K. Marks. 1987. *Results of Final Testing for Significance at the Ink Bayou Site (3PU252), Pulaski County, Arkansas.* Arkansas Archeological Survey, Fayetteville. Submitted to the Arkansas Highway and Transportation Department, Little Rock.

Weissner, P. 1977. *Hxaro: A Regional System of Reciprocity for Reducing Risk among the !Kung San.* Ph.D. dissertation, Department of Anthropology, University of Michigan. University Microfilms International, Ann Arbor.

Welch, P. D. 1986. *Models of Chiefdom Economy: Prehistoric Moundville as a Case Study.* Ph.D. dissertation, Department of Anthropology, University of Michigan. University Microfilms International, Ann Arbor.

Williams, N. F. 1959. Mineral Resources of Arkansas. *Arkansas Geological and Conservation Commission, Bulletin 6.* Little Rock.

Wobst, H. M. 1977. Stylistic Behavior and Information Exchange. In *For the Director: Research Essays in Honor of James B. Griffin*, edited by C. Cleland, pp. 317–342. Museum of Anthropology, University of Michigan, Anthropological Papers 6.

Wolf, E. R. 1966. *Peasants*. Prentice-Hall, Englewood Cliffs, New Jersey.

Wolf, E. R. 1982. *Europe and the People without History*. University of California Press, Berkeley.

# Part IV

# Innovation and Style in Projectile Points

Lithic studies, including those in this volume, typically involve assemblages of artifacts, though they occasionally focus on a specific cross-cutting attribute such as raw material or chipping technology. Concentration on one typological entity within an assemblage can easily become a particularistic endeavor. However, when a specific type is assessed both within and outside its immediate context and is considered within sufficiently broad systemic frameworks, it can provide new insight into human behavior.

A stone tool type that has frequently captured research interest is the projectile point, for at least three fundamental reasons. First, typological points, at least in America, are usually bifacially shaped with great care and are therefore end-products of a multitude of manufacturing decisions and an impressive amount of human effort. These decisions and efforts frequently leave vestiges on the finished object that can be directly translated into cultural information, a contention supported by several recent experimental and archaeological studies (Callahan 1979; Frison and Bradley 1980; Young and Bonnichsen 1984).

Second, the projectile point was not functionally unidimensional. That is, a typological point was not necessarily employed prehistorically to tip a projectile, but often served to fulfill a variety of purposes—an issue that has been verified through several studies of use-wear (e.g., Ahler 1971; Wylie 1975; Odell 1988). This situation has occasioned changes in nomenclature to better reflect its multivariate utility, as in "projectile point/knife" (Faulkner and McCollough 1973) or "projectile point/ hafted cutting tool" (Ahler and McMillan 1976); or to remove functional suggestions entirely, as in "hafted biface" (Emerson and McElrath 1983; Fortier 1983) or just "biface" (Shott 1993). The quality of multidimensionality renders the projectile point intriguing—not just because a point might occasionally have served as a drill bit, but because its use was culturally determined and reflective of its role within its cultural environment. From such a flexible and adaptable tool, much information can be extracted.

And third, the history of the functional projectile point as part of a complex of weapons delivery systems is still open for debate. In North America this debate is especially interesting, as positions have crystallized around two scenarios concerning the inception of the bow-and-arrow from a spear/dart/atlatl technology. A "Late-Comer" camp considers the formal shaped projectile point the true representative of such weaponry and judges the inception of bow-and-arrow technology to have occurred ca. AD 200–700 in most areas of North America (Hester 1973; Blitz 1988; Seeman 1992; Shott 1993). In contrast, an "Early Bird" camp depends on nontypological data from archaeological context to push early bow-and-arrow technology back to 1000–2000 BC (Aikens 1970; Webster 1980; Odell 1988; Patterson 1993). This debate is continued, implicitly, in Shott's contribution to this volume.

The multivariate nature of projectile points has rendered them susceptible to various manipulations, interpretations, and research questions. For example, why does a projectile point look the way it does: because people made things a certain way at a particular time, because they had specific tasks in mind for specific shapes, or because they altered the form of their points through multiple episodes of use and resharpening (Flenniken 1985)?

Nowhere has the debate over the determinants of form (and its corollary, the use of typology) raged more passionately than in the American West. Here Jeffrey Flenniken has argued that percussive use and resharpening of point types such as the Elko corner-notched have changed significant numbers of them into other point types such as Rosegate, a contention hotly denied by David Thomas. In this volume Michael Rondeau evaluates both sides of the issue and introduces data of his own. His discussion clarifies some of the issues involved and provides insight into their resolution.

John Rick is concerned less with the mechanism of projectile point stylistic change than with its consistency and the societal levels at which it occurs. Rick's data set consists of well provenienced points from two deeply stratified preceramic sites in central Peru. Through a process that he terms "strategized induction," he establishes four levels of stylistic variability. These are ultimately associated with different levels of mobility, exploitation, social identity, and long-term processes of cultural adaptation. Rick's approach and his interpretations of chronological patterns are both innovative and provocative.

Also provocative is Michael Shott's discussion of projectile points from the American Midwest. Shott begins by debunking the most fundamental assumption on which conventional lithic typology is based, i.e., that culture is episodic and characterized by long periods of equilibrium disturbed by short bursts of innovation—reflected, of course, by changes in projectile point styles. In contrast to normative views, Shott's approach emphasizes the behavioral practices to which the projectile weaponry was responding, practices that influenced the sizes and shapes of the constituent barbs and tips. His data show a continuous trend of declining projectile point measurements, a result that fits a non-episodic model of culture change. Shott's explanation of these trends using a Diet Breadth model taken from Optimal Foraging Theory elicited some of the liveliest debate at the symposium.

## REFERENCES

Ahler, S. A. 1971. *Projectile Point Form and Function at Rodgers Shelter, Missouri.* Missouri Archaeological Society, Research Series, No. 8.

Ahler, S. A., and R. B. McMillan. 1976. Material Culture at Rodgers Shelter: A Reflection of Past Human Activities. In *Prehistoric Man and His Activities: A Case Study in the Ozark Highland,* edited by W. R. Wood and R. McMillan, pp. 163–199. Academic Press, New York.

Aikens, C. M. 1970. *Hogup Cave.* University of Utah, Anthropological Papers, No. 93.

Blitz, J. H. 1988. Adoption of the Bow in Prehistoric North America. *North American Archaeologist* 9:123–145.

Callahan, E. 1979. The Basics of Biface Knapping in the Eastern Fluted Point Tradition: A Manual for Flintknappers and Lithic Analysts. *Archaeology of Eastern North America* 7:1–180.

Emerson, T. E., and D. L. McElrath. 1983. A Settlement-Subsistence Model of the Terminal Late Archaic Adaptation in the American Bottom, Illinois. In *Archaic Hunters and Gatherers in the American Midwest,* edited by J. Phillips and J. Brown, pp. 219–242. Academic Press, New York.

Faulkner, C. H., and C. R. McCollough. 1973. *Introductory Report of the Normandy Reservoir Salvage Project: Environmental Setting, Typology, and Survey.* Normandy Archaeological Project, Vol. 1; Department of Anthropology, University of Tennessee, Report of Investigations, No. 11.

Flenniken, J. J. 1985. Stone Tool Reduction Techniques as Cultural Markers. In *Stone Tool Analysis: Essays in Honor of Don E. Crabtree,* edited by M. Plew, J. Woods and M. Pavesic, pp. 265–276. University of New Mexico Press, Albuquerque.

Fortier, A. C. 1983. Settlement and Subsistence at the Go-Kart North Site: A Late Archaic Titterington Occupation in the American Bottom, Illinois. In *Archaic Hunters and Gatherers in the American Midwest,* edited by J. Phillips and J. Brown, pp. 243–260. Academic Press, New York.

Frison, G. C., and B. A. Bradley. 1980. *Folsom Tools and Technology at the Hanson Site, Wyoming.* University of New Mexico Press, Albuquerque.

Hester, T. R. 1973. Chronological Ordering of Great Basin Prehistory. *Contributions of the University of California, Archaeological Research Facility,* No. 17.

Odell, G. H. 1988. Addressing Prehistoric Hunting Practices through Stone Tool Analysis. *American Anthropologist* 90:335–356.

Patterson, L. W. 1993. Current Data on Early Use of the Bow and Arrow in Southern North America. *Ohio Archaeology* 43:21–24.

Seeman, M. F. 1992. The Bow and Arrow, the Intrusive Mound Complex, and a Late Woodland Jack's Reef Horizon in the Mid-Ohio Valley. In *Cultural Variability in Context: Woodland Settlements of the Mid-Ohio Valley,* edited by M. Seeman, pp. 41–51. MCJA Special Paper No. 7, Kent State University Press, Kent, Ohio.

Shott, M. J. 1993. Spears, Darts, and Arrows: Late Woodland Hunting Techniques in the Upper Ohio Valley. *American Antiquity* 58:425–443.

Webster, G. S. 1980. Recent Data Bearing on the Question of the Origins of the Bow and Arrow in the Great Basin. *American Antiquity* 45:63–66.

Wylie, H. G. 1975. Tool Microwear and Functional Types from Hogup Cave, Utah. *Tebiwa* 17:1–31.

Young, D. E., and R. Bonnichsen. 1984. *Understanding Stone Tools: A Cognitive Approach.* Center for the Study of Early Man, University of Maine at Orono, Peopling of the Americas, Process Series, Vol. 1.

# Chapter 8

# When Is an Elko?

MICHAEL F. RONDEAU

## ABSTRACT

Perennial disagreements in the assignment of individual specimens to projectile point types and of general types to chronological placement are symbolized by the Flenniken-Thomas debate on the temporal utility of Great Basin point types. Microscopic inspection of use breakage and rejuvenation is used to evaluate the morphological stability of one collection of Elko Corner-notched points. Some claims from both sides of the debate are supported. It is concluded that the detailed analysis of additional collections could make the debate irrelevant and signal more useful avenues for projectile point research.

## INTRODUCTION

The contribution of projectile point analysis to the study of behavioral systems has often been neglected by conventional research. This lack is evident in the debate on the temporal utility of Great Basin projectile point types, with its perfunctory consideration of diverse ecological contexts (Flenniken and Novic 1984; Thomas 1984, 1986a, 1986b; Flenniken and Raymond 1986; Flenniken and Wilke 1986; Bettinger, O'Connell, and Thomas 1991; Wilke and Flenniken 1991).

The purpose here is not to review in detail or otherwise dwell at length upon the Flenniken-Thomas debate, but to argue that a detailed analysis of point modifications, a reading of the flake scars, should lay this debate to rest. In support

---

**MICHAEL F. RONDEAU** • 10 Alvares Court, Sacramento California 95833.

*Stone Tools: Theoretical Insights into Human Prehistory*, edited by George H. Odell. Plenum Press, New York, 1996.

of this purpose, it is submitted that these point types retain evidence of the influence of breakage and rejuvenation events on their morphological stability. This hypothesis is tested by the analysis of a selected Elko Corner-notched collection.

The second purpose here is to present for testing a condensed model of hunter-gatherer projectile point context. Point context refers to the set of factors that influenced the manufacture, use, curation, rejuvenation, and discard of projectile points. This preliminary model is offered as part of a working hypothesis regarding one possible mechanism for the change of projectile point forms through time. Findings suggest that the form and production of projectile points may elucidate the relationship of mobility strategies to lithic reduction modes. The prediction that projectile point assemblages can be found that will support opposite sides of the Flenniken-Thomas debate is offered.

## THE ARGUMENT

The concept of point rejuvenation as a primary engine of change in the form of specific projectile points has been offered by Flenniken and others (Flenniken and Novic 1984; Flenniken and Raymond 1986; Flenniken and Wilke 1986; Wilke and Flenniken 1991). From this idea has followed the argument that the rejuvenation of use-damaged points has resulted in the morphological transformation of certain Great Basin point forms into other types. This claim has been rejected by others (Thomas 1984, 1986a, 1986b; Bettinger, O'Connell, and Thomas 1991). The Flenniken argument has not led to the demonstration that any archaeological sites have been misplaced in time by the use of Great Basin point types. The rejection of this argument, on the other hand, has not shown that projectile points have changed form as a result of breakage and rejuvenation. Regardless of these opposing views, the initial Flenniken concept bears closer examination for its potential to suggest a mechanism by which a gradual shift in the mental template of point makers took place through time.

The persistence of projectile point forms through time and across space can be argued as evidence for the operation of traditions in the application of mental templates. Not only did the manufacturers of points have a finished form in mind, but this conceptualized form was subject to greater or lesser degrees of flexibility, given the context of its manufacture. This context consisted of influences such as the size and quality of available toolstone, the available tools with which to work the stone, and available materials to which the finished points would be hafted. Therefore, the initial mental template, or conceptual form sought at the time of manufacture, cannot be assumed to have been an inflexible ideal, but was more likely a practical guideline.

It may be argued that the mind set of the point user held an even more flexible concept of projectile point form. This flexibility may have been greatest during the repair of impact-damaged points, to regain their functional utility. A greater range of morphological variability as a result of point rejuvenation would be expected.

Extrapolating from the Flenniken argument, it may be concluded that there are a number of ways in which each of the point types tends to fracture. Each type may be said to have a breakage pattern that subsumes a set of common fragment types. Of these fragments, some can be reworked into serviceable projectile tips. The form of these breaks, hafting requirements, toolstone quality, and available flaking techniques can be posited as constraints on the ways that each point type could be rejuvenated. These limits can be expected to result in a series of forms diagnostic of the constraints noted above.

However, the fact that a suite of point forms has the potential to be created by replicative studies does not necessarily signal that these were significant, or that they even existed in prehistory. To date, much of the Flenniken-Thomas debate has involved arguments about whether or not the transmutation of point types occurred frequently, or at all. Flenniken's perspective of change in point forms seems to be too synchronic. The body of extant data argues that the change in forms occurred gradually over a much greater time span.

Since the points themselves constitute the evidence for changes in form over both the short run of individual specimen use-lives and the longer term history of type-lives, they may also retain evidence suggesting the mechanisms that contributed to those changes. The question is whether or not the short-term flexibility in mental templates involving the rejuvenation of points influenced gradual, longer-term changes in initial mental templates.

## SOME POTENTIAL INFLUENCES ON HUNTER-GATHERER POINT FORMS

There has been little study of the ways in which the different Great Basin point types, within different ecological contexts, were actually rejuvenated. There was variability in the geographic range of any given type during its type-life. The ecological context for the use, curation, rejuvenation, and discard of those points could have been influenced by several factors. Seasonal and longer-term climatic influences on carrying capacity may be argued to have had both short- and long-term effects on mobility strategy. Shifts in mobility strategy might then be expected, given other factors, to have influenced rates of curation and therefore rejuvenation.

Curation, as used here, is the duration of projectile point use-life, commencing with completion of projectile manufacture and ending with loss or discard of the point. Within use-life are subsumed transport, use, damage, rejuvenation, and rehafting events. The curation rate of a projectile point, therefore, may be characterized as the number of use-life events to which it was subjected. Evidence for the rate of curation may be suggested by the degree of size diminution resulting from accumulated impact damage and rejuvenation events, as well as the number of attributes retained from those events. Increases in curation rate, and therefore the potential number of use-damage and rejuvenation events, argue for the existence of potentially greater morphological variability in the archaeological record.

Testing for a relationship between point form and ecological context appears warranted. Such testing would need to deliver the technological life history of the point specimens, if questions about the influence of curation rates on their final form are to be addressed. A condensed model is offered below to structure the projectile point context for such testing.

## HUNTER-GATHERER PROJECTILE POINT CONTEXT

Point form may be characterized as mutable during its artifact life, from toolstone vagaries, use-damage, maintenance, and potential reuse. From manufacture to discard there is a context of interplay between environmental limitations and the behavioral systems that operated within those limitations. Thus, for any single region and temporal period, some variation in the morphology of a point type may be expected.

Variation in typologically diagnostic point attributes may first be influenced at manufacture by the lithic landscape, which includes differential sizes, workability, and availability of toolstone. Second, the food producing capacity of the area may influence mobility behavior and therefore curation and subsequently rejuvenation rates. These elements of the regional resource base may vary independently of each other and through time to complicate the prehistoric picture.

It is suggested for hunter-gatherers that initial shifts in point forms were most likely to have occurred in resource-poor regions. These were regions limited in both toolstone and subsistence resources. Such circumstances could have encouraged increased mobility and point curation. Thus it can be expected that point collections will have the strongest potential to exhibit broad morphological variation in regions with marginal resources.

In contrast, one might expect that areas of relatively sedentary occupation with abundant high quality toolstone should exhibit significantly less curation and rejuvenation of points. The size of the points discarded may also be notably larger than in more marginal localities. There is, however, the possibility that a greater range of influences on point forms may emerge with more sedentary settlement patterns. While this deserves more study, especially in relation to agricultural societies (Rondeau 1979), those inquiries are beyond the scope of this paper.

Beyond the two alternatives above, the potential for other causes of point form change is recognized. The introduction of the bow-and-arrow is only one additional example. Projectile point collections exhibiting different ranges of morphological variation for a single type may be anticipated according to this model. The prediction is offered that some collections resulting from different contexts should, to varying degrees, support both sides of the Flenniken-Thomas debate.

The general ideas outlined above are intended to guide one aspect of hunter-gatherer research. They are not intended as an interpretive model, but rather as a model suitable for testing and modification as the data warrant.

## TEST CASE BACKGROUND

The study area for testing this model is located in the central Sierra Nevada mountains. These mountains exceed 400 miles in length, average about 60 miles in width, and trend northwest to southeast. In cross-section they have been characterized as a great wedge pointing westward (Alt and Hyndman 1975), the crest being near the steep eastern face, and easily 90 percent of the surface area lying on the western slopes. The crest of the Sierra Nevada forms the western margin of the west-central Great Basin.

Both ethnographically and prehistorically, these mountains were a transition zone between what has been called the Great Basin and the California culture areas. Previous to European settlement, peoples on both sides of the crest exploited the Sierra Nevada seasonally. Subsistence resources included seasonal deer migrations into the uplands, high elevation spawning runs, and a variety of plants that became available at progressively higher elevations as spring and summer progressed. Trade spanned the crest in both directions (Davis 1961), and people from both sides of the crest are said to have sometimes wintered on the opposite side of the crest. Heavy winter snowfall from Pacific storms occurred frequently, and it can be argued that this acted as a seasonal barrier and a constraint on the use of higher elevations.

For the western slopes of the central Sierra Nevada, permanent villages were usually located below 4,000 feet to avoid the winter snow line (Barrett and Gifford 1933). However, considerably higher elevations were occupied during harsh winters by limited numbers of people, as illustrated by accounts of the Washo who attempted to feed the Donner party (Elston et al. 1977).

Paleoenvironmental reconstructions and prehistoric cultural sequences for the higher elevations of the Sierra Nevada remain formative. There are ample indications that past conditions were at times dryer, wetter, cooler, and warmer than today. However, a mosaic of more subtle variation, and some suggestion of microclimates, is apparent. This paper will forego a discussion of the paleoenvironment, since the timing of these shifts and the temporal placement of Elko assemblages with respect to them remains to be refined for the Sierra Nevada (Rondeau 1982).

The study collection comes from Alp-152 site, which lies at an elevation of 6600 feet on the western slope of the central Sierra Nevada in the Gabbott Meadow locality of the Highland Creek drainage, a tributary of the North Fork Stanislaus River. Alp-152 is situated on a terrace above the upstream end of Gabbott Meadow. In 1986, slightly less than 55 cubic meters were excavated at Alp-152 (Peak and Neuenschwander 1991). Archaeological deposits reached a meter in depth. Artifacts of flaked and ground stone were recovered, though no other archaeological materials were preserved.

Current forest vegetation includes white fir, Jeffrey pine, and lodgepole pine. The riparian community includes quaking aspen, black cottonwood, willow, and white alder (Peak and Neuenschwander 1991). Gabbott Meadow is a grassland largely devoid of trees.

The test locality meets the criteria of marginality in that it generally lacks good quality raw materials for flaking. Available toolstone occurs in limited quantities. Obsidian was imported primarily from the western Great Basin. X-ray fluorescence analysis of over 300 specimens from the region and a detailed inspection of eight site collections indicates that ca. 99% of the obsidian originated at the Bodie Hills source (Peak and Neuenschwander 1991; Rondeau and Rondeau 1989), which is slightly less than 50 air miles southwest of Gabbott Meadow in the western Great Basin. On foot, given the rugged terrain, the distance is much greater.

Given the environmental setting and ethnographic land use patterns, it is a reasonable expectation that forays into this locality were seasonally limited. This environment provided a short-duration resource base that required a fairly high level of seasonal mobility.

## GENERAL TEST RESULTS

The technological examination of points to determine the role of curation in the creation of final form requires that other components of the flaked stone assemblage be used as controls over this test. These controls include the relative contributions of local tool stones, the level of biface manufacture verses projectile point maintenance, and the general behavior patterns that accounted for point deposition.

At Alp-152 the analysis of the recovered collection included 26,318 pieces of debitage, 126 points, 19 unfinished bifaces, and four cores (Rondeau and Rondeau 1989). Obsidian was the dominant flaked stone in all artifact categories except cores. The presence of only four cores suggests a limited use of local toolstones. They included two of quartz, and one each of basalt and andesite.

The fact that alternatives to obsidian were limited is also indicated by the quantities of the second and third most common materials found in the debitage: quartz, with 336 (1.4%); and welded tuff, with 331 (1.3%). The quartz was worked by bipolar percussion. Debitage of welded tuff was mainly produced during biface manufacture. Other igneous rock types, chalcedony, chert, and metavolcanics had a very minor presence in the debitage.

Evidence of percussion biface manufacturing is limited. The unfinished bifaces were divided between obsidian (n = 13) and welded tuff (n = 6). A range of early-to-late manufacturing stages was evident for both materials. While welded tuff biface debitage was limited, obsidian comprised 97.1% of all debitage at Alp-152. This flaking waste included 6666 flakes (26.1%) that retained a bifacial edge, 331 small pressure flakes, and 91 (0.3%) notching flakes, one exceeding a centimeter in width.

The high percentage of obsidian biface edged flakes argues for the finishing and rejuvenation of points as having been the dominant contributors to the obsidian debitage. The parallel pressure flakes are thought to have been produced during the working of projectile point blade elements. This flake scar pattern is evident to varying degrees on some Elko Corner-notched points from this site and on others from the region. The notching flakes appear to have had two possible origins: the creation and the refurbishing of bases on Elko Corner-notched points.

Evidence among the debitage for obsidian biface percussion flaking was extremely limited, with the exception of a single excavation unit (there is always one!). Even so, the debitage data indicate that the volcanic glass flaking waste resulted mainly from point finishing and rejuvenation.

At Alp-152 the projectile point specimens (n = 126) were largely fragmentary. Only three were whole. Elko Corner-notched point fragments were the most common (n = 34), but there were also two Elko Eared points. Late prehistoric points included three Cottonwood Triangular and two Desert Side-notched points. One other point was assigned to a non-diagnostic, concave-base morphology, and the rest were undiagnostic fragments.

While a range of events may account for the presence of points (e.g., loss, caching, interment as grave goods), the fragmentary nature of nearly all specimens (97.6%) argues for intentional discard. The most common fragment forms were basal elements (n = 45, 36.6%). The presence of basal fragments has been suggested to be a result of discard, with the broken pieces having been removed from the haft during the retooling of projectiles (Keeley 1982).

Some of the 21 (17.1%) edge fragments may also have been returned to the site within the haft wrapping. Others, among which were 11 tip fragments, 10 tip or barb fragments, and six barb fragments, may have entered the archaeological record during carcass refuse disposal (Robertson 1980). A small number of these fragments may have been produced by manufacturing or rejuvenation errors. The possibility that some barbs may have been intentionally removed during refurbishing of point margins was not indicated in this assemblage.

The larger fraction specimens (n = 39, 31.7%), i.e., those retaining the main body or blade element, included 15 with the tip missing, 13 with a barb missing, nine missing the base, and two with an edge missing. The slightly larger number of basal pieces, when compared to blade element pieces, does not appear significant. Some blade elements may have been lost or totally destroyed off-site during hunting. Thus bases may have been more likely than other parts of a point to have been returned in the haft. Points retaining significant portions of their blade elements probably were more often subjected to refurbishing and reuse, resulting in their removal from the site. The highly reduced state of most larger pieces from Alp-152 (Figure 1) argues for this treatment. When compared with the condition of the points presented by Thomas (1981), the Alp-152 specimens suggest intensive curation.

## SPECIFIC TEST RESULTS

To determine the influence of curation on final point form, three lines of evidence were considered: size, morphology, and technological attributes. The size test follows Thomas (1981), who differentiated Elko Corner-notched points from Rosegate points by the former having basal widths of 10 mm or greater. This size test was used with the understanding that Thomas (1981:37) has cautioned that his point key may be of diminished utility with increasing distance from Monitor Valley. However, the expanding base formed by the notches of Elko Corner-notched points has been recognized as a fragment form diagnostic of this point type in the western Great Basin (Bettinger 1981) and the northern Sierra Nevada (Rondeau 1982).

Experimental manufacture of this point type has established that the expanding base element can be snapped off during manufacture (Titmus and Woods 1986). Experimental shooting of this point type has indicated that impact damage was the most likely cause of such basal snaps (Flenniken and Raymond 1986; Titmus and Woods 1986). Flenniken and Raymond (1986) found that a diminution of the expanding base occurred as a result of refabrication following impact damage.

If such refabrication occurred at Alp-152, then the expanding base fragments should exhibit a size range of pieces trending larger than the basal elements of large fraction specimens (those that retain a majority of the blade element). This was expected to be supported by the data if those more complete specimens owed their condition, in part, to basal repair. However, evidence for the snapping off of previously refabricated basal elements was also anticipated. If these were also present, some overlap in the size ranges of the two fragment categories might be expected.

Maximum width of the expanding base, as noted above, was chosen as the measure of basal size. This study used 29 specimens from Alp-152. Fifteen belonged to the relatively whole, large fraction category (Table 1), and 14 were expanding base fragments (Table 2). The expanding base fragment widths had a mean of 15.4 mm (15.6 mm with the incomplete specimen removed). The

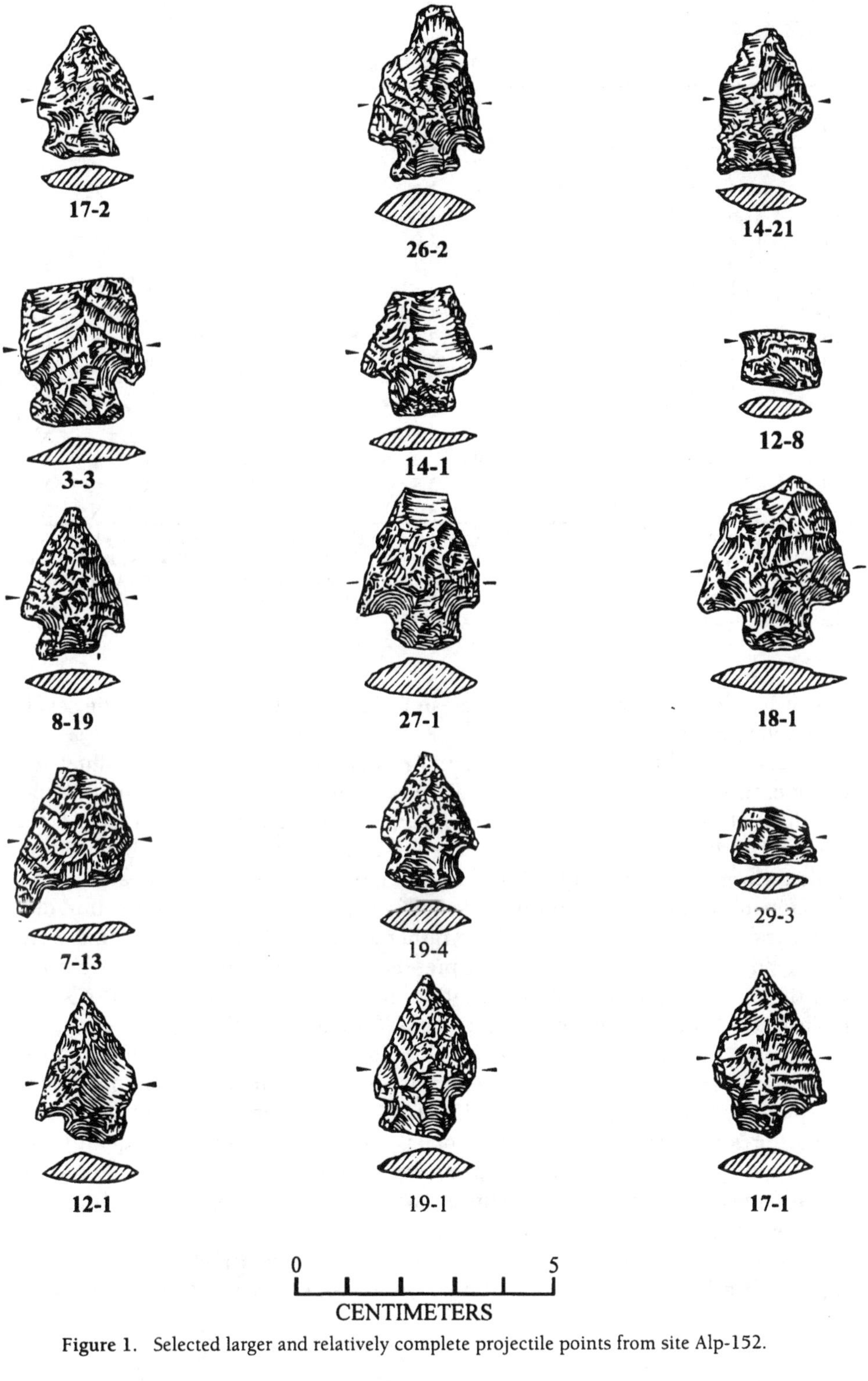


Figure 1. Selected larger and relatively complete projectile points from site Alp-152.

**Table 1. Elko Corner-Notched Large Fraction Point Data**

| Cat (#) | BW (mm) | Lg (mm) | Wd (mm) | Th (mm) | Wt (g) | HBW (u) | Ib | IB | Rb | RB | IR | Ab | AB |
|---|---|---|---|---|---|---|---|---|---|---|---|---|---|
| 7-13 | 12.2 | 28.2 | 20.9 | 3.6 | 1.7 | 2.6 | X | X | X | — | | X | — |
| 8-19 | 11.9 | 26.9 | 19.4 | 5.0 | 1.8 | 2.9 | X | X | X | | X | X | |
| 12-1 | 11.6 | 26.5 | 19.0 | 5.7 | 1.8 | 1.0/1.7 | X | X | B | — | X | — | — |
| 14-1 | 12.2 | 22.8 | 21.7 | 4.4 | 1.6 | 2.0 | X | X | X | X | X | X | X |
| 14-21 | 15.0 | 27.0 | 17.9 | 5.3 | 2.1 | 2.2 | X | X | X | X | X | - | X |
| 17-1 | 10.1 | 28.6 | 22.0 | 5.3 | 2.5 | 2.8 | | X | B | N | X | X | - |
| 17-2 | 13.8 | 20.1 | 18.6 | 4.9 | 1.5 | 1.3/1.6 | | X | B | X | X | X | X |
| 18-1 | 13.5 | 30.7 | 29.3 | 6.4 | 4.6 | 3.3 | X | X | B | X | X | X | X |
| 19-1 | 14.1 | 28.7 | 18.5 | 5.7 | 2.3 | 1.3/1.7 | X | X | X | N | X | X | X |
| 19-4 | 15.1 | 25.0 | 18.6 | 5.6 | 1.8 | 1.7/1.7 | X | X | B | | X | X | |
| 25-2 | 16.6 | 20.3 | 23.4 | 4.9 | 2.1 | 2.6 | X | | — | | | | |
| 26-2 | 12.7 | 31.9 | 21.6 | 7.1 | 3.6 | 2.6 | X | | B | N | | | X |
| 27-1 | 13.0 | 29.4 | 22.7 | 7.1 | 3.9 | 3.2 | X | X | B | N | X | X | X |
| 29-1 | 17.1 | 26.0 | 22.8 | 4.7 | 2.8 | 3.8 | X | | B | | X | X | X |
| 30-1 | 9.7 | 24.1 | 17.2 | 4.0 | 1.3 | 4.1 | X | | X | X | X | X | X |

Cat = Catalog Number BW = Basal Width Lg = Length Wd =Width Th = Thickness Wt = Weight HBW = Hydration Band Width Ib = Blade Impact Damage IB = Base Impact Damage Rb = Blade Rejuvenated RB = Base Rejuvenated IR = Impact Damage Over Rejuvenation Ab = Asymmetrical Blade Associated With Rejuvenation AB = Asymmetrical Base Associated With Rejuvenation X = Attribute Present B = Includes Rejuvenation Over Missing Barb N = Includes New Notches — = Point Element Insufficient or Missing

relatively complete pieces showed a mean of 13.6 mm (14.5 mm with the seven incomplete measurements removed).

The expanding base fragment sample contains the three largest values for this measurement. The relatively whole piece sample contains the smallest six, but three of these are incomplete. The incomplete measurements on pieces from both categories were caused by a damaged or missing corner on each base. These broken corners on the expanding bases suggest that some rejuvenation, and therefore diminution of basal width, probably occurred without refabrication of a new basal element. Fourteen of these rather small basal corner fragments were recovered from Alp-152. Even so, the presence of the expanding base fragments, and the pattern of basal width diminution, both argue that there was also some refabrication of the same, only smaller, basal form at Alp-152.

Since the intent of this study was to determine the effects of extended curation on point form, it was necessary to determine the relationship of point morphology to the technological evidence of impact damage and rejuvenation. It is worth restating that increased curation is expressed by an increased quantity of damage events in the use-life of a specimen. Further, most of the morphological variation was created by impact damage, with rejuvenation operating within the confines of the surviving fragment.

The 15 suitable large-fraction specimens were inspected under incandescent light with a binocular Bausch & Lomb microscope at a power of up to 25X.

Table 2. Elko Corner-Notched Expanding Base Fragment Data

| Cat (#) | BW (mm) | HBW (u) |
|---|---|---|
| 1-6 | 17.9 | 2.4 |
| 9-14 | 15.0 | — |
| 12-8 | 15.3 | 1.4 |
| 12-24 | 15.2 | 2.5 |
| 17-4 | 13.3 | — |
| 18-9 | 17.9 | — |
| 19-10 | 14.7 | — |
| 20-2 | 19.7 | — |
| 20-35 | 14.2 | 2.2 |
| 22-34 | 17.2 | 2.1 |
| 25-5 | 13.8 | 1.7 |
| 25-35 | 14.0 | 2.4 |
| 29.2 | 13.3 | — |
| 29-3 | 16.7 | 3.0 |
| 35-3 | (a) | 1.7 |

Cat = Catalog Number BW = Basal Width HBW = Hydration Band Width
[a]Data Unavailable

Thirteen retained impact damage to the blade, 11 to the base. Fourteen showed rejuvenation of the blade. The final specimen was too damaged to allow a determination. Nine had rejuvenated bases, and four showed new notching scars. Twelve had been impact damaged after having been repaired from earlier damage. Twelve had asymmetrical blade morphologies after rejuvenation. Nine had asymmetrical basal elements after rejuvenation. Eight had bifacial blade edge rejuvenation replacing a broken barb. While these changes in form were sufficient to increase morphological variability within the point type, they were not sufficient to suggest a change of type assignment.

## TEMPORAL CONSIDERATIONS

That rejuvenation of these points may indicate that they were used during a later period also needs to be addressed. According to the Thomas (1981) basal width criteria, these Elko points did not become Rosegates during their use-life. Even so, scavenging and reuse of these points might have occurred during a later period.

Available typological evidence strongly suggests that at Alp-152 the Elko period accounted for the bulk of its occupation. Peak and Neuenschwander (1991) followed Thomas (1981) in suggesting that the main use of Alp-152 was between

3300 and 1250 B.P. The radiocarbon assays provided dates of 370±75 B.P., 3180±90 B.P., and 4060±90 B.P.

It was thought that obsidian hydration band width analysis might provide a more direct approach to dating the use and rejuvenation of these points. The 15 large-fraction specimens and nine expanding base fragments were analyzed. The first 15 and two of the latter nine had been cut previously and sourced by X-ray fluorescence. All were sourced to Bodie Hills. Seven additional basal fragments were visually sourced to Bodie Hills. Four of the large-fraction specimens were cut a second time in an attempt to find band width variability across rejuvenated and earlier point surfaces.

Band widths ranged from 1.0 to 4.1 microns (Tables 1 and 2). At much lower elevations (e.g., 2000 feet) on the western slopes of the central Sierra Nevada, the hydration range noted above for Bodie Hills volcanic glass would span both the late prehistoric periods marked by Desert Side-notched and preceding Rosegate points, and the earlier Elko Corner-notched points. However, these lower elevations have moisture and temperature regimes considerably different from the Gabbott Meadow area. Unfortunately, not enough study has been done in the region at higher elevations to know if a three-micron range is reasonable for the Elko period. The effect on obsidian hydration rates of microenvironments at widely different elevations in the Sierra Nevada are poorly understood at this time.

Three of the four double-cut specimens exhibited a second band width, but none were sufficiently divergent to indicate two different periods of use (Table 1). None of the single-cut specimens exhibited two band widths. The obsidian hydration data do not indicate a delayed reuse of the points. It should be noted that the band width readings for the Bodie Hills debitage show a similar range (Peak and Neuenschwander 1991).

## CONCLUSIONS

The testing of the analytical approach of reading flake scars to identify evidence for breakage and rejuvenation of projectile point forms indicated that rejuvenation could clearly be identified. This is not to say that all breakage and rejuvenation events could be identified, nor that it is even necessary to do so. A preliminary demonstration of the utility of this approach has been achieved. Further testing is required for other time periods, point types, and regions, since there is no guarantee that this analytical approach will be useful in all cases.

The testing of a preliminary ecological model of hunter-gatherer projectile point context also received support in that the Elko Corner-notched points from Alp-152 exhibit evidence of extensive modification. The extent of modification suggests an extended use-life, which fits well with the expectation of a high curation rate. However, Alp-152 is only one site in the area. A reminder to resist the single-site-as-region interpretation is warranted, particularly since there is a greater variability in the form of Elko points in the region than was found at Alp-152.

An Elko-to-Rosegate transition is not suggested by the diminution of basal width at Alp-152. Only one specimen (30–1) has a basal width of less than 10 mm, and it has the thickest hydration band width of all. Size dictates that these specimens were dart points. They exceed width and thickness parameters for Rosegate points in the region. This does not mean that such points could not have been reused on arrows. Further, the Elko/Rosegate transition is partly attributed in the region to the transition from dart and throwing stick to bow-and-arrow. The role of projectile launchers in the change of point forms is a special case and is not tested by the model proposed here.

How well the results support or refute arguments in the Flenniken-Thomas debate is not a concern of this paper. However, perennial disagreements in the assignment of specific artifacts to point types, and of types to temporal placement, are symbolized by the debate. This debate suggests that uncertainties in the control of form continue to plague researchers. These uncertainties are a predictable result of the failure to consider context and to identify the results of curatorial variability. To not be concerned with why points change does not obviate the fact that they did. There were places and times when these changes occurred, probably not in a common and widespread scenario as championed by Flenniken, but perhaps rather where influences converged to encourage the retention of a modified point type.

If this is accepted, what is the utility of points as temporal markers in transitional areas and during transitional times? How might these areas and times be identified? How do they relate to long-term influences? Does point rejuvenation relate to a general trend of point size diminution across millennia? Did initial mental templates change because it was observed that smaller points also worked well, particularly since smaller points tend to be more expedient in terms of manufacturing time and more flexible in raw material procurement requirements? Were there periods when other factors encouraged the retention of forms that already existed as rejuvenated pieces? Would the diminution in big game availability, regardless of cause, be such an example? If a general trend in point size diminution is recognized, what do the exceptions, those of smaller-than-typical size for a type, tell us? Are these exceptions the result of the sizes of locally available toolstone? Do these exceptions indicate increased curation rates? Would the potential for shifts in manufactured point forms be more likely among highly dispersed groups of hunter-gatherers?

This test case can offer no answers for why point types changed, but it does argue for the need for studies to determine how points changed during their use-lives and ultimately across their type-lives. It is only in the aggregate that answers may be articulated. Yet it should be clear that there is more to learn from projectile points than: when is an Elko?

## ACKNOWLEDGMENTS

Data on the archaeology of Alp-152 were provided by Ann Peak and Neil Neuenschwander, for which I am grateful. Ann Peak also supported this effort by allowing the use of illustrations from Peak and Neuenschwander (1991). Thanks go to Alan Bryan, John Dougherty, George Odell, Michael Shott, and Georgie Waugh, who provided useful comments on a draft of this paper.

## REFERENCES

Alt, D. D., and D. W. Hyndman. 1975. *Roadside Geology of Northern California.* Mountain Press Publishing Co., Missoula.

Barrett, S. A., and E. W. Gifford. 1933. *Miwok Material Culture. Indian Life of the Yosemite Region.* Bulletin of Milwaukee Public Museum 2(4). Yosemite National Park Association, California.

Bettinger, R. L. 1981. *Archaeology of the Lee Vining Site, FS # 05–05–51–219 (CA-Mno-446), Mono County, California.* Inyo National Forest, Bishop.

Bettinger, R. L., J. F. O'Connell, and D. H. Thomas. 1991. Projectile Points as Time Markers in the Great Basin. *American Anthropologist* 93:166–172.

Davis, J. T. 1961. *Trade Routes and Economic Exchange among the Indians of California.* University of California, Archaeological Survey Reports 54, Berkeley.

Elston, R. G., J. O. Davis, A. Leventhal, and C. Covington. 1977. *The Archaeology of the Tahoe Reach of the Truckee River.* Northern Division of the Nevada Archaeological Survey, University of Nevada, Reno.

Flenniken, J. J., and A. L. Novic. 1984. Morphological Projectile Point Typology of the Great Basin: Replication, Experimentation and Technological Analysis. Paper presented at the Forty-Ninth Annual Society for American Archaeology Meetings, Portland.

Flenniken, J. J., and A. W. Raymond. 1986. Morphological Projectile Point Typology: Replication, Experimentation and Technological Analysis. *American Antiquity* 51:603–614.

Flenniken, J. J., and P. J. Wilke. 1986. The Flaked Stone Assemblage from Hogup Cave, Utah: Implications for Prehistoric Lithic Technology and Culture History in the Great Basin. Paper presented at the Twentieth Biennial Great Basin Anthropological Conference, Las Vegas.

Keeley, L. H. 1982. Hafting and Retooling: Effects on the Archaeological Record. *American Antiquity* 47:823–833.

Peak, A. S., and N. J. Neuenschwander. 1991. *Cultural Resource Studies, North Fork Stanislaus River Hydroelectric Development Project, Vol. V.* Northern California Power Agency, Roseville.

Robertson, J. 1980. *Chipped Stone and Socio-Cultural Interpretations.* M. A. Thesis, Department of Anthropology, University of Illinois, Chicago.

Rondeau, M. F. 1979. *Projectile Point Analysis for the Kahorsho Site, NA 10,937, Central Arizona.* M. A. Thesis, Department of Anthropology, California State University, Sacramento.

Rondeau, M. F. 1982. *The Archaeology of the Truckee Site, Nevada County, California.* California Department of Food and Agriculture, Sacramento.

Rondeau, M. F., and V. L. Rondeau. 1989. *Technological Investigations of Flaked Stone Assemblages from Eight High Sierran Sites, Alpine and Tuolumne Counties, California.* Rondeau Archeological, Sacramento.

Titmus, G. L., and J. C. Woods. 1986. An Experimental Study of Projectile Point Fracture Patterns. *Journal of California and Great Basin Anthropology* 8:37–49.

Thomas, D. H. 1981. How to Classify the Projectile Points from Monitor Valley, Nevada. *Journal of California and Great Basin Anthropology* 3:7–43.

Thomas, D. H. 1984. Diversity in Hunter-Gatherer Cultural Geography. Paper presented at the Forty-Ninth Annual Meeting of the Society for American Archaeology, Portland.

Thomas, D. H. 1986a. Contemporary Hunter-Gatherer Archaeology in America. In *American Archaeology: Past and Present,* edited by D. J. Meltzer, D. D. Fowler, and J. A. Sabloff, pp. 237–276. Smithsonian Institution Press, Washington D. C.

Thomas, D. H. 1986b. Points on Points: A Reply to Flenniken and Raymond. *American Antiquity* 51:619–627.

Wilke, P. J., and J. J. Flenniken. 1991. Missing the Point: Rebuttal to Bettinger, O'Connell, and Thomas. *American Anthropologist* 93:172–173.

## Chapter 9

# Projectile Points, Style, and Social Process in the Preceramic of Central Peru

JOHN W. RICK


## ABSTRACT

This paper argues that, to move forward in research on stone tool style, we must supplement theoretical discussion with examination of large data sets for stylistic patterning. Using a large projectile point collection excavated from cave sites in the high altitude region of central Peru, I demonstrate the presence of robust patterning at a number of levels of stylistic resolution. I use hierarchy of stylistic subdivisions within the projectile point class to evaluate key variables at a number of stylistic scales, including measures of diversity and continuity. The analysis produces evidence of stylistic foci and transitions between them that reflect social process: the evolving social conditions which affect the material output of identity. I argue that there is evidence of complex residential and interactional relationships across time and among the occupants of these Andean sites.


## INTRODUCTION

In this paper I present a data-driven analysis of style in tools of the Peruvian preceramic period, in the belief that we need to look to our data for a structure of


**JOHN W. RICK** • Department of Anthropology, Stanford University, Stanford, California 94305.



*Stone Tools: Theoretical Insights into Human Prehistory*, edited by George H. Odell. Plenum Press, New York, 1996.

style that will confine the range of interpretations made. I will first comment on the current state of theory related to stone tool style, and make observations on the necessity of deriving some of our models directly from data. The nature of archaeological data helps designate the analytical strategies most likely to offer insights into stylistic behavior. An extended examination of a large diachronic data set will follow, emphasizing the search for robust patterning related to social process.

In spite of recent interest in issues of style, archaeologists know amazingly little about isolating and analyzing stylistic information in stone tools. In fact, the theoretical basis for stone tool style has been more frequently emphasized than the direct, problem-oriented consideration of style in archaeological stone tools. In some ways this is healthy, because theoretical considerations can presumably help avoid the naive interpretations of a purely inductive approach. Also, cautionary tales coming from the ethnoarchaeological literature of style (e.g., Weissner 1983) warn us that simplistic assumptions about stone tool style, such as directly equating tool types with residential groups, are unlikely to be correct. Similarly, studies of ethnographic collections of projectiles suggest that more stylistic information may be concentrated in perishable arrow shafts than in stone points (Sinopoli 1991).

A familiar debate of the past decade has alternately argued that style is consciously or unconsciously imparted; that style communicates or is not primarily involved in communication; and that style and function are or are not overlapping (for recent versions, see Binford 1989; Sackett 1990; Weissner 1990). The conversations about these alternate positions have been useful, but have not been followed by stone tool stylistic research dealing primarily with the archaeological record. In fact, the rarity of such research (but see Close 1989 for the latest in a series of data-driven articles by this author) argues that the recent theoretical literature has been either difficult to apply to analysis of stone tool style, or discouraging to researchers (e.g., Clark 1989:33).

If the stone tool style researcher is discouraged by the lack of operational theoretical models for stylistic interpretation, it may be because style is one of the most difficult subjects that exists for erecting a deductive research framework. Predictions about where and how style is visible or what patterns it takes are linked by relatively weak logical chains to past phenomena. Style is unlikely to have strong universal structure, any more than that social conditions would be uniform across broad extensions of space and time. There are undoubtedly many ways that style originated, persevered, was transformed, and disappeared in classes of tools. Where to look for style and how to interpret it are conditioned as much by pragmatic conditions as by theoretical positions. In essence, the theory of style gives us ways to think about the subject, but we also need practical ways to localize style, and to decide on the basic parameters of any stylistic research program.

The study of style has been subject to what Wobst (1978) calls the "tyranny of the ethnographic record." Archaeologists have often dealt with style as if it were a snapshot of a moment of frozen social time—in essence, that it identifies the location or identity of individuals or groups. We are caught up in trying to do "paleoethnography," believing that style can establish prehistoric group identity. But given the long time scale and coarse temporal resolution of most of the archaeological record, only rarely will individuals or groups be stylistically visible, even if they were imprinting a clearly identifiable style on their material culture. It is rather the differentiation or change exhibited by style over substantial spatial or temporal dimensions that is likely to be of utility in archaeology.

What happens to stylistic entities across time, and how does this variation relate to that of neighboring entities? While there is nothing new about this perspective, I want to underline that a goal more consistent with the strengths of archaeological data is to use the dimension of time to view social process, rather than to attempt to isolate a group or individual at any point in time. We may never know with confidence the social correlates of a specific style, and thus the entities will often remain shady characters at best. Instead, we can hope to see longer-term trends which reflect *social process*.

Social process, in the narrow sense in which I use it here, refers to the evolving social conditions which affect the material output of identity, conscious or otherwise. Social process can be seen in the archaeological record of changing identity in time and space. The record of identity can include the number of levels of identity, the units of identity, the strength of their definition, their location and duration, and the processes which affected these dimensions.

An important related issue is the appropriate scale, and the concordance between analytical techniques and phenomena that we are hoping to monitor and eventually explain. A focus on social process requires the identification of stylistic units in the archaeological record. Much effort has been spent attempting to isolate stylistic attributes or types that can be used to identify prehistoric groups. Pursuit of specific groups is methodologically and conceptually problematic, except in rare cases. In reality, there is no obvious way to know that any particular aspect of material culture is related to any specific scale of patterned prehistoric behavior. Thus a stylistic classification might relate to identity-based patterning at the level of the individual, family, local group, or regional population.

In sum, my position on style is that:

1. We need to start examining extensive data sets for patterning that is arguably stylistic. Theory-building alone is not sufficient to understand the nature of stylistic behavior, or to provide reasonable models for style in prehistory.
2. We need to recognize the strengths of archaeology by emphasizing spatial and diachronic vantages. The structure of patterning across time is one of the most important clues to the processes behind stylistic change.
3. We need a zoom lens for examining the stylistic record, capable of encompassing a number of levels of scale. Since we do not have obvious

ways to know the scale of the prehistoric identities that produce patterns in artifacts and patterned behavior, we should condition our methods to allow varying scales of archaeological patterning to potentially register a range of scales of prehistoric behavior.

In keeping with these observations, this study will analyze an abundant record of stylistic data from preceramic Peru. I will examine summary measures of stylistic variability at a number of levels of typological scale over a lengthy period of time.

## THE DATA

The data for this examination of stone tool style and social process consist of a series of complete projectile points recovered from excavations in two intensively occupied preceramic cave sites from the high altitude puna region of Central Peru (Figure 1). The first site, Pachamachay, was investigated from 1973 to 1976, and a long, artifact-rich stratigraphic record was recovered from modest-sized excavations (Rick 1980). The second site, Panaulauca, is located 27 km from Pachamachay, and was excavated from 1979 through 1986 by a multidisciplinary team (Bocek and Rick 1984, Rick n.d.). For brevity, I will refer to Pachamachay as *Pax*, and Panaulauca as *Pan*. A larger sample was recovered at Pan, but excavations of major extension could not be carried through to basal levels due to civil unrest which effectively closed the area to research in 1986.

Both caves are believed to have been residential bases for hunter-gatherers during the preceramic period of about 9000–1700 B.C. (Bocek and Rick 1984). Both contain extremely abundant lithic and faunal remains concentrated in cultural sediments of 2.5–4 m maximum depth that literally choke the mouths of these small caves. The succeeding ceramic period deposits witness the use of some domestic camelids and probably local cultigens, but also show notably lighter occupation intensity. Changes in functional tool class proportions suggest that these were specialized and short-term occupations, compared with the intensive and broad-based preceramic use of the sites.

Plant and animal remains from the Pan deposits have been extensively analyzed from a variety of viewpoints, and are documented in Pearsall (1988, 1989, n.d.) and Moore (1988, 1989). These analyses are too detailed to describe here, but two salient points emerge. First, the plant remains are consistent with, but do not by themselves prove, year-round occupation of the site (Pearsall n.d.:425–439), while the animal remains show some seasonality in early and late occupation periods (Moore 1989:395–397). The complementing seasonality of plant and animal assemblages provides a reasonably strong argument for year-round occupation of the site during the portions of the preceramic considered here, and for most of the ceramic periods. Second, the diversity of animals taken

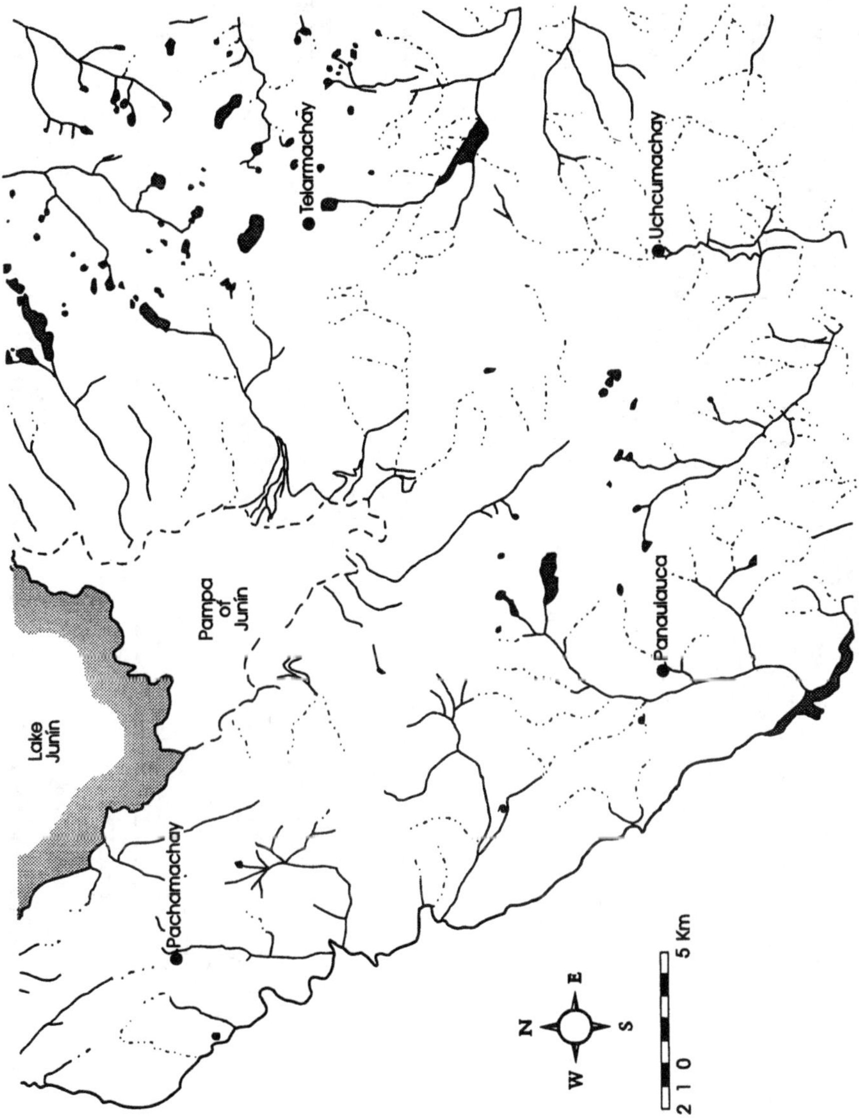


Figure 1. Map of the Puna of Junín, showing important preceramic and formative cave sites.

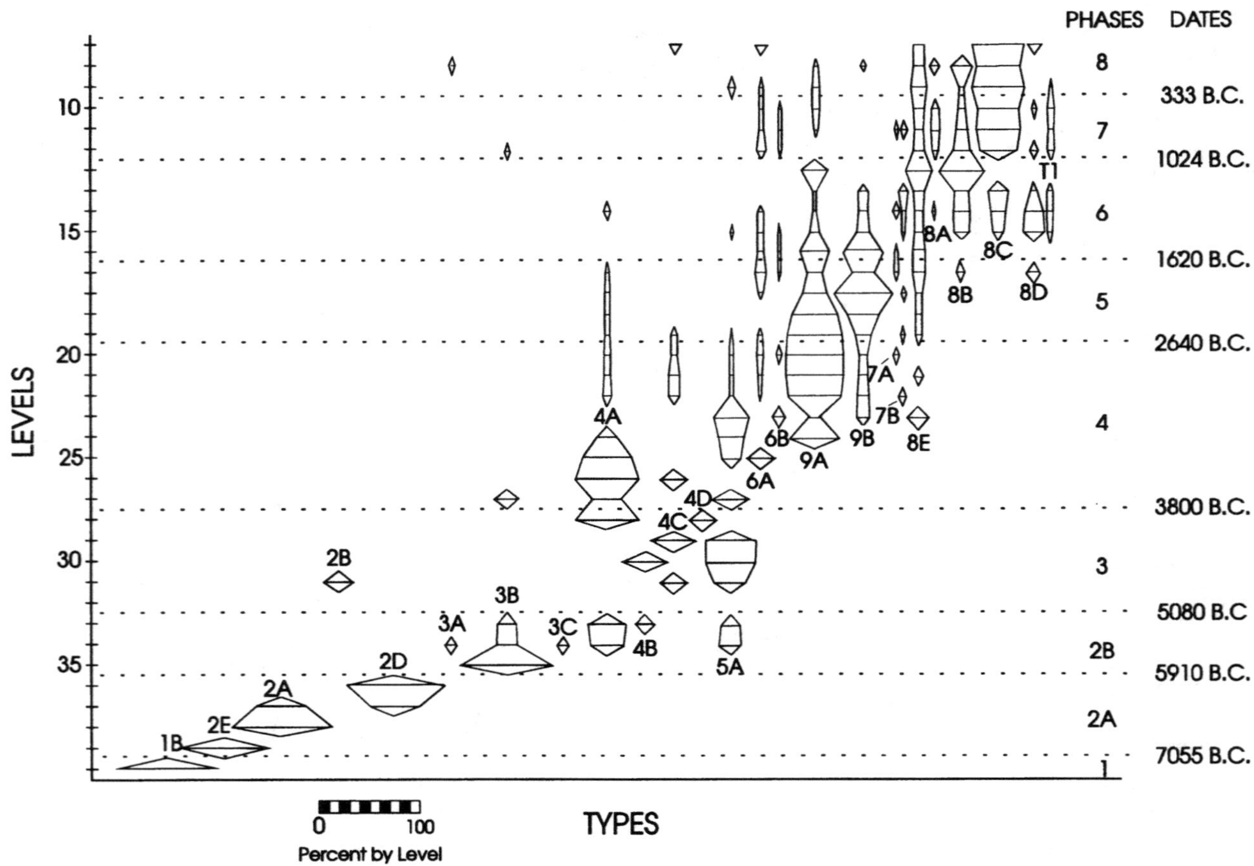


**Figure 2.** The projectile point type relative frequencies from Panaulauca, across the entire stratigraphic sequence. The data for this study are related to the upper part of the chart, levels 22–27.

by the inhabitants of Pan is quite low, with camelids comprising approximately 90% of the assemblage for all but the earliest preceramic period; the majority of the remaining animals consist of deer. Thus all game animals were effectively of similar body size.

The projectile points of Pan, in conjunction with those of Pax (Rick 1980), offer an exceptional possibility to observe stylistic variability in stone tools fairly continuously across as much as 10,000 years. These tools are both abundant and diverse, and in many cases have been made to quite specific forms which can reasonably be called stylistic. The multiplicity of contemporary projectile point forms (Figure 2) finds no obvious explanation in the realm of functional variability, since there are few game species, and only one major prey size. I have proposed that puna stylistic variability was most easily observed within functional classes, and have noted that the only clear case of stylistic variability was in class 3, the projectile points. Projectile points may not be the only objects manifesting style, but they are the most easily documented ones. Similarly, their extensively retouched character makes isochrestic style more easily detected, although I do not contest the argument that more extensive modification is not equivalent to a capacity to contain style (Sackett 1990).

I created a two-level hierarchy of stylistic forms (Rick 1980:145–146)—the first being the *type group*, reflecting coarse morphological divisions; and the finer, second level being the *type*, representing finer divisions of form within the type groups (Figure 3). Type was the original unit of earlier Pax analyses, and was defined largely on the conjunction of different tip, midsection, and base outline forms. Additional attributes of importance, although not necessarily diagnostic, include edge treatment (expedient/wavy, straightened/trimmed, beveled, coarsely serrate, finely serrate, ground) of different edge segments; ratio of thickness to width; and cross-section. The following is an example of a type description:

> Type 5A. Rounded base, straight-sided triangular tip, joined in midsection by a slight shoulder between a wider base and narrower tip. Mostly heavily ground on basal edges, with very smooth basal outline; tip edges wavy with no attempt to trim to straightness after the last round of intrusive flaking; very low thickness/width ratio; thin and symmetrical diamond-shaped cross-section.

Type groups are groups of types that show coarse morphological similarities, reflecting major patterns of projectile point temporal variation. Type Group 5 includes wide and thin, bipointed, or nearly bipointed forms. Eight type groups were defined, within which were contained a total of 32 types (Figure 4).

The primary Pan cave mouth excavation area yielded 2043 finished projectile points sufficiently unfragmented to allow stylistic classification, while additional specimens came from excavations in the talus slope and cave interior. Points from these smaller excavations are not included in the sample used here, because imperfections in level correlations will introduce temporal error. The main exca-

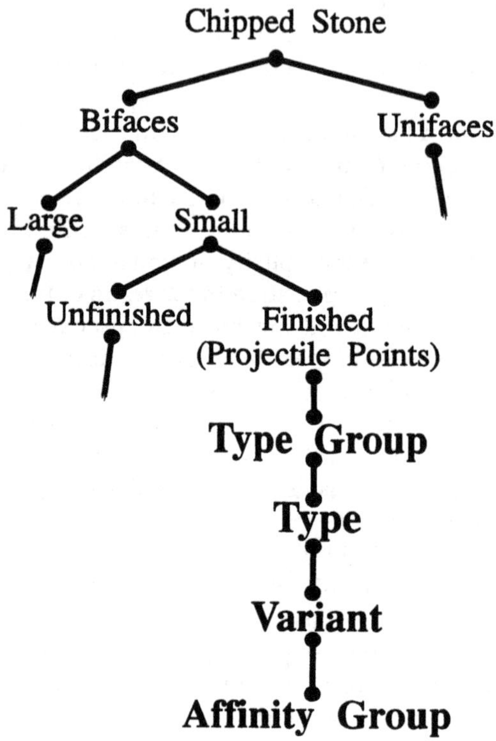


**Figure 3.** Schematic diagram of projectile point classification within the overall lithic typology of Junín materials.

vation is a large and well-stratified sample excavated to mid-preceramic level 23, below which only a small test was made. Finished projectile points were distinguished from a preform class by having sharp points, relatively regular and non-wavy edges, haftable bases, and profile straightness and symmetry. I cannot argue for a perfect separation between finished and unfinished classes, inevitably relying on an intuitive assessment of all relevant characteristics, including those mentioned. Preforms are about half as common as finished points, although this varies widely in different site contexts.

My main objective is to explore the stylistic variability of the projectile points and its relationship to social change at the site. This is a very fragile analysis, in that any major stratigraphic mixing or point misclassification will diminish its potential significantly. Similarly, only with the largest possible samples will cultural patterning overshadow stochastic variability. Thus I will primarily rely on Pan projectile points from layer 7 (the latest intact stratum) to layer 22 of the large excavation surface, a sample of 1505 points. This covers the approximate period

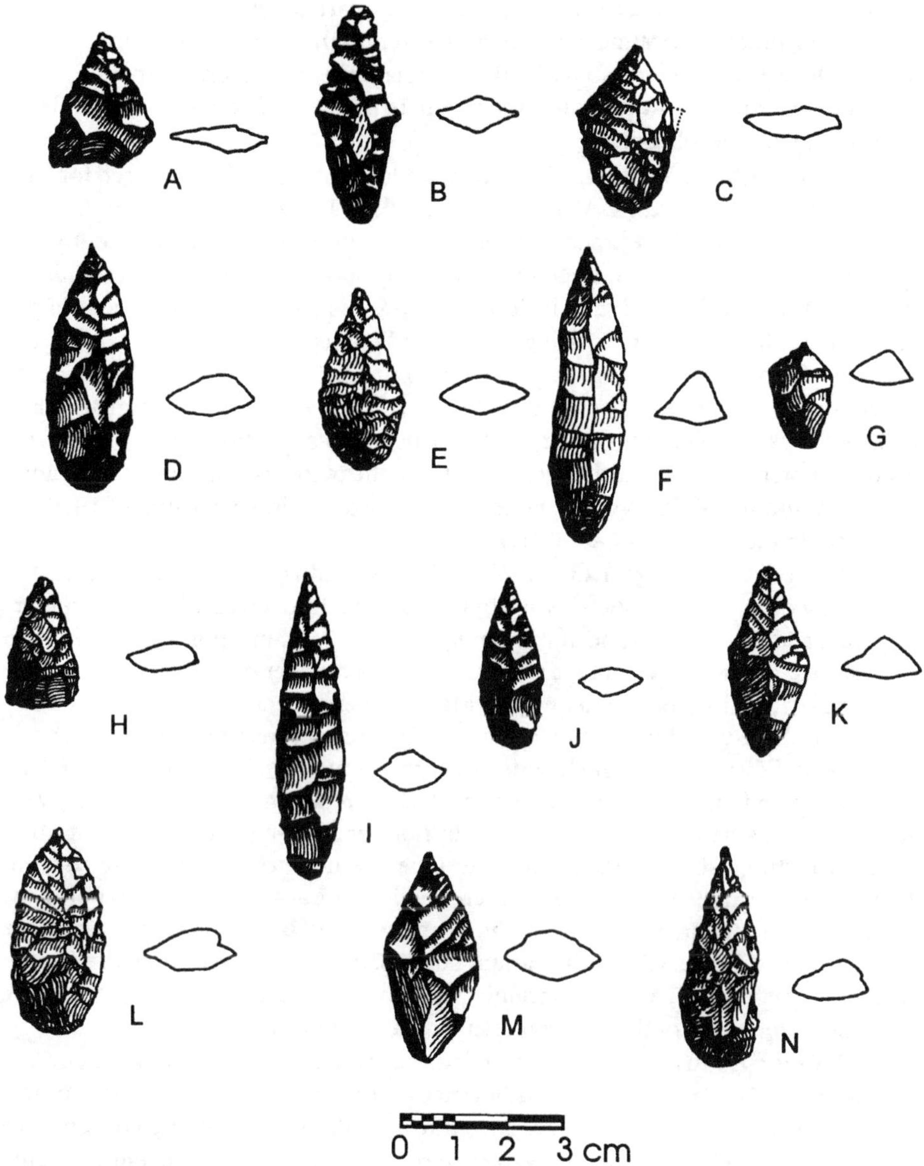


**Figure 4.** Selected projectile points representing type groups and subdivisions of type group 9 from Panaulauca Cave, Junín, Peru. A: Type Group 1; B: Type Group 2; C: Type Group 3; D: Type Group 4; E: Type Group 5; F: Type Group 6; G: Type Group 7; H: Type Group 8; I: Type 9a Variant 1; J: Type 9 Variant 2; K: Type 9A Variant 5; L: Type 9B Variant 1; M: Type 9B Variant 2; N: Type 9A Variant 4.

of 3300 B.C. to A.D. 1200, although the great majority of this record is dated prior to 700 B.C. Pan levels 22–16p span the latter part of the middle preceramic through terminal preceramic, while level 16c represents the ceramic horizon, and later levels are mostly formative. I will also relate the Pan record to that of levels 3–19 in Pax, which span the same approximate period and contain 892 whole or nearly whole, typeable points.

The type groups and types of the Pan samples conform to those defined for Pax, with some major exceptions. All eight type groups of Pax were clearly recognized at Pan. A large number of late preceramic projectile points from Pan fall outside any Pax type group. These have been given the new type group 9—containing 764 points, or more than half of the total sample. Type group 9 has a series of distinctive shapes based on outline, combined with highly trimmed points and rudimentarily executed bases. These points stand in strong contrast to the stubby or elongate, leaf-shaped, exceptionally thick but well-trimmed type groups 6 and 7, which predominate in Pax at the same period. This large new type group, nearly totally lacking at Pax, argues for a strong, potentially stylistic distinction between the two sites in the later preceramic period. A number of Pax types were not observed at Pan, leaving a total of 18 of the Pax types in the main level 7–22 Pan sample.

The relatively large Pan sample, with its limited time range, offers the possibility of further stylistic subdivision. Thus, two finer levels of stylistic hierarchy were added—*variants* and *affinity groups* (Figure 3). Conceptually, variants can be regarded as members of a type which differ on a number of characteristics other than those used in the type definition. Most of these characteristics are at a finer scale than those used for types—particular edge or point treatment, profile characteristics, and distinctive retouch patterns, to mention a few. Whereas type subdivisions are paradigmatic (to use the definitions of Whallon 1972)—all types were based on the same variables—variant definitions are tree-type in that the variables used to distinguish variants within one type are not necessarily those used in breaking down another type. These differentiations are at a finer level of typological detail than is found in most archaeological treatments of projectile points. The large number of specimens allows the variant subdivision, but did not force this extension of the typology. Thirty-six variants, with counts ranging from 15 to 229, were defined within types of the late preceramic and ceramic periods.

Affinity groups are even finer stylistic distinctions within variants and are based on *similarities,* rather than differences, to other points. In practical terms, effectively identical points were grouped into a single affinity group; in the absence of any such matching points, a given specimen would compose the entire group. Of the 151 thusly defined groups, nine have only one specimen, and the largest group has 29; most fall between five and 15 members. Affinity groups were formally described, but often the differences between them were non-categorical—a more or less acute angle between two straight edges, a similar length:width ratio, and a particular retouch strategy for producing a needle-like tip could justify a grouping. They are regarded as an experimental attempt to rapidly distinguish extremely fine levels of morphological similarity.

Some of the variability in variant or affinity groups may have been caused by different levels of reduction or rejuvenation (see Rondeau, this volume). It is difficult to evaluate this possibility, since rejuvenation could create almost any imaginable form a knapper might choose. The forms of variants and affinity groups do not create continua of increasing shortness, edge bevelling, or other characteristics known from rejuvenation. These points are made in rather unfragile materials, and are mostly rather short and often thick; arguably, they are relatively unsusceptible to in-use fragmentation or edge loss. My evaluation is that, while rejuvenation may have occurred, it is not a major contributor to the variability on which this typology is based; certainly it could not account for the number of categories observed here. Also, the variability does not seem to be functional in nature, although I note that Vaughn (1985) has suggested that a small number of projectile points from nearby Telarmachay (Lavallee et al. 1985) shows evidence of varied use.

This hierarchy of stylistic levels (Figure 3), while based on systematic differences or similarities, cannot be considered entirely objective. As with most classifications, different characteristics could be emphasized, yielding a very different result. I cannot conclusively demonstrate that a subdivision at the type level should not have been at the variant level. I might not even be able to achieve an exact replica of my typology, if deprived of my notes. I will argue, at the least, that the typology was consistently applied, since I did all classifications myself within about a 10 year period. Although it might seem trivial, I am convinced that the hierarchy running from type groups to affinity groups uniformly sorts formal differences running from profound to slight.

## SOURCES OF VARIABILITY IN THE STYLE SEQUENCE OF PANAULAUCA

In essence, we would like to know why stylistic entities begin, end, and change in relative frequency; and what the meaning of contemporary stylistic multiplicity is. The standard explanation for these patterns is that certain forms of material culture start through innovation or adoption. Their frequency of use in a culture increases, then decreases, and eventually the form is abandoned. This explanation is tenable when only a few stylistic types of functionally identical items are produced in a given moment, in that individuals within a small group could all conceivably participate in a stylistic tradition with more than one variant per functional form. If the stylistic forms found in a given moment are myriad, however, it becomes increasingly likely that some of this diversity reflects stylistic differences between contemporaneous individuals or groups. I call such loci of style *production units or entities.*

Components are not moments in time, however, as they usually imply significant duration. Thus the question remains whether the multiple stylistic entities were co-residential or sequential. Also, those who produced a stylistic class may not have been those who used and/or deposited the artifacts in their archaeological context. Many site formation processes have further distorted the record. Stratigraphic mixing within a site increases the temporal range of styles, which increases stylistic multiplicity at any point in time. Thus there are many sources of variability in the stylistic record, and it is unlikely that the influence of social, exchange, or site formation processes can be neatly segregated. Similarly, while exchange of stylistic forms may be understood through raw material studies, it is improbable that artifacts can be conclusively tied to places of origin, much less their original production units. Thus the record of movement, origin, disappearance, increase, or decrease of those production units or their outputs may be heavily modified by cultural and natural transformations of the archaeological record, requiring vigilance on the part of the analyst.

## Measures of Stylistic Variability

Put straightforwardly, the analysis of style in Pan projectile points is the assessment of the relationships among various artificial taxonomic units, which reflect variability not assignable to dimensions other than style. If each of the hierarchical tiers of the projectile point classification is monitoring stylistic information, then we have an instrument akin to a zoom lens, capable of observing a scale of entities going from fine to coarse. But what measures can be observed across time in this framework? Frequencies can be used to seriate classes in time, but while changing frequencies of *functional* artifact classes have intrinsic, if complex, meaning, it is difficult to interpret the meaning of proportions of stylistic classes. Thus diachronic summary measures that de-emphasize the specific consideration of class frequency are necessary.

## Diversity

Since the multiplicity of stylistic classes should be related to the diversity of production entities, the stylistic diversity of layers in Pan is important. Diversity is generally considered to be made up of richness, or number of classes found in a component, and evenness (Figure 5), a calculation of how even the frequencies of these classes are (Kintigh 1984).

### *Richness*

Richness is related to sample size, since large samples usually have a greater number of classes of items than small ones. Kintigh (1984, 1989) has developed a method of estimating expected sample size-corrected diversity through assem-

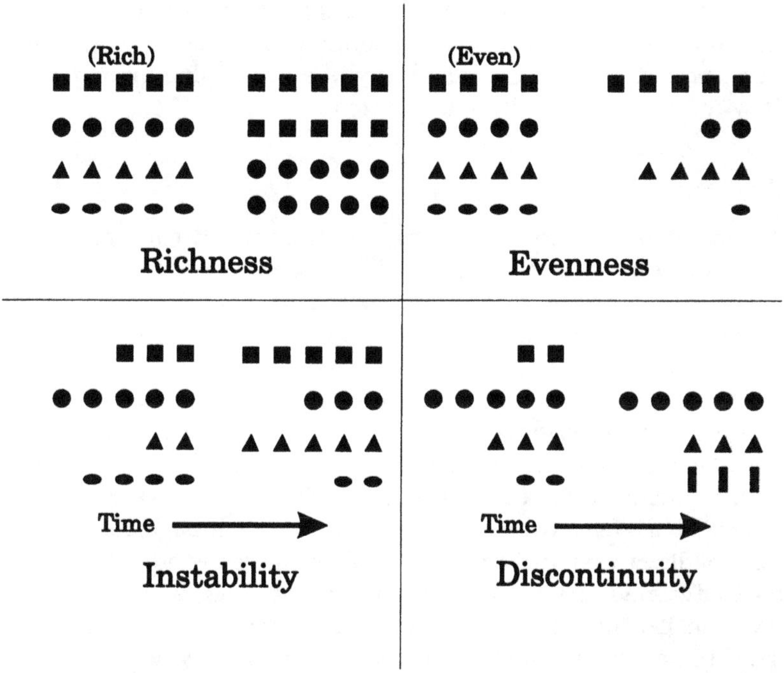


Figure 5. Graphic depiction of assemblages, illustrating the measures used in this study.

blage simulation, which may have some advantage over regression methods. Kintigh's (1988) Monte Carlo methods for assemblage simulation not only give the mean expected richness, but also provide standard deviations around that mean. Richness *per se* is of limited interest, compared to how much richer or poorer than expected the component is. Thus I use the number of standard deviations the actual richness of Pan style classes are above or below the simulated mean.

An important factor in evaluating the richness of seriatable levels is that simulation, as currently implemented, gives the expected richness *as if there were no temporal restriction to the style classes*. The limited temporal range of classes in a seriation intrinsically reduces class richness for a given level. Pan projectile points are highly time-specific, and thus simulation does not yield realistic estimates of expected richness. Therefore, while Pan projectile point richness is understandably much lower than the simulated means, it can be contrasted between levels as a relative measure. Note, however, that mixing a level with its temporal neighbors will increase the richness of a component in the same way that a longer-lasting component will have a greater richness of time-sensitive artifacts than a component of short duration.

This factor has not been considered in most applications of diversity analysis, and some unexpectedly rich sites or components may reflect lengthy or mixed time periods. In Pan the possibility of stratigraphic mixing must also be dealt with, as it could greatly affect richness.

### *(Un-)Evenness*

Options abound for computing the intuitively simple concept of evenness, i.e., the degree of balance in proportional representation of classes. Many have adopted the Shannon-Weaver information statistic **J** or **H/Hmax**, but this is strongly criticized by Bobrowsky and Ball (1989:6–7) as being overly dependent on sample size and species richness. They suggest the variance of the proportional abundance of classes as a simple and direct measure of evenness. I use the coefficient of variation (standard deviation/mean) of non-zero class frequencies as an equivalent yet standardized measure, which yields intuitively satisfactory evenness measurements. Because this is based on variance, however, larger values indicate greater *unevenness* among the classes. Even though this unevenness statistic is not directly related to richness, the two measures are not independent. As finite-sized assemblages become richer, their unevenness is intrinsically limited by the ever-smaller numbers of items distributed among the increasing subdivisions. Thus some attention must be paid to the predictably negative correlation between these two measures.

## Continuity

An interesting measure of stylistic variability is the degree to which stylistic class proportions change through time. This can be measured in two ways: proportion continuity and the continuity of categories or classes (Figure 5).

**Proportion (dis)continuity (Instability).** The degree to which a layer matches the proportions of the immediately previous level was calculated as the Euclidian distance between the two sets of proportions. A layer's continuity is relative to the preceding component, so the 17 layers in the Pan data set yield 16 values. This is a distance, so this measure is actually *dis*continuity or, as I prefer to term it, instability.

**Class continuity.** Continuity was also evaluated on a presence/absence basis using the Jaccard coefficient, which gives a high value to layers with many classes in common. Differences between components are thus based on the disappearance or appearance of classes, rather than proportional change as with Euclidian distance. Proportion and class continuities are clearly related, so instability and continuity can be expected to be highly (negatively) correlated. Each Pan level from 7 to 21 has a value rating its *similarity* with the immediately previous level.

## STRATEGY OF ANALYSIS

The basic data consist of a matrix of four stylistic measures (richness, unevenness, instability, and continuity), for four hierarchical typological classes (type group, type, variant, and affinity group), across Pan levels with adequate sample size. An important part of the analytical strategy is to compare the pattern of each measure across the resolution range of the classification. I will concentrate on robust data patterns, defined as trends that are replicated in multiple style classes or across different measures; or on single or multi-variable patterns that show clear, non-random patterning in time sequence. Such strength of patterning is necessary for making interpretations, because the numbers of style classes present are sometimes small (especially for type group), or the number of items within the style class may be quite small (especially for affinity group). Thus, consistent patterning involving independent measures or temporal neighbors increases the likelihood that non-random variability is being portrayed. I also use the convergence of pattern in the Pan measures with those of other data sets to help restrict the interpretations for explaining the strong patterns mentioned above, including:

1. The density (quantity/$m^2$) of lithic waste and stone tools (#/$m^2$) in the Pan site, as an approximate measure of occupation intensity (Figure 6).
2. Similar stylistic measures for the projectile points of Pax, over the same time range. Due to smaller sample size, only measures for type group and type are available for the Pax sample.
3. The match or mismatch between the specific types found in the two sites.

### Stratigraphic Mixing and Point Reuse

I previously mentioned that this analysis could be complicated by points shifted from their levels of origin by various processes. At Pan no process that could stratigraphically concentrate given point styles exists, other than rapid and permanent deposition of points in the strata of their manufacture. Thus temporal style distributions should only increase in range, either by intruding points to lower levels, or extruding them to superior levels. While no Pan projectile point classes exist throughout levels 7 to 22, most are found over a large portion of the sequence. This is probably caused by a combination of post-occupation dispersion processes and the long duration of the classes. For instance, there seems to be a notable reuse of early preceramic projectile points in late phases at Pax (Rick 1980:163). Pan shows a similar pattern, with scattered points of predominantly early preceramic origin showing up in much later strata. Scavenging of casually-encountered surface materials or stratigraphic mixing may be responsible for these patterns. These two factors have quite different implications, because the deep stratigraphic mixing necessary to produce this pattern would inevitably be disruptive to this analysis of style. While we did not observe stratigraphic disruption, it might have occurred in unexcavated parts of the site.

To explore patterns resulting from stratigraphic mixing, a simple simulation was programmed. It showed that upward or downward mixing of originally temporally restricted materials is unlikely to leave items separated by many levels from their place of origin. Thus single Pan projectile points far outside of the primary temporal distributions of their class were probably scavenged and reutilized, rather than simply extruded by stratigraphic mixture. Points were considered extrusive and were eliminated from this analysis if they were single specimens separated by two intervening levels from their main earlier distribution, or if two specimens of a class occurred in an upper level separated by four intervening vacant levels. In this way six points (0.4%) were deleted from the type group sequence, 12 (0.8%) from the type, 19 (1.3%) from the variant, and 52 (3.5%) from the affinity group tier.

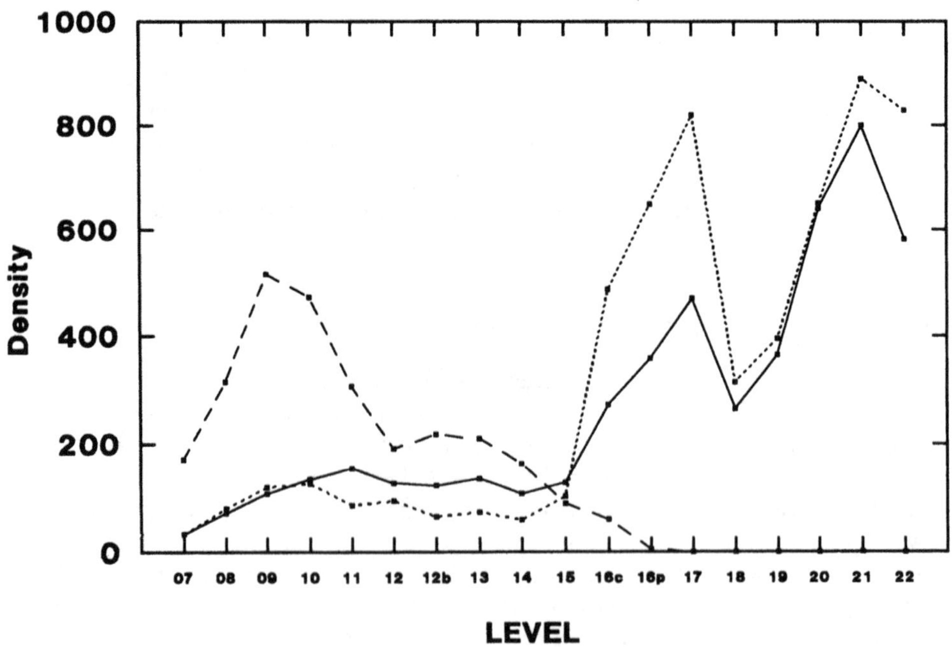


**Figure 6.** Density of cultural materials in later strata of Panaulauca Cave. Solid line is tools, and scale at left is quantity per $m^3$. Finely dashed line is lithic waste, and scale at left is in units of 10 g per $m^3$. Coarsely dashed line is ceramics, and scale is in units of 4 g per $m^3$.

## ANALYSIS OF PANAULAUCA STYLE PATTERNS

### Richness

Richness generally increases across time, although type group and, to a lesser degree, type deviate from the trend (Figure 7). It is not surprising that type group richness does not increase with time, since there are few type groups in this coarse class—a maximum of seven and a minimum of two. And, with their longer duration, type groups fit diversity expectations, having the richness expected of their sample size. The number of type groups is thus relatively invariant across time, which is in contrast with the finer classes. The discrepancy is the greatest in the earliest levels under consideration: the finer classes have very low richness, highly distinct from type group.

While this is true for all three finer classes, it is most pronounced for variants; there exist very few variants, considering the sample sizes. This suggests a high

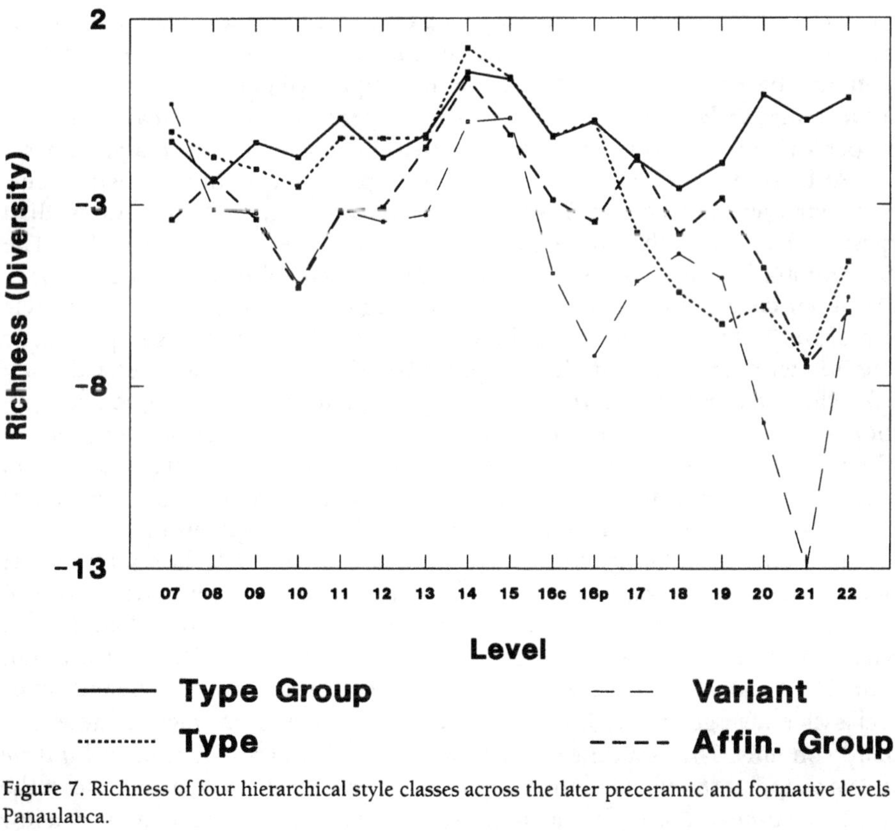


Figure 7. Richness of four hierarchical style classes across the later preceramic and formative levels of Panaulauca.

degree of standardization and production at the variant level: there were a few highly productive entities responsible for variants, while those producing affinity groups and types were slightly closer to diversity expectations. In general, the variant remains the poorest of the style classes, although it is strongly differentiated only in the late preceramic, and at other times is simply among the least rich classes.

Type plays a dynamic role across the richness sequence. While it is in the range of magnitude of the finer classes during most of the preceramic portion of the sequence (levels 22–16p), it distinctly tracks type group in the ceramic periods (levels 16c-7). Given the richness-decreasing effects of time sensitivity in the points, it is likely that ceramic period types and type groups are at least as diverse as expected from sample size. On the other hand, affinity group and variant, tracking each other through the ceramic period, are consistently poorer. In a sense, the richness graph has shifted from considering variant to be poorest, to considering both variant and affinity group to be poorest, and shifting type up to join type group. Either the stylistic typology is biased in the later periods, finely classifying what really are coarse differences, or there has been a change in the behavior of style. In general there is increased diversity of classes, each with few points, but affinity groups and variants hang on to vestiges of the high frequencies and poorness of class. This is especially the case around level 10, which is dominated by a distinctive triangular point complex (Figure 4H). The apparent shift to a finer scale of stylistic differentiation suggests that this increased production per style was at a smaller scale entity than in earlier, late preceramic times.

At three specific times in the site there is particularly sharp stylistic focus, i.e., when there are notably fewer and better represented stylistic units than expected. Two are in the preceramic, and one is in the mid-late formative. The most profound of these comes around level 21, corresponding to the end of a peak in material density in the site (Figure 7). It registers in three of the four style classes, but is seen most clearly at the variant level. The next, less sharp, stylistic focus is in terminal preceramic level 16p, again corresponding to a material density peak. This time only two style classes—variant, and to a lesser degree, affinity group—are involved, while type joins type group at a much higher level of richness. The last focus comes at level 10, and is again on a less marked material density peak. It clearly involves variant and affinity group, and may slightly deflect the type group and type curves, although not truly affecting their magnitude.

Lying between the stylistic foci are times of relatively rich classes with fewer points per class on average. The classes all converge on a common richness—no particular scale of entity is now playing a predominant role. Beyond that, the two stylistic recesses are rather different. The earlier one, around level 18, is not marked by a high average richness. The later stylistic recess is quite different, since all classes converge on a richness peak, by far the greatest average richness this record contains. This is a time when class richness is as great as expected *if there were no temporal specificity* to the point classes. This is, in effect, a collapse of the seriation: a time of overlap of the three predominant type groups, coming at a low point of material density just after the terminal preceramic peak. This low intensity occupation witnesses the presence of more stylistic entities than the site has ever seen, although none of them were deposited in any quantity in the site, nor were they present for very long. The restricted timing of this phenomenon shows that this is not likely to have been a product of stratigraphic mixing, since it would have affected preceding and following levels. In terms of distributions, it is sharply defined and radical.

## Unevenness

Layers with low richness have a greater potential than layers of high richness to be uneven for any given sample size (a sample of 10 items found in 10 categories cannot by definition have *any* unevenness). Pan levels of low richness also tend to be uneven, as seen by the downward-trending graphs through time (Figure 8),

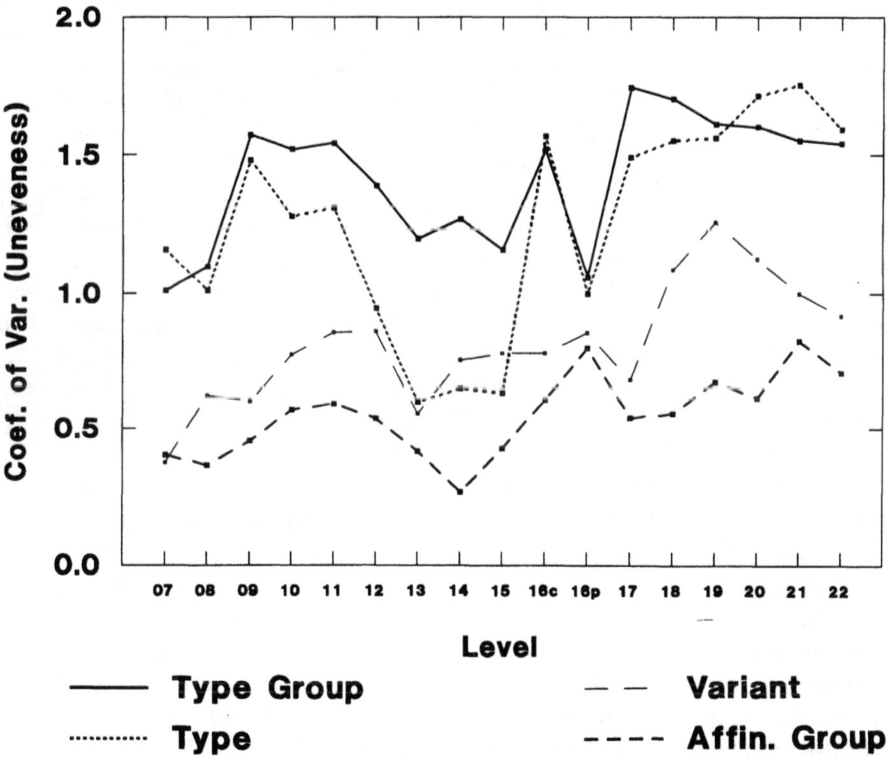


Figure 8. Unevenness of four hierarchical style classes across the later preceramic and formative levels of Panaulauca.

and by the strong negative correlation (r=-.810) between richness and unevenness. Thus the general trend, and to some degree the specific configuration of unevenness, is controlled by richness, but this measure still contains important information.

The low richness spike of level 21 is not matched by a peak of unevenness, but rather by a broad, if high, plateau. Variant, by far the poorest of the tiers at this time, is yet to reach maximum unevenness. Thus the relatively few variants in this level are not highly unbalanced in quantity, suggesting that a number of variant-scale entities are contributing to this major stylistic focus. The next class-poor style focus, level 16p, is matched by small variant and affinity group unevenness peaks, but corresponds to much steeper unevenness *dips* for type group and type. Level 10, the center of the late richness low, rides the middle of a long-duration unevenness high. These are strikingly variable patterns, and while confirming general tendencies to unevenness at times of richness, they show that much remains to be explained.

The style classes show a magnitude ranking that follows the classification hierarchy from most uneven (type group) to least uneven (affinity group). This is expected, as unevenness is limited by the ever smaller samples within class subdivisions. The graphs of the tiers roughly follow each other, and the overall decline of unevenness across time is the smoothest trend of any of the four variables recorded. The greatest exception to this is a strong trough in the type graph from levels 13 to 15—obviously a time of exceptional evenness within the existent types. Type is the richest of the tiers during this period, but not by nearly as much of a gap as the evenness distinction. The exceptionally great richness, and particularly evenness, seen in these levels is expressed most strongly at the type level, suggesting a major process at the scale of type. Alternatively, type unevenness is briefly very like that of lower tiers, as if the hierarchical relationships between the tiers are being compressed. A broader range of stylistic entities seems to be sharing the site. This coincides with other observations that the site at this time shifts away from being the base of low mobility hunters toward being an habitually used secondary site for more mobile groups, possibly of herders (Rick n.d.).

To get beyond the general tendency of unevenness to inversely track richness, the four classes were averaged within both measures, the averages regressed against each other, and the residuals examined. All three major style foci are represented by low points, indicating that these levels are more *even* than expected from the negative linear relationship of richness and unevenness. So, while unevenness is generally found at times of low richness, a small number of classes continues to be surprisingly well represented at times of the style foci. This unevenness limit is difficult to explain—it could be a background of displaced projectile points, movement or interaction creating a standard background noise of diversity, or multiple stylistic entities in Pan. In the case of variant, the style foci each have one variant that has double or near double the number of any other variant, so it seems that dominance of a single classification remains, but with a notable background of intermediate frequencies.

## Instability: Euclidian Distances

Although the graph of instability shows parallel tendencies among the classes, affinity group stands out as distinct from the rest (Figure 9). It follows a generally upward trend across time, without serious reversals, and is generally insensitive to the sharp peaks and valleys shown by the other tiers. At the finest stylistic resolution, layers became increasingly dissimilar with time.

At the other extreme, type group shows the closest proportions between consecutive levels—not surprising, because few type groups are involved, and these are reasonably constant in the later part of the overall sequence. Levels 15, 16c, and 16p, however, are exceptions: each of these levels represents a sharp break with the previous one, a trend reflected in all the style classes. The actual change

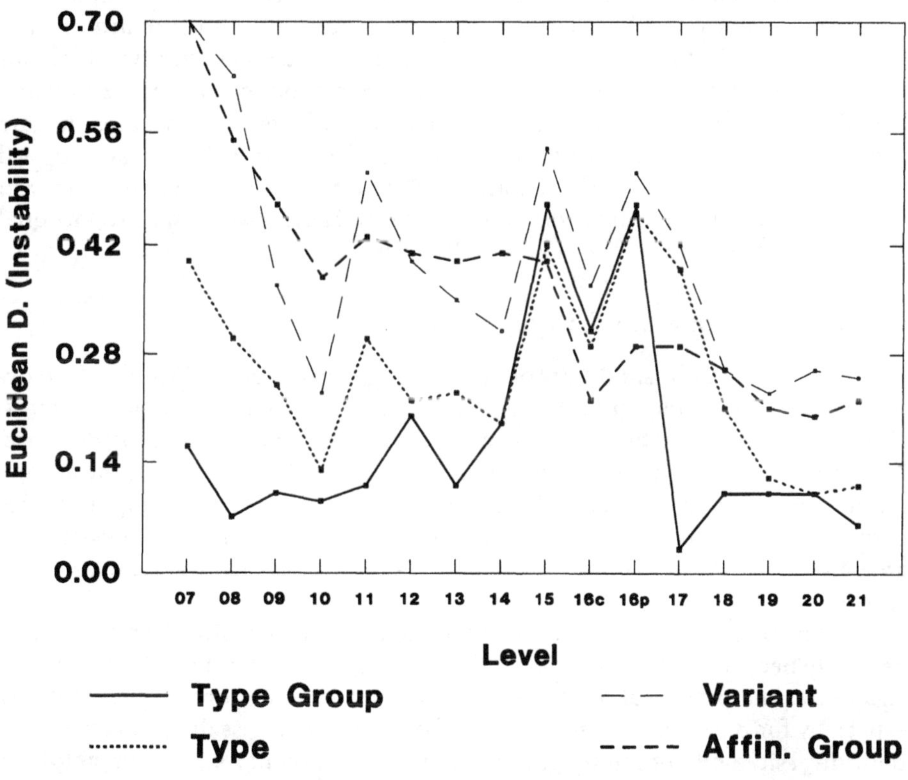


**Figure 9.** Instability of four hierarchical style classes across the later preceramic and formative levels of Panaulauca.

is a sharp shift from dominant proportions of type group 9 to type group 8 in level 16p, then a drop in type group 8 and a return of type group 9 in level 16c, and a final change to a lasting dominance of type group 8 for the remainder of the sequence.

Thus the preceramic-ceramic transition is a time of considerable instability. Long before the major type group transition occurs, there was a "destabilizing" period in which the intermediate-scale style classes have already begun to change. This may have been caused by social factors such as increased mobility, social fluidity, local economic restructuring, or trade. Whatever the reason, there seems to have been a major flux just before the shift in type group proportions. Interestingly, the reflection of these processes in the fine-scale affinity group proportions is very muted. Thus level 16p witnesses the sudden appearance of different large-scale stylistic entities from those that previously existed in Pan. The affinity group record suggests that at this time there is less displacement of some small-scale, perhaps local and long-term, entities.

With level 15, however, the record is utterly clear—now *all* four tiers in all four variables show the same directionality: sharply increased richness, sharply increased evenness, and a sharp break in continuity of both proportions and categories. To affect all these variables and stylistic levels so radically and simultaneously, the stylistic entities must have suffered a major change in identity and organization. While the proportions of functional tool classes at this transition are completely stable (Rick n.d.), the density of lithic materials in the site has dropped roughly an order of magnitude (Figure 6). The meaning of these phenomena will be considered later, but note that the instability chart also shows proportional continuity during the later, level 10, style focus.

## Continuity: Jaccard Similarity

The striking low continuity of the affinity groups (Figure 10) is undoubtedly due to the more common level mismatches in these highly subdivided groups. Only in level 16p does the gap between the higher tiers and affinity group decrease. At this time of strongly decreased class continuity for the higher tiers, the affinity group value remains constant. Thus, while new forms are appearing, old ones continue as well, particularly in the affinity groups. Of eight types present, five are new for the site and none disappear; thus there is no richness depression for type at this time.

The other three tiers are very close to each other in magnitude, and all classes generally decrease across time—one of the strongest long-term trends in the data. Again, the sharpest break is at level 16p, when continuity drops quite strongly, especially for type. The 16p style focus is based on a richness drop in variant, but both diversity and continuity graphs show that 16p is not a focus arising from continuity. Rather, it is, to a large degree, a new assemblage, perhaps established at the variant level by a relatively small number of new entities. Affinity group

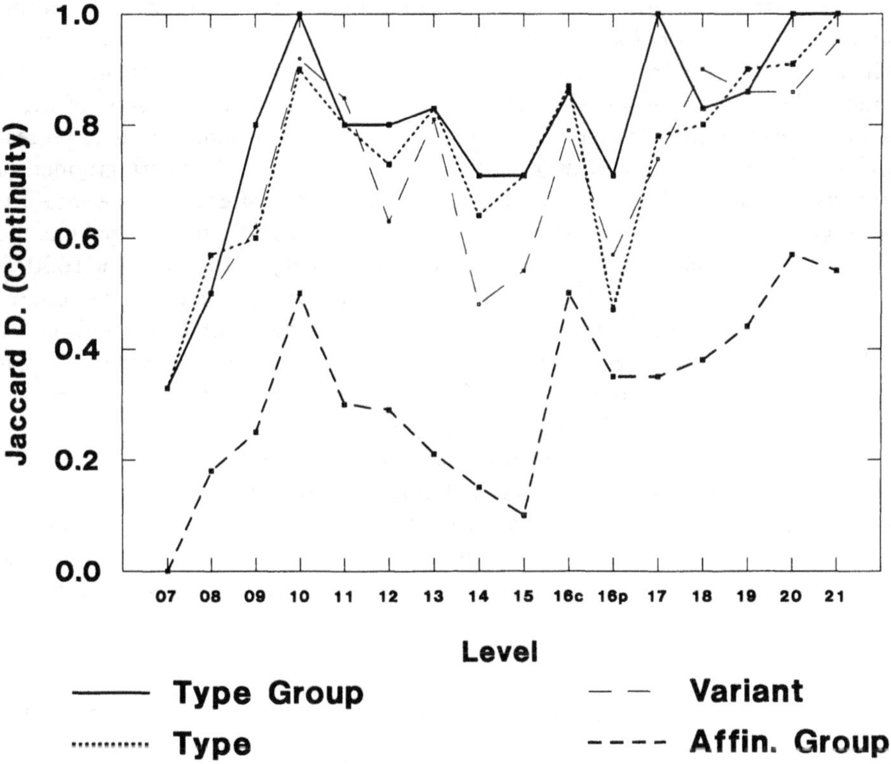


**Figure 10.** Continuity of four hierarchical style classes across the later preceramic and formative levels of Panaulauca.

continuity, like instability, does not show a sharp change at this time, but it has previously dropped, perhaps anticipating the trend seen in the other classes.

Uniformly high levels of class continuity are found in stratigraphic levels 20–21, 16c, and 10, the first and last clearly conforming to the first and last stylistic foci respectively. Thus these two foci are clearly part of long-term trends, which reach their pinnacles of correlated poorness, unevenness, stability, and continuity, and then fade away. The level 16c class continuity peak conforms to an instability trough and an unevenness peak for type and type group. While 16p has low variant richness, arguing for a style focus, the continuity break and mixed record of unevenness are at variance with this interpretation. Thus levels 16p and 16c are a complex and different phenomenon, a radical stylistic transition. Level 16c resembles 16p more than 16p-17 or 15–16c. Level 16p sees the introduction of a new series of entities, but 16c returns relatively close to level 17 proportions and makeup. Then level 15 returns to the newer pattern of 16p, after which there is considerably greater continuity.

The last item of analysis for projectile point style is the residual information found after accounting for the strong relationship between class-averaged instability and continuity (r=-.88). Regression assumptions of normal distribution and homoscedasticity (constant variance) are confirmed in regression plots of these variables, leaving little evidence of autocorrelation or other limitations of regression in this instance. Surprisingly, when the regression residuals are graphed, a fairly regular sinusoidal time series is apparent, especially after moderate smoothing (Figure 11). This cycle of alternation between periods of proportional change and periods of broken class continuity has a periodicity in the range of 1000 to 1600 years, but too few cycles are present to argue for a particular wave length. This graph is provocative, hinting at a regularity and pattern unlike those observed in the remaining data. This is a very small residual from a highly correlated pair of variables with their major axes of variability annulled by the regression. Instability and continuity both show a long-term pattern of reducing similarity, which from about level 16p to level 10 has a plateau, if not a slight reversal. The residual curve of Figure 11 shows that the similarity between levels is most strongly influenced by class continuity at the peaks of the curve, and primarily reflects instability of proportions in the troughs. This regular alternation could relate to a pattern of stylistic change underlying more obvious long-term trends.

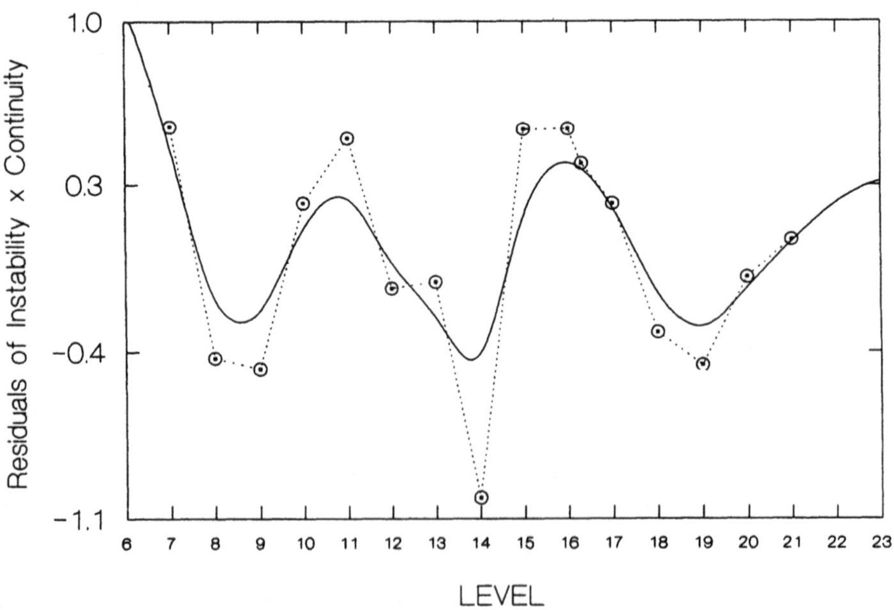


**Figure 11.** Residual of instability X continuity regression, by level. Dotted line follows actual data, solid line is distance weighted least squares smoothing.

Periods emphasizing proportional change may be thought of as somewhat more gradual, and those emphasizing class change suggest times of radical transitions.

This cycling of the similarity residuals appears to be related to the three stylistic foci recognized in richness and other graphs. The location of these foci on the residual cycles is complicated by the fact that level 21, one of the foci, is our earliest level with similarity data. If at level 21 the residual curve is near its peak, then the stylistic foci all occur near the peaks of maximum categorical (as opposed to proportional) change. This pattern is logical in that poorness comes immediately after the loss of entities; proportional change is then associated more with times of increasing richness, or stylistic multiplicity. The peaks in occupation intensity come at times of greatest stylistic coherency (especially at the finest tiers of style definition in the style foci of levels 10 and 21), following immediately upon times of relatively abrupt change in style.

## Relationships between Panaulauca and Pachamachay

Only *type* richness and the temporal match between specific types will be compared between the two sites, because finer tiers are not available for Pax. The temporal correlation of levels between the two sites is a complex issue, and an exact correspondence is not likely. For purposes of simple graphing, however, Pan levels 7 to 22 can be linked on a one-to-one basis with Pax levels 3 to 19. Radiocarbon dates support this matchup, and the general contemporaneity of these stratigraphic sequences is sufficient for examining broad patterns.

The richness curves from the two sites are similar in pattern and magnitude, with several exceptions (Figure 12). Pan shows a slightly richer record, especially in the early ceramic levels 13 to 15. Both sites have their lowest richness scores in the late preceramic period, and generally increasing richness at the ceramic horizon, with constant, if elevated, richness at the end of the sequences. The late style focus of Pan level 10 is sharply apparent on the Pax graph, even though it is primarily registered in finer-scale classes in Pan. This focus at both sites consists of the same single type and its subclasses, so the same phenomenon is being seen.

To explore the relationship between the types found in the two sites, type frequencies from each level were assembled as records in a single data base, and multidimensional scaling was performed on the proportion-based Euclidean distances between pairs of levels. This provides a map of the type similarity relationships between the levels of the sites (Figure 13). The map shows three different regions: a group of late levels from both sites combined, and two groups of earlier levels in which the sites are segregated. These early levels are highly distinct between sites, but by the end of the sequence the sites' types are quite similar. In fact, Pan levels 12 to 15 and 16p are reasonably close to contemporaneous levels at Pax, notably more so than levels 16c and 17 to 22. Thus there is an increasing stylistic resemblance between the two sites across time, with the exception of a *decrease* in similarity from 16p to 16c. The late preceramic stylistic

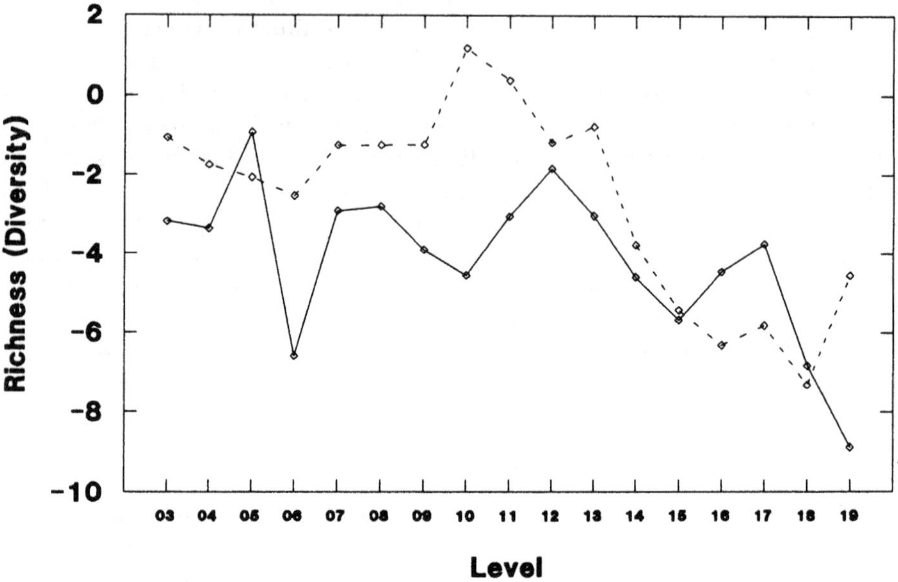


Figure 12. Richness of the Panaulauca and Pachamachay type classes. Lower scale shows Panaulauca levels corresponding to Pachamachay levels 3-19.

distinction between these sites is quite great, with many types and even type groups not shared between the two. This indicates that the late preceramic of the area is not necessarily characterized by few stylistic entities, but rather by entities that differ between sites. The increasing richness of types across time within sites, however, is accompanied by increased sharing of stylistic entities between sites.

Although its significance is not totally clear in the reduced multidimensional space of Figure 13, it appears that the changes in type frequencies at Pax follow a simple linear tendency through time in this chart. Pan, however, shows a strong inflected pattern, which ends in a parallel linear tendency congruent to Pax. Thus the approximation of these two sets of levels arguably shows more change in Pan than in Pax; in process, Pan becomes like Pax, rather than the other way around.

## SUMMARY AND INTERPRETATION

Measures of projectile point style have proved a rich, if complicated, source of information. Many of the robust data patterns are unlikely to have been caused by chance, and must relate to influential processes of some kind. Stylistic analysis should reveal something about the relations, numbers, and kinds of groups in existence. Unfortunately, stylistic variability is subject to the homogenizing or

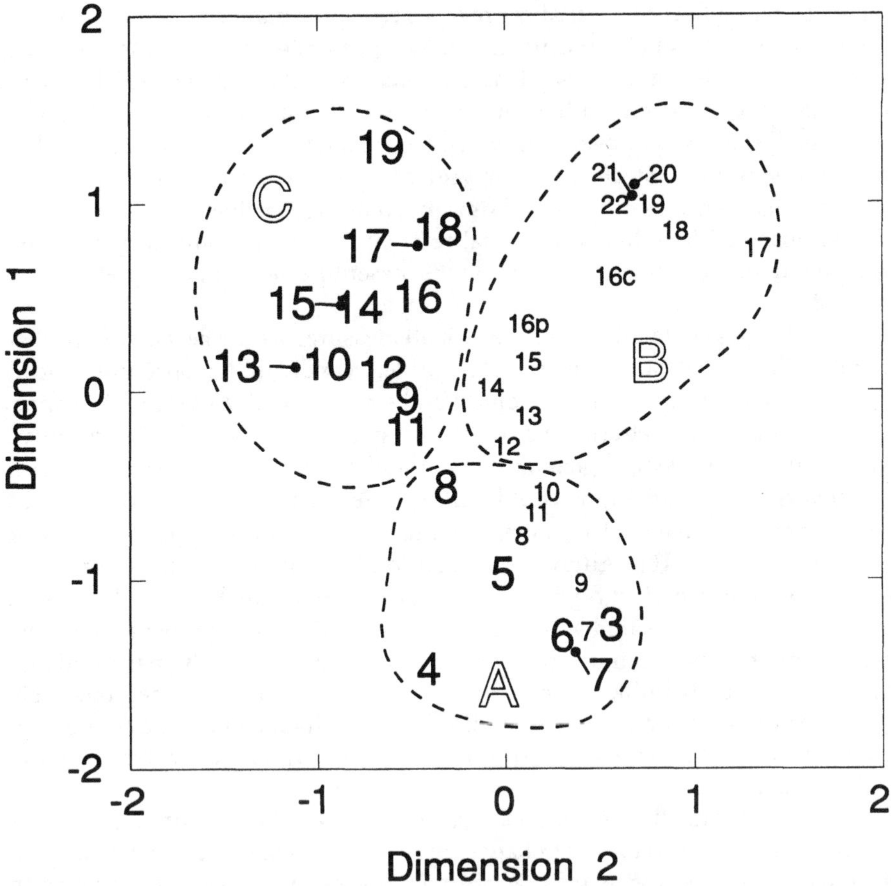


**Figure 13.** Plot of dimension map from two-dimension multidimensional scaling of Panaulauca levels 7–22 (small numbers) and Pachamachay levels 3–19 (large numbers). Ellipses surround A) late levels from both sites with great similarity, B) earlier segregated levels from Panaulauca, and C) earlier segregated levels from Pachamachay.

sorting processes, cultural and natural, that affect other types of data. Even if we were confident that our stylistic units reflected social differences, we would have no obvious way of understanding the degree of social differentiation, nor the scale of the groups being monitored.

Some time ago I suggested that Pax projectile point types might correspond to local co-residential stylistic production units, roughly equivalent to bands (Rick 1980:314–316). The addition of two more tiers of finer stylistic differentiation in projectile points appropriately complicates the issue. From additional information accumulated since, I recognize that *some* stylistic forms seen at Pax have too wide a distribution to reflect the spatial range of a single preceramic band. I now prefer to think that the tiers of the stylistic hierarchy represent stylistic production units of varying scale. By units of stylistic production, I do not necessarily mean individuals or groups, but rather whatever unit, delimited in time and space, within which there was stylistic similarity. Direct interpretations of exactly what style units were must await a tighter grid of local, excavated sites capable of mapping stylistic entities in space and time, but it does seem likely that the stylistic variation observed here has some social referent. Even with only two sites, the data demand an interpretation, which I will present in the form of a preliminary history of stylistic units.

The changes centered on level 16p for all measures are the best case of robust patterning in these data. But while it is an ideal candidate for observing social process, alternative explanations should also be entertained. At present I cannot, however, support any alternatives to the idea of strong social discontinuity. Stratigraphic mixing would generally *dilute,* rather than create, such a pattern. While inversion of stratigraphy could indeed have created the return of previous stylistic forms seen in level 16c, it could not have created the original sharp break to new forms seen in 16p. Although a scenario of discontinuities in occupation, combined with near-perfect flipping of strata, could generate this record, it seems extremely unlikely compared with explanations based on social process. That the transition might have been a product of functional change in the nature of site occupation is equally unlikely, because occupation intensity remained relatively high during the transition, and the stylistic classes which come to dominate Pan during subsequent times were those present in Pax during both intensive and more sporadic occupation.

In summarizing the stylistic analysis, I will use a frankly interpretive perspective, explaining most variability as changes in style-producing units. Archaeological style units can start or stop, expand or contract, suggesting an associated process in the production unit. Disappearance of a style from the site must mean that the unit has left, stopped existing, or changed in style. Initial appearance of a style can mean the complementary activities of a unit's arrival, creation of the style, or style adoption. Increase or decrease in a style's proportion could mean the shrinkage or growth of the production unit, a change in amount of time the unit functioned or was present, or a change in the unit's production rate. All of these imply that production and deposition occur in the site where the unit existed.

As for the life history of the points—who made them and where—I rely on two arguments. The first is that in Pan, raw materials are predominantly local, especially in the time range considered. Many of these materials are not found at Pax, so the trading or movement of points between these nearby sites was not extensive. Second, much of the time Pan and Pax are quite different on the various stylistic indicators, something not likely to occur if large numbers of points were moving between these two sites. Thus assumptions of in-site origin are not unreasonable for the large majority of the points. However, available evidence does not rule out some intrasite exchange of points.

The long-term trends in Pan and Pax style have a clear origin. The increase in richness and instability and the decrease in unevenness and continuity that generally characterize Pan from late preceramic times on are products of the decreasing temporal periods in which style-producing entities were present in this site. The late preceramic was a time of high occupation intensity in all known preceramic sites from the Junin puna of central Peru. At this time stylistic entities were few, of long duration, and of unbalanced proportions in Panaulauca; and, to the degree to which it has been explored here, Pax. The extremely dense late preceramic deposits in these cave sites were produced by relatively fewer entities than was the case in later periods, and these entities had greater continuity in time. Given the strong stylistic differentiation between adjacent residential bases in the late preceramic, these phenomena argue for a high degree of style localization and a low degree of movement of the entities tied to specific styles.

The most parsimonious explanation is that local style production units, themselves probably constituting elements of social groups, were relatively immobile and constant in their occupation of specific sites. This pattern has implications for the nature of the stylistic divisions under analysis. Given the overall dense occupation of the Junin plateau at this time and the relatively close spacing of stylistically distinct, contemporaneously occupied residential bases (Figure 1), stylistic distinctions are unlikely to have been caused by a lack of interaction of style-producing entities. Even if they were relatively sedentary and economically independent, the residents of adjacent bases must have been in contact with their neighbors, and aware of the stylistic characteristics of their projectile points. Thus Junin stylistic differentiation seems at least partially reactive to, rather than simply a passive product of, communication and contact. It seems possible that late preceramic peoples were consciously and intentionally producing style in their tools, although this extension of the argument is tentative, at best.

The above argument may be seen as style in the service of an ulterior argument—that is, my prior position that small-group, non-complex Junin hunter-gatherers had achieved low mobility in the preceramic period (Rick 1980). While I believe the style record supports effectively sedentary occupations in the late preceramic, the patterns of style go far beyond such simplistic cover statements. The strongly patterned details of the Panaulauca sequence argue for a number of processes during the transition away from low richness and high continuity.

Notable are the three stylistic foci—times when an unexpectedly small number of stylistic entities are present in the site. The method of examining a hierarchical typology has shown that these foci are sharply defined only at a relatively fine scale in the classification, and that the scale of definition may change between foci. The variant level shows the greatest variability in richness, as well as the lowest values. Variant probably is monitoring the greatest flux in related entities, and if there ever was a local residential social unit tightly related to projectile point form, it was probably stylistically differentiated at about the variant level. The operative stylistic unit may have been somewhat larger-scale in the late preceramic level 21 focus, and become increasingly smaller in scale through the level 10 focus, at which time the smallest scale class, affinity group, seems to have been as responsible as variant for the existence of the focus. Our typological instrument is much too uncalibrated to be confident in this conclusion, but it hints that as the duration and intensity of site occupation decreased into the ceramic periods, the stylistic foci were not only less pronounced, but were produced by smaller entities.

The foci themselves, and particularly the cyclical pattern in which foci originate at times of stylistic discontinuity and fade during periods of proportional change, demand explanation. A simple interpretation is that they represent short, intense occupations of the site, founded by carriers of different style. Given the continuous stratigraphic record of dense remains in which even light-density levels are very artifact-rich, I believe that longer-term processes are involved. My preferred interpretation is that these cycles represent long-term processes of establishment, continuance, and rejuvenation of stylistic entities. In effect, a local entity is founded, diversifies, disperses, and is eventually replaced by another focused entity—often one related to, and perhaps descended from, earlier entities. I am not convinced that all such cycles reflect the actions of individuals or groups, but they may represent long-term social processes relating to identity. The observed cyclicality of proportional and categorical change were not likely caused by chance or by some undetected site formation process, but both possibilities should be further explored.

The sharply demarcated unevenness, instability, and continuity transitions that accompany the level 16p style focus, and the sharp changes in specific type groups, need special attention. The level 16p to 15 period is one of sharp changes, suggesting that radical social processes were occurring. The stylistic pattern shows a dramatic first appearance of some types previously present in neighboring Pax, the subsequent return of the distinctive Pan type group 9 (which was almost totally lacking at Pax), and its eventual (post-16c) near-disappearance. It is tempting to argue that the long-term residents of Pan were replaced by other local social units after destabilization of some sort. Given the demise of the strongest stylistic focus and the incursion of stylistic forms known earlier elsewhere, some change in the occupying population seems likely. There is, in fact, minimal stylistic continuity across the 16p-15 watershed.

The 16p richness depression is caused by the presence of a small number of finer-level entities *which are drawn from a larger number of larger-scale entities.* There is increasing unevenness among affinity group and variant, reflecting the dominance of a relatively few small-scale entities, while the type and type group classes become decidedly even. At the same time, elevated instability and decreased continuity tell us that new entities are present (particularly in the variant and larger-scale classes), although there is significant continuity at the affinity group scale. Large-scale entities display discontinuity, while some smaller-scale entities remain. The fact that the focus is composed of few small-scale units drawn from multiple large-scale ones suggests a hybrid character of the entities using Pan at this time.

Given the radically decreasing density of materials in the site at this time, however, it would be presumptuous to suggest a simple displacement of an original population by another group. In fact, it seems likely that whoever continued to live in Pan had a very different occupation pattern, with greater mobility and perhaps less stability in group membership. This is suggested by the increased richness and evenness of most tiers of the style record, and by the relatively great inter-level differences in proportions of the lower tiers of style. Notably, however, the coarsest tiers of style returned to relatively similar level-to-level proportions, suggesting larger-scale social stability. The easiest explanation is that more groups spent less time in Pan, and probably none occupied the site on a very continuous basis during ceramic times.

This pattern may have been partially broken by the style focus of level 10, when there was a brief period of continuity, and even a moderate increase in artifact densities. This may not have been a simple return to later preceramic conditions, however. The type which dominates Pan is common to Pax, and may represent a very different level of entity—it may be widespread throughout Junin, and maybe even in other areas of Peru. Thus, it may well represent a systematic utilization of the cave by part of a much larger-scale entity, as I originally inferred (Rick 1980:325–326).

A final pattern of importance is what I term *sequencing*, which refers to the evidence of a process that starts at one level of the typological tiers, and progressively affects others across time. There are two notable cases of this, one being the earlier instability in smaller-scale entities prior to the large-scale ruptures of levels 15–16p. This suggests a social process in which the duration of occupation by local entities decreased, anticipating a major social change.

A second case is the elevated unevenness at the variant and affinity group tiers in level 16p, followed by marked unevenness at the type group and type tiers in level 16c. In fact, unevenness trends of type and type group were uniformly anticipated by those of variant and affinity group, and within those pairs there is a tendency for the finer level to show the trend sooner. This is perhaps the clearest example of time sequencing, but even in this case, there is a possibility that the pattern is due to chance. Taken at face value, this pattern suggests that dominance of stylistic entities starts at the smallest scale and graduates to larger entities through time. This could reflect the growth and preferential survival of certain strains of a style cohort. Could it be that stylistic entities become increasingly coherent, starting from small scale distinctions and gradually expanding into coarser levels of differentiation, with lower levels gradually diversifying again? Large traditions in lithic industries would thus be derived from particularly successful small trends at a finer scale. This pattern is enticing, because it is one that is reasonably expected from a variety of biological and social processes applicable to small human groups, ranging from differential group survival and reproduction through establishment of dominant style by preferential adoption.

## CONCLUSION

This analysis has ranged from the observation of strong patterns in typological entities to speculative interpretations of possible social processes behind the patterns. These patterns and explanations may not all be valid, but I hope to have shown that time, typological scale, and space can be used creatively to distinguish long-term patterns. This approach to stylistic analysis is particularly suited for the strengths and weaknesses of archaeology.

Stylistic units in the archaeological record should be defined at variable scales, since we cannot easily know the scale of the prehistoric entities that might be stylistically detectable. Long-term perspectives can help find patterns and assign meanings to them, a process that can help us understand the long-observed temporal patterning of styled things. Battleship curves are one such acknowledged pattern, but one that has relatively little explanatory potential. Analysis of long-term relationships between multiple levels of scale of stylistic classes for multiple measures like diversity or continuity can refine our knowledge of stylistic patterning. Regularities in patterning at the local or regional level could lead to important interpretations of social processes behind the archaeological observations. This latter step, however, will undoubtedly be a long-term process itself, demanding extensive negotiation among those who aspire to the challenge.

## ACKNOWLEDGMENTS

This paper has benefited from a number of rounds of sage advice. The research and early versions were birthed during a fellowship I held at Clare Hall in Cambridge, England. A reduced version was presented at the 1992 meetings of the Institute of Andean Studies in Berkeley, and informally presented versions have been inflicted on numerous colleagues. Students in a number of Stanford classes have puzzled over the included intricacies. Most importantly, the extensive discussions in and subsequent to the Tulsa Lithics Conference have been very influential. I will not single out any of the many individuals who have been helpful, and, most importantly, I do not mean to impute that the contents of this work have passed their final muster. I do thank them all, and acknowledge essential funding for the field work from the National Science Foundation and Stanford University.

## REFERENCES

Binford, L. R. 1989. Styles of Style. *Journal of Anthropological Archaeology* 8:51–67.

Bobrowsky, P. T., and B. F. Ball. 1989. The Theory and Mechanics of Ecological Diversity in Archaeology. In *Quantifying Diversity in Archaeology*, edited by R. D. Leonard and G. T. Jones, pp. 4–12. Cambridge University Press, New York.

Bocek, B. R., and J. W. Rick. 1984. La época precerámica en la puna de Junín: Investigaciones en la zona de Panaulauca. *Chungara* 13:109–128.

Clark, G. A. 1989. Romancing the Stones: Biases, Style and Lithics at La Riera. In *Alternative Approaches to Lithic Analysis*, edited by D. O. Henry and G. H. Odell, pp. 27–50. Archaeological Papers of the American Anthropological Association No. 1.

Close, A. E. 1989. Identifying Style in Stone Artefacts: A Case Study from the Nile Valley. In *Alternative Approaches to Lithic Analysis*, edited by D. O. Henry and G. H. Odell, pp. 3–26. Archaeological Papers of the American Anthropological Association No.1.

Kintigh, K. W. 1984. Measuring Archaeological Diversity by Comparison with Simulated Assemblages. *American Antiquity* 49:31–63.

Kintigh, K. W. 1988. *The Archaeologist's Analytical Toolkit*. Published by the author, Tempe.

Kintigh, K. W. 1989. Sample Size, Significance, and Measures of Diversity. In *Quantifying Diversity in Archaeology*, edited by R. D. Leonard and G. T. Jones, pp. 25–36. Cambridge University Press, New York.

Lavallée, D., M. Julien, J. Wheeler, and C. Karlin. 1985. *Telarmachay: Chasseurs et Pasteurs Préhistoriques des Andes - I*. Institut Français d'Etudes Andines, Paris.

Moore, K. M. 1988. Hunting and Herding Economies on the Junín Puna. In *Economic Prehistory of the Central Andes*, edited by E. S. Wing and J. C. Wheeler, pp. 154–166. B.A.R. International Series 427, Oxford.

Moore, K. M. 1989. *Hunting and the Origins of Herding in Peru*. Ph.D. dissertation, Department of Anthropology, University of Michigan, Ann Arbor.

Pearsall, D. M. 1988. Interpreting the Meaning of Macroremain Abundance: The Impact of Source and Context. In *Current Paleoethnobotany: Analytical Methods and Cultural Interpretations of Archaeological Plant Remains*, edited by C. A. Hastorf and V. S. Popper, pp. 97–118. University of Chicago Press, Chicago.

Pearsall, D. M. 1989. Adaptation of Prehistoric Hunter-gatherers to the High Andes: The Changing Role of Plant Resources. In *Foraging and Farming: The Evolution of Plant Exploitation*, edited by D. R. Harris and G. C. Hillman, pp. 318–332. Unwin Hyman, London.

Pearsall, D. M. n.d. Prehistoric Economy and Subsistence. In *Hunting and Herding in the High Altitude Tropics: The Early Prehistory of the Central Peruvian Andes*, edited by J. W. Rick, pp. 399–448. Unpublished manuscript, Stanford University.

Rick, J. W. 1980. *Prehistoric Hunters of the High Andes*. Academic Press, New York.

Rick, J. W. n.d. *Hunting and Herding in the High Altitude Tropics: The Early Prehistory of the Central Peruvian Andes*. Unpublished manuscript, Stanford University.

Sackett, J. R. 1990. Style and Ethnicity in Archaeology: The Case for Isochrestism. In *The Uses of Style in Archaeology*, edited by M. W. Conkey and C. A. Hastorf, pp. 105–112. Cambridge University Press, New York.

Sinopoli, C. M. 1991. Style in Arrows: A Study of an Ethnographic Collection from the Western United States. *Michigan Discussions in Anthropology* 10:63–87.

Vaughn, P. 1985. Analyse Tracéologique. In *Telarmachay: Chasseurs et Pasteurs Préhistoriques des Andes - I*, edited by D. Lavallée, M. Julien, J. Wheeler, and C. Karlin, pp.401–424. Institut Français d'Etudes Andines, Paris.

Whallon, R. E. 1972. A New Approach to Pottery Typology. *American Antiquity* 37:13–33.

Wiessner, P. 1983. Style and Social Information in Kalahari San Projectile Points. *American Antiquity* 49:253–276.

Wiessner, P. 1990. Is There a Unity to Style? In *The Uses of Style in Archaeology*, edited by M. W. Conkey and C. A. Hastorf, pp. 105–112. Cambridge University Press, New York.

Wobst, H. M. 1978. The Archaeo-ethnology of Hunter-gatherers or the Tyranny of the Ethnographic Record in Archaeology. *American Antiquity* 43:303–309.

*Chapter* **10**

# *Innovation and Selection in Prehistory*

## A Case Study from the American Bottom

MICHAEL J. SHOTT


## ABSTRACT

In the American midcontinent, first millenium A.D. point types show a single continuum of metric variation, not a time-series of distinct forms. This view changes our understanding of the technological transition from dart to arrow, which may have begun earlier and unfolded over a longer time than commonly believed, and involved for some time simultaneous use of both weapon types. Reduction in point size may have persisted even after the arrow was adopted, possibly in response to broader subsistence changes. The Diet Breadth model suggests one explanation for the continuous decline in arrow-point size. Whatever its cause, this time-dependent trend can improve our ability to measure time as the continuum it is.


## INTRODUCTION

This paper addresses two related questions on the Woodland and later prehistory of the American Bottom region of Illinois: 1) the timing and nature of

**MICHAEL J. SHOTT** • Department of Sociology and Anthropology, University of Northern Iowa, Cedar Falls, Iowa 50614–0513.

*Stone Tools: Theoretical Insights into Human Prehistory*, edited by George H. Odell. Plenum Press, New York, 1996.

the transition from dart to arrow as weapons delivery systems; and 2) the nature and explanation of time-dependent variation in the metric attributes of first dart, and then arrow, points[1] in that region. The paper briefly addresses a related issue, the chronometric potential of continuous variation in points and, by implication, other artifact classes. To raise these issues, it first considers in general the kinds and causes of variation or change in material culture.

## CONTINUOUS AND EPISODIC CHANGE IN MATERIAL CULTURE

Archaeologists of all theoretical persuasions use patterns in the changing size and form of artifacts to impose order on the material record. Projectile points are among the most popular of artifact classes used in this way, as diagnostic time-markers. This much is commonplace, an element of practice internalized by all archaeologists. Nevertheless, it is worthwhile to emphasize an apparently self-evident principle established by the use of points and other artifacts as time-markers, i.,e., change in artifact size and form occurs as a function of time.

Conventional usage assumes, for all practical purposes, that variation in material culture is episodic, characterized by long periods of stasis punctuated by occasional furies of rapid change. Specific types are identified with chronological phases, and thereby acquire status diagnostic of those phases. Obviously, such use of size and form variation yields a view of time as a series of discrete, successive phases.

Nowhere is this view more common nor more problematic than in lithic studies. Although we have always recognized metric and formal variation in artifact classes, we have more often taken it at face value than we have sought to explain it. To be sure, some changes are explained on functional grounds, such as one subject of this study: the presumed shift in eastern North America from atlatl and dart to bow and arrow some time during the first millennium A.D. In most cases, however, the shifting forms of points are assumed to reflect vagaries of style or the replacement of one culture by another. In short, even now we tend to regard points largely as markers of cultural norms.

Normative views of cultures stress the modal properties of their components, including their artifacts. To establish our time-space frameworks, we find it useful to define the markers of successive phases in ways that emphasize their differences (Dunnell 1992:218). It is a poor point diagnostic of Phase A that is difficult to distinguish from Phase B's marker. This practice colors our perception of the past; it encourages us to maximize the variation between the properties of what we consider successive phases and their cultures, and to minimize the variation within them (Frankel 1988). That is, we view the past as a succession of discrete phases whose characteristics are given by ideal artifact types and whose boundaries are defined by major, episodic change. Obviously, how we perceive the past influences how we explain it.

This view of culture change can be needlessly self-limiting. Type definitions, for instance, can be notoriously problematic. Archaeologists often disagree on what constitutes a particular type, as well as on the assignment of specimens to types. Your Madison may be my Hamilton, your Clovis my Folsom. Archaeologists invest a significant amount of time and effort agonizing over such matters, effort that might be better spent on more productive activities. We regard the imprecision of type definitions and the difficulties of assignment as disagreeable consequences of the need to impose chronological order on the archaeological record. We may even view them as products of lax standards of prehistoric artesanry. Less often do we regard them as a property requiring or deserving explanation, still more rarely as a source of exceptionally fine chronological resolution of change in the past.

## Some Perceived Causes of Variation

A great deal of variation in artifact size and form is relatively subtle, however; more a case of continuous or gradual change than of episodic revolutions in style or function. This property of material culture variation is more faithful to our understanding of time as a continuum, but it is difficult to reconcile with our prevailing treatment of time in the archaeological record.

Even when continuous variation is observed, it is often explained by reference to one of several factors: style in its various guises; individual variation in motor skills, ability and other properties; laxness or low standards of conformity to mental ideals or norms; drift from those norms as a function of time or space; intrinsic mechanical and other properties of raw material; and measurement error as a fundamentally extraneous source of variation. All of these factors are important in some cases, and all but raw material and measurement error are of legitimate anthropological interest.

## Performance Requirements as Causes of Variation

There exist, however, other equally important and anthropologically significant factors that can produce continuous variation in points and other artifacts. These are manufacturing and functional or performance criteria. With respect to projectile points, these include ease, replicability, and accuracy of manufacture (Guthrie 1983; Knecht 1991); ease of mounting (Fischer 1989); ballistic performance requirements; penetration; changes in prey size (Churchill 1993, Shott 1990:28); and general design criteria relating to resistance to breakage from mechanical stress.

Among ballistic performance criteria, Christenson (1986a:115, 1986b) proposed three especially important ones for projectiles: accuracy, range, and thrust (see also Fischer 1989 and Schiffer 1979). Schiffer (1979:363) identified additional performance criteria such as firing rate, ease of use in dense cover (see also Hall 1980:438) and in ambush, and risk of accidental injury in use. Fischer

(1989:37) and Guthrie (1983) listed penetration and wound potential as further important performance criteria.

Guthrie argued that thickness among metric attributes most strongly determines penetration and that, obviously, the relationship is inverse. But very thin points may be too fragile to be effective (Guthrie 1983:282–285). By Guthrie's reasoning, the narrow margin between the thinness desired for penetration and the thickness required for structural integrity of the point can produce a high rate of point failure in either manufacture or use. As a corollary of penetration requirements, therefore, point design should facilitate rapid and economical manufacture; that is, points must be produced reliably and cheaply (Guthrie 1983:290; Schiffer and Skibo 1987:600). Fischer (1989:38) regarded the time-series of Mesolithic and Neolithic forms he documented "in terms of continuous adjustments aiming at the optimum interplay between" the performance requirements of penetration, cutting ability, and accuracy.

Guthrie (1983:289–290) also emphasized the importance of haft design and hafting to optimal projectile functioning. Thus, performance criteria can impose structural constraints on haft dimensions, perhaps stronger constraints than those operating upon blade dimensions. At any rate, Knecht's (1991:136) study of Upper Paleolithic bone points suggests that blade length varied more widely, and therefore more freely, than did haft dimensions.

It is a truism (Guthrie 1983:273) that hunting practices influence point size and form, but it is one that archaeologists still deride at times (e.g., Yerkes 1992:69). Such attitudes exact a steep price: they risk the neglect of an important source of knowledge about the nature of prehistoric hunting practices, and the nature and rate of change in them through time. If change in hunting practices is continuous in nature, it may produce change of a similar nature in point dimensions over time, change that reflects functional or performance requirements, not style or drift.

Thus subtle and continuous variation in point size and form can document comparable variation in manufacturing, and especially performance, criteria. Viewed in this way, points do not merely stand as gross markers of discrete chronological phases; they register, in their temporal patterns of change, the continuously changing nature and magnitude of performance requirements. Variation in points can also register the processes by which economic changes or technological innovations are accommodated to existing traditions and practices. This possibility exploits the diachronic quality of archaeological data and interpretations based on them. Studies of the adoption or diffusion of innovations are hampered by the relatively short time frame accorded researchers in other social sciences (Doolittle 1984:125). Archaeological data formed over long periods of time can overcome such deficiencies. Finally, to the extent that continuous, if subtle, variation characterizes points, these patterns of change can be used to measure time in its inherently continuous form, rather than as a series of discrete phases. Ceramics of various kinds have been used in this fashion to produce what

Braun (1985) calls "absolute seriations" (Binford 1962; Braun 1985, 1987; Plog and Hantman 1990), but the chronological potential of points, though substantial (Fischer 1989:37; cf. Dunnell and Feathers 1992:26; George and Scaglion 1992), has not been assayed.

## CONTINUOUS VARIATION: AN AMERICAN BOTTOM CASE STUDY

The fact that continua of change in size and form have been documented widely in the archaeological record has been largely overlooked. Examples range from the European Paleolithic and Mesolithic (Fischer 1989:39, Fig. 10; Knecht 1991:134) to comparable periods (Deller and Ellis 1988:255) and mid-prehistoric ones (Forbis 1960:86–94; Fawcett and Kornfeld 1980) in North America. Significantly, there are numerous examples from Woodland and late prehistoric contexts on this continent. Blitz (1988:131) sees "developmental continua" in late prehistoric triangular points. Others have reported similar findings in eastern North America (Schambach 1982:173; Milner 1984:92;Lynott 1991; George and Scaglion 1992). And Seeman (1992:42) spoke of a "smooth evolution of one modal projectile point into another through a sequence of . . . named categories" during the Middle and Late Woodland periods in the Ohio Valley. As Lynott (1991) noted, explaining these continua is not easy if we assume that changes in material culture are episodic.

Such a combination of change in both the size and form of points characterized a study of Late Woodland assemblages in the upper Ohio Valley (Shott 1993). Unfortunately, the relatively small size of the assemblages limited the potential and scope of that study. In contrast, the tremendous volume of fieldwork in the American Bottom of Illinois in the 1970s and 1980s, most conducted by the University of Illinois on the renowned FAI-270 project (Bareis and Porter 1984), brought to light a substantial body of data from Late Woodland assemblages. There we have the necessary combination of large, systematically described assemblages and a well established local chronological sequence to study the dart-to-arrow transition and to gauge relatively subtle and continuous variation in points over a substantial time interval. In addition, the American Bottom as a study area minimizes the effects of several sources of variation discussed earlier: space, since a reasonably small and well defined area is identified; measurement error, since a common set of attributes was coded in different assemblages by a small and relatively close-knit set of researchers; and raw material, since only a few major sources were used (Kelly 1984; Moshage 1987).

The following analysis first treats the ordering in time of American Bottom assemblages, an obvious prerequisite to any study of chronological change. It then considers the two questions raised at the outset, offering possible explanations for the continuous patterns of variation observed in the data.

## Chronology

American Bottom Late Woodland occupations can be dated in several ways and to varying degrees of precision. At the coarsest scale, they can be ordered by phase affiliation in the Rosewood-Mund-Patrick-Sponemann sequence (Kelly et al. 1984; Fortier et al. 1991). Such ordering is adequate for certain purposes, but there is ambiguity in phase assignments because radiocarbon dates from components of a single phase often span much greater time intervals (Shott 1992:204). Alternatively, several FAI-270 investigators (e.g., Milner 1984:163–170; Kelly et al. 1987:430, 1990:284) have proposed chronological trends through the Woodland and Mississippian in ceramic attributes, structure size and shape, and other traits. These include trends in the relative frequency of vessel classes and decorative elements (Kelly et al. 1990:Table 9.3), analogous to patterns that Plog and Hantman (1990) have observed in American Southwestern sites. Such trends can be used to order site occupations in time and perhaps to establish chronological order among sites themselves. Although such internal "seriations" (Kelly et al. 1990:284) often are consistent with the ordering produced using radiocarbon dates, they are ignored in this analysis. To reduce the risk of circular reasoning, chronological orderings here are based purely on the single criterion of radiocarbon dating. The ambiguity and imprecision of radiocarbon dating are well known, but there are ways to determine to an acceptable degree of probability whether a set of assays is consistent with a single occupation of relatively short duration (Shott 1992).

Orderings by averaged and rounded radiocarbon dates, without regard to phase assignment, appear in Tables 1 and 2. Sites are ordered in Table 1 as indivisible units, ignoring some complexities of occupational histories. Alpha 1 and the Late Woodland occupations at Range are omitted from this ordering because the dispersion of their dates is inconsistent with a single occupation. The Dohack and Holdener sites yielded valid averaged results, but information on their projectile points is unavailable. The Cramer and Columbia Farms sites have projectile points, but no dates.

The second ordering appears in Table 2, by various combinations of original and averaged dates. For this ordering, some sites were subdivided into clusters thought by their excavators to represent distinct periods of occupation. Clusters were dated either by their single original assay (e.g., Cluster IV at George Reeves) or by the averaging of multiple assays. Even with this treatment, pooled dates within clusters do not always yield concordant results (e.g., Cluster 1 at Fish Lake). Such clusters are omitted from the table and from analysis. It bears emphasizing, however, that the dates assigned to some sites are unchanged in the second ordering. The Late Woodland occupation at Mund, for instance, is considered on contextual grounds an essentially single-component site (Fortier et al. 1983:389), a judgment supported by the results of date averaging. Thus Mund appears in Table 2 exactly as it does in Table 1, as a single occupation associated

**Table 1. Chronological Ordering Using Sites as Units**

| Site | Averaged date B.P. |
|---|---|
| Leingang | 1580 ± 52 |
| George Reeves | 1538 ± 50 |
| Mund | 1426 ± 65 |
| Steinberg | 1394 ± 70 |
| Sponemann | 1248 ± 45 |
| Fish Lake | 1244 ± 50 |
| Range (Dohack phase) | 1015 ± 50 |

with the same date. Other occupations that yielded a statistically significant averaged date lacked measurable points (e.g., Community 4 at Sponemann), and so were omitted from the table.

The second chronological ordering introduces a measure of inconsistency to the data. Lone assays from occupations are taken at face value, but averaged multiple assays from the same occupation presumably yield more accurate, or at least statistically reliable, results. In addition, the relative order of sites by averaged date is not necessarily consistent with site ordering produced by phase assignments. These inconsistencies are acknowledged but considered no further. Obviously, all presumed projectile points are included in analysis when sites are treated as units, while only specimens from the appropriate subdivisions of multiple-occupation sites are analyzed when separate occupations are the unit of observation.

**Table 2. Chronological Ordering Using Occupations as Units**

| Site or occupation | Averaged date B.P. |
|---|---|
| George Reeves IV | 1620 ± 70 |
| Leingang | 1580 ± 52 |
| George Reeves II | 1496 ± 61 |
| Mund | 1426 ± 65 |
| Steinberg | 1394 ± 70 |
| Fish Lake 3 | 1360 ± 130 |
| Sponemann 1 | 1240 ± 70 |
| Sponemann 3 | 1205 ± 66 |
| Range 2 R-1 | 1195 ± 76 |
| Fish Lake 2 | 1176 ± 86 |
| Range 2 R-3 | 1016 ± 64 |
| Range (Dohack phase) | 1015 ± 50 |
| Range 2 R-2 | 883 ± 73 |

## The Dart-To-Arrow Transition

The first question raised in this paper concerns the timing and nature of the dart-to-arrow transition in the study area. To address it requires some way to distinguish dart and arrow specimens. Ordered by their time of occurrence, it should be possible to determine when the transition occurred, and whether it involved the abrupt and near-complete replacement of darts by arrows.

Unfortunately, complete projectiles are rarely recovered from archaeological sites. Projectile points, however, are systematically related to the size of the composite weapon of which they formed a part (Thomas 1978; Christenson 1986a, 1986b), making it possible to deduce, at least within broad limits, the relevant properties of the whole from the archaeologically common part.

Design correlates of projectile point size are widely recognized. Archaeologists have registered a gratifying degree of concordance on the modal dimensions of dart and arrow points. Arrows rarely exceed 1 cm in neck width and 0.5 cm in thickness, while darts measure circa 1.4 cm in neck width (Thomas 1978:469; Fawcett and Kornfeld 1980:72; Patterson 1985:82; Lorentzen 1989). These findings concur with Forbis's (1960) interpretations of trends at Old Woman's Buffalo Jump, where specimens having neck width values near 1.3 cm are considered darts, those with values near 0.8 cm, arrows. Spencer's (1974:Table 5) experimental study suggests that dart points can exceed 3 cm in maximum width and 0.6 cm in thickness. Compared to size, design correlates of point form have been neglected. Fischer (1989:38), however, has identified longitudinal symmetry as a requirement for accurate projectiles.

Such information is useful, but the best way to distinguish archaeological specimens is by reference to ethnographic ones of known function. Thomas used this reasoning in his study of projectile size and form, in an attempt to distinguish arrow from dart points (Seeman 1992; Shott 1993). From a sizable arrow sample (n=132) and a much smaller one for darts (n=10), Thomas (1978:470) defined a function to distinguish the forms and to establish classification functions for both that can be used to assign specimens of unknown function to the proper category. The classification functions are as follows:

Function 1. Darts:

$$C = 0.188\ ML + 1.205\ MW + 0.392\ MT - 0.223\ NW - 17.552$$

Function 2. Arrows:

$$C = 0.108\ ML + 0.470\ MW + 0.864\ MT + 0.214\ NW - 7.922$$

where C = the function result expressed numerically
ML = maximum length
MW = maximum width
MT = maximum thickness, and
NW = neck width.

Metric attributes of points can be inserted into each function, the higher C value indicating the correct classification. The use of Thomas's function is described elsewhere (Shott 1993), but it requires emphasis here that it has yielded an accuracy rate of only 70% for dart points; that is, fully 30% of actual darts are incorrectly classified as arrows. This deficiency may reflect the extremely small sample size for darts more than any other factor. To correct it, more data on ethnographic and archaeological hafted dart points are being collected from museums. Until that effort is completed, however, the original data and the functions defined from them are used. Seeman further suggests that the function may not apply to points made during a period of transition from dart to arrow, arguing that "Thomas's comparative sample was drawn from contexts well after any initial introduction or invention, and his classification functions may misclassify points associated with initial adoption" (Seeman 1992:42).

However, a different data set suggests how the classification functions may be able to identify such transitional specimens. Gould's (1966) Tolowa informants identified a series of archaeological points from a late prehistoric site in northern California. The known data points in this case are not museum specimens but Tolowa standards or norms; normative data, of course, retain *some* value in archaeological research. Taking at face value the judgment of informants, archaeological points were identified as definitely arrows, definitely darts or harpoons, or intermediate between the two (Gould 1966:56–57). Data from these specimens were classified using a variant of Thomas's original function that omitted the attribute of neck width, since most lacked the notching that defines necks. Although intermediate specimens were few in number, they yield a mean absolute difference between dart and arrow score that is much smaller than the corresponding value for specimens identified as darts and arrows. That is, intermediate specimens yield equivocal results, in which dart and arrow classification scores are very similar. At least provisionally, then, it is worth considering archaeological specimens that yield similar dart and arrow scores as intermediate or, more appropriately in a diachronic sense, transitional specimens. To extend the reasoning, a convergence in dart and arrow scores for archaeological specimens may identify them as marking the transition from dart to arrow technology. It may reflect a process by which a basic dart form or model is transformed ("translated" in Hall's [1980:437] terms) for use as arrow points in an emerging new technology.

Nevertheless, there are perils in this approach. First, these results ignore the effects of point resharpening and reduction on classification results, which can greatly reduce original dimensions of darts, especially length, to the degree that they might be mistaken for arrows (Shott 1993). Lacking data on original size of stemmed and notched points from caches or workshops, degree of reduction in archaeological specimens cannot be estimated. Possible resharpening effects are acknowledged here but are not considered further. Second, point size alone, without regard to the point's place within the design of a larger projectile, can produce misleading results. Few archaeologists would argue that Early Archaic or Late Archaic hunters used the bow and arrow (cf. Odell 1988; Amick 1994), yet some points of those periods are sufficiently small to perhaps be classified as arrow points. Such possibilities are in fact intriguing, if uncertain on present evidence. Alternatively, the functions can be used to classify specimens dating to well after the bow-and-arrow adoption, but this does little more than prove the obvious. Perhaps their best use—always emphasizing the provisional character of inferences drawn from them—is during the period of transition, working back in time from specimens "known" (more accurately, assumed [Shott 1993]) to be arrows to those of uncertain function. That basic approach is followed here.

## Results

Classification of points by time-ordered site yields the results shown in Table 3, which are similar for the two classification functions. In both, darts form a significant minority in early assemblages (except for the George Reeves site), before dropping precipitously in later occupations. The absence of classified darts from the Fish Lake 3 component onward among time-ordered components is slightly deceptive. Such specimens were recovered from those sites, but by chance none was found in the dated components.

The bow-and-arrow is thought to have been introduced to the American Bottom during the Patrick Phase, circa 1350–1150 B.P. (Kelly et al. 1984:122). Classification results suggest that it appeared substantially earlier, during Rosewood times, later rising to predominance during the Mund and Patrick Phases. George Reeves notwithstanding, the transition from dart to arrow seems to have been first gradual and only later abrupt. Significantly, the appearance of arrows did not at once spell the demise of darts, which persisted in substantial numbers for roughly 150 years. Darts then were retained in small quantities for specialized use or by conservative or disadvantaged members of American Bottom societies. Possible reasons for retention of darts are discussed elsewhere (Shott 1993). Alternatively, they were entirely abandoned, the few later-occupation specimens perhaps so classified only in error.

Dunnell and Feathers (1992) reached a similar conclusion regarding the replacement of sand-tempered by shell-tempered pottery elsewhere in the Mississippi Valley. They argued (1992:35) for a gradual transition rather than an abrupt replacement, concluding that both kinds of temper may have been used concurrently "for a significant period of time." As in this case, detailed study resolved

Table 3. Results of Classification by Time-Ordered Site

| | Date | σ | Function 1 | | Function 2 | |
|---|---|---|---|---|---|---|
| | | | Dart | Arrow | Dart | Arrow |
| Site | | | | | | |
| Leingang | 1579 | 52 | 3 | 2 | 3 | 2 |
| George Reeves | 1538 | 50 | 1 | 14 | 3 | 12 |
| Mund | 1426 | 65 | 9 | 16 | 6 | 19 |
| Steinberg | 1394 | 70 | 1 | 1 | 1 | 1 |
| Sponemann | 1248 | 45 | 1 | 19 | 0 | 20 |
| Fish Lake | 1244 | 50 | 1 | 14 | 0 | 15 |
| Range (Dohack) | 1015 | 50 | — | — | 1 | 33 |
| Component | | | | | | |
| George Reeves IV | 1620 | 70 | 1 | 3 | 2 | 2 |
| Leingang | 1580 | 52 | 3 | 2 | 3 | 2 |
| George Reeves II | 1496 | 61 | 0 | 3 | 0 | 3 |
| Mund | 1426 | 65 | 9 | 16 | 6 | 19 |
| Steinberg | 1394 | 70 | 1 | 1 | 1 | 1 |
| Fish Lake 3 | 1360 | 130 | 1 | 3 | 0 | 4 |
| Sponemann 1 | 1240 | 70 | 0 | 2 | 0 | 2 |
| Sponemann 3 | 1205 | 66 | 1 | 8 | 0 | 9 |
| Range 2 R-1 | 1195 | 76 | — | — | 0 | 3 |
| Fish Lake 2 | 1176 | 86 | 0 | 2 | 0 | 2 |
| Range 2 R-3 | 1016 | 64 | — | — | 0 | 1 |
| Range (Dohack) | 1015 | 50 | — | — | 0 | 15 |
| Range 2 R-2 | 0883 | 73 | — | — | 0 | 6 |

Function 1 is Thomas's original function. Function 2 omits neck width from consideration.

into a continuous process of transition what appeared on first inspection to have been a fundamentally discontinuous transformation.

## Explaining the Apparent Transition

In his survey of Woodland cultural developments in Illinois, Hall (1980:437) argued that the later Late Woodland "Stemmed and notched points of small size could be interpreted as an example of the translation of larger Woodland atlatl 'dart' (spear) points into the smaller size suitable for the arrow, a shift of dimensions within the older pattern." This comment echoes sentiments registered earlier in Cole and Deuel's (1937:55) critical evaluation of west-central Illinois Maples Mills points as "an attempt to reproduce the small projectile or arrow points of Mississippian dimensions by a technique unsuitable for the result sought." They argued, in effect, that an existing custom of point form was adapted without drastic alteration to the metric requirements imposed by a changing hunting technology.

This argument finds support from other fields. Doolittle (1984:134) regarded much technological change as an incremental process in which gradual innovation is "the product of in-the-field trial and error." This view is shared by Sahal (1981), whose lengthy study is especially apposite in the present context. Hall's (1980) "translation" argument essentially holds that a preexisting design influences the direction of innovation, and persists long after the optimal design defined by changing needs and constraints is altered. Thus:

> very often there emerges a pattern of . . . design as an outcome of prolonged development effort. The pattern in turn continues to influence the character of subsequent technological advances long after its conception. Thus innovations generally depend on bit-by-bit modifications of a design that remains unchanged in its essential aspects over extended periods of time. (Sahal 1981:33)

Ultimately, knappers in the American Bottom and across eastern North America shifted from stemmed or notched to triangular points. Though no more than speculation at this point, perhaps the morphological change—but not the metric change, since size differences between the forms are slight—occurred abruptly, once the limits of the prexisting form were reached under conditions of changing performance requirements (Sahal 1981:74–75).

## Documenting Continuous Variation

The second question raised in this paper follows from the first. It concerns the nature and explanation of time-dependent variation in the metric attributes of projectile points.

Mean metric attributes and their corresponding coefficients of variation (cv) by time-ordered sites appear in Table 4. Most show a clear chronological trend of declining values, but the relationship is fairly weak for maximum length and shoulder width. Other attributes are more closely controlled by haft requirements. The use of points as normative markers tends to obscure such patterns of continuous variation. In effect, the nature of time-dependent variation in this study is more continuous than episodic.

As Christenson (1986b) and Thomas (1978:467) suggest, haft attributes like neck and base width most closely articulate a point with the larger apparatus of

**Table 4. Projectile Mean Metric Attributes by Site**

| | | | Length | | | | Width | | | | | | | |
|---|---|---|---|---|---|---|---|---|---|---|---|---|---|---|
| Site | Date | σ | Stem | CV | Max. | CV | Shoul. | CV | Neck | CV | Base | CV | Thickness | CV |
| Leingang | 1580 | 52 | 1.41 | .07 | 6.02 | .16 | 2.73 | .13 | 1.89 | .08 | 2.14 | .07 | 1.03 | .16 |
| George Reeves | 1538 | 50 | 1.32 | .30 | 4.26 | .24 | 1.87 | .22 | 1.45 | .28 | 1.80 | .23 | 0.66 | .27 |
| Mund | 1426 | 65 | 1.26 | .17 | 5.33 | .14 | 2.03 | .12 | 1.45 | .11 | 1.67 | .15 | 0.90 | .17 |
| Steinberg | 1394 | 70 | 2.35 | .37 | 4.50 | .04 | 2.15 | .16 | 1.60 | .06 | 1.60 | .00 | 0.90 | .00 |
| Sponemann | 1248 | 45 | 0.61 | .38 | 2.92 | .37 | 1.37 | .24 | 0.84 | .29 | 1.20 | .36 | 0.42 | .50 |
| Fish Lake | 1244 | 50 | 0.64 | .19 | 2.74 | .27 | 1.24 | .21 | 0.70 | .21 | 1.00 | .24 | 0.34 | .23 |
| Range (Dohack) | 1015 | 50 | — | — | 2.91 | .21 | 1.41 | .18 | — | — | — | — | 0.39 | .29 |

**Table 5. Correlation of Attributes with age B.P. for All Specimens**

| | By site | | By component | |
|---|---|---|---|---|
| Attribute | r | p | r | p |
| Length, stem | .50 | .31 | .62 | .07 |
| Length, maximum | .82 | .02 | .82 | .00 |
| Width, shoulder | .79 | .03 | .74 | .00 |
| Width, neck | .90 | .01 | .85 | .00 |
| Width, base | .97 | .00 | .82 | .01 |
| Thickness | .79 | .03 | .74 | .00 |

which it formed a part. Figure 1 plots neck and base width against time-ordered site. The correlation is especially strong for base width (Table 5); regressions are as follows:

$$\text{neck width} = -2.80 + .003\ (\text{Age B.P.})$$

$$\text{base width} = -2.42 + .003\ (\text{Age B.P.}).$$

In neither case do residuals pattern clearly, although there is a slight tendency for neck width values to increase with age. On balance, the data show a clear and highly significant size reduction with time in haft attributes, especially neck and base width. Base width in particular yields very clear results.

Table 6 shows metric data by time-ordered components. Results are similar; neck and base width again correlate very strongly with age (Figure 2) such that:

**Table 6. Projectile Mean Metric Attributes by Component**

| | | | Length | | | | Width | | | | | | | |
|---|---|---|---|---|---|---|---|---|---|---|---|---|---|---|
| Site | Date | σ | Stem | CV | Max. | CV | Shoul. | CV | Neck | CV | Base | CV | Thickness | CV |
| George Reeves IV | 1620 | 70 | 1.60 | .16 | 5.15 | .15 | 2.04 | .07 | 1.51 | .10 | 1.93 | .16 | 0.74 | .07 |
| Leingang | 1580 | 52 | 1.41 | .07 | 6.02 | .16 | 2.73 | .13 | 1.89 | .08 | 2.14 | .07 | 1.03 | .16 |
| George Reeves II | 1496 | 61 | 1.20 | .12 | 4.43 | .04 | 2.07 | .06 | 1.77 | .07 | 1.90 | .07 | 0.87 | .06 |
| Mund | 1426 | 65 | 1.26 | .17 | 5.33 | .14 | 2.03 | .12 | 1.45 | .11 | 1.67 | .15 | 0.90 | .17 |
| Steinberg | 1394 | 70 | 2.35 | .37 | 4.50 | .04 | 2.15 | .16 | 1.60 | .06 | 1.60 | .00 | 0.90 | .00 |
| Fish Lake 3 | 1360 | 130 | 0.55 | .16 | 2.90 | .35 | 1.36 | .21 | 0.76 | .28 | 0.67 | .67 | 0.34 | .15 |
| Sponemann 1 | 1240 | 70 | 0.45 | .11 | 2.60 | .08 | 1.25 | .04 | 0.75 | .07 | 0.95 | .16 | 0.30 | .00 |
| Sponemann 3 | 1205 | 66 | 0.62 | .35 | 3.11 | .41 | 1.44 | .25 | 0.90 | .27 | 1.30 | .32 | 0.47 | .45 |
| Range 2 R-1 | 1195 | 76 | — | — | 2.57 | .17 | 1.22 | .24 | — | — | — | — | 0.33 | .27 |
| Fish Lake 2 | 1176 | 86 | 0.65 | .08 | 2.90 | .00 | 1.20 | .08 | 0.65 | .08 | 0.85 | .18 | 0.30 | .33 |
| Range 2 R-3 | 1016 | 64 | — | — | 2.70 | .00 | 1.40 | .00 | — | — | — | — | 0.40 | .00 |
| Range (Dohack) | 1015 | 50 | — | — | 2.91 | .21 | 1.41 | .18 | — | — | — | — | 0.39 | .28 |
| Range 2 R-2 | 0883 | 73 | — | — | 2.78 | .17 | 1.47 | .16 | — | — | — | — | 0.40 | .17 |

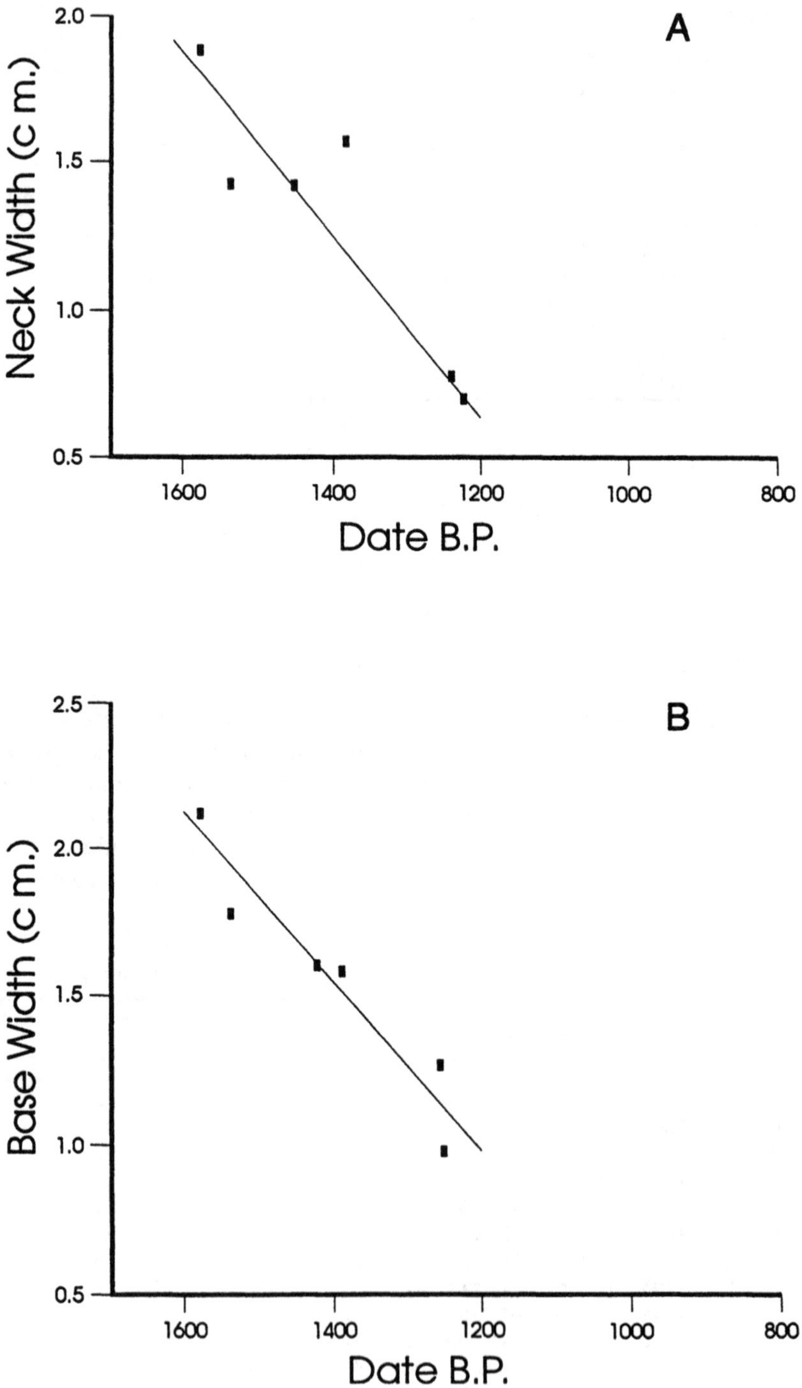


**Figure 1.** Neck and base width against time-ordered sites.

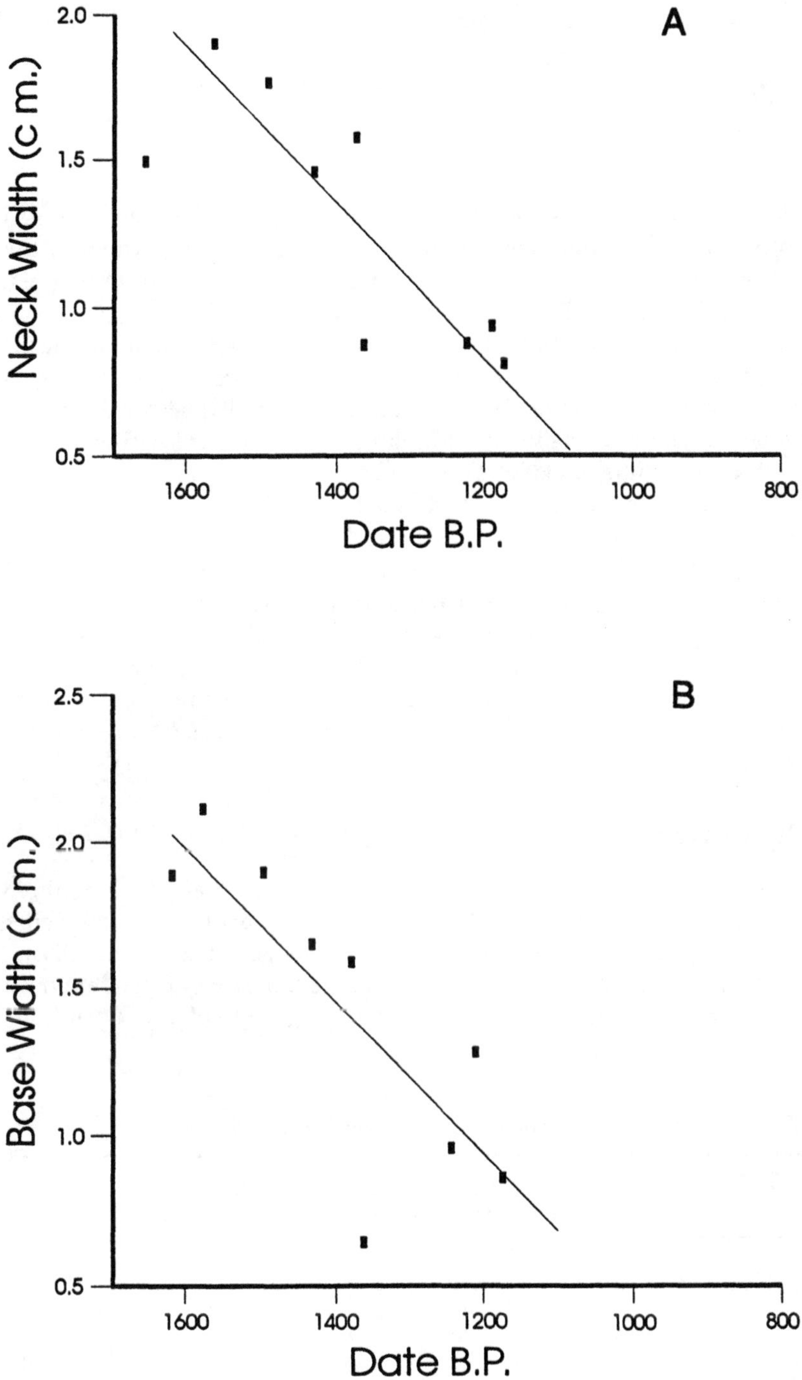


**Figure 2.** Neck and base width against time-ordered components.

neck width = -2.33 + .003 (Age B.P.)

base width = -2.30 + .003 (Age B.P.).

Again, residuals show no clear pattern. And again base-width results are especially strong, with the exception of a pronounced outlier formed by the Fish Lake 3 component. In this data set, neck width results are also strong, with no extreme outlier or general pattern of divergence.

These metric trends might be regarded as the simple product of the technological transition from dart to arrow discussed above. Surely the transition explains some of the variation, but perhaps not all of it. Even after their adoption, arrows varied continuously in size through time, as did darts earlier.

Using the results of classification, arrow points can be separated from dart points. Descriptive statistics for arrows are listed in Tables 7 and 8 for sites and components, respectively. Although these data exhibit a trend similar to the combined data set, the correlation between width attributes and age is slightly lower (Table 9). Cross-plots of neck and base width against time closely resemble those for the combined data sets, so only the results by time-order sites are presented (Figure 3). Again, however, analysis reveals a strong correlation between age and point size. Although the sample of darts is much smaller, a similar pattern characterizes it. Shoulder and neck width correlate especially strongly with age in this class.

## An Alternative View

On the limited available evidence, the data can be interpreted in other ways, as well. There is a gap at ca. 1300 B.P. in the distribution of neck width values and perhaps in base width, as well (Figures 1–3), that suggests two distinct size groups. Separately, the groups do not necessarily yield the same time-dependent trend that the entire data set does together. (In fact, the groups of lower values in the

**Table 7. Arrow point mean metric attributes by site**

| | | | Length | | | | Width | | | | | | | |
|---|---|---|---|---|---|---|---|---|---|---|---|---|---|---|
| Site | Date | σ | Stem | CV | Max. | CV | Shoul. | CV | Neck | CV | Base | CV | Thickness | CV |
| Leingang | 1580 | 52 | 1.40 | .07 | 6.10 | .23 | 2.25 | .02 | 1.75 | .03 | 2.00 | .00 | 1.20 | .17 |
| George Reeves | 1538 | 50 | 1.12 | .29 | 4.04 | .22 | 1.78 | .20 | 1.34 | .31 | 1.61 | .25 | 0.66 | .29 |
| Mund | 1426 | 65 | 1.26 | .17 | 5.24 | .16 | 1.91 | .10 | 1.46 | .12 | 1.73 | .14 | 0.94 | .12 |
| Steinberg | 1394 | 70 | 1.50 | | 4.70 | | 1.80 | | 1.50 | | 1.60 | | 0.90 | |
| Sponemann | 1248 | 45 | 0.61 | .08 | 2.87 | .08 | 1.35 | .04 | 0.82 | .07 | 1.13 | .09 | 0.41 | .12 |
| Fish Lake | 1244 | 50 | 0.65 | .18 | 2.61 | .22 | 1.16 | .16 | 0.66 | .14 | 0.92 | .18 | 0.34 | .20 |
| Range (Dohack) | 1015 | 50 | | | 2.88 | .22 | 1.34 | .17 | | | | | 0.39 | .04 |

**Table 8. Arrow Point Mean Metric Attributes by Component**

| | | | Length | | | | Width | | | | | | | |
|---|---|---|---|---|---|---|---|---|---|---|---|---|---|---|
| Site | Date | σ | Stem | CV | Max. | CV | Shoul. | CV | Neck | CV | Base | CV | Thickness | CV |
| George Reeves IV | 1620 | 70 | 1.50 | .07 | 5.25 | .07 | 1.95 | .02 | 1.35 | .04 | 1.60 | .12 | 0.75 | .17 |
| Leingang | 1580 | 52 | 1.40 | .07 | 6.10 | .23 | 2.25 | .02 | 1.75 | .03 | 2.00 | .00 | 1.20 | .17 |
| George Reeves II | 1496 | 61 | 1.20 | .12 | 4.43 | .04 | 2.07 | .06 | 1.77 | .07 | 1.90 | .07 | 0.87 | .06 |
| Mund | 1426 | 65 | 1.26 | .17 | 5.24 | .16 | 1.91 | .10 | 1.46 | .12 | 1.73 | .14 | 0.94 | .12 |
| Steinberg | 1394 | 70 | 1.50 | — | 4.70 | — | 1.80 | — | 1.50 | — | 1.60 | — | 0.90 | — |
| Fish Lake 3 | 1360 | 130 | 0.55 | .16 | 2.32 | .12 | 1.17 | .09 | 0.60 | .12 | 0.80 | .10 | 0.32 | .12 |
| Sponemann 1 | 1240 | 70 | 0.45 | .11 | 2.60 | .08 | 1.25 | .04 | 0.75 | .07 | 0.95 | .16 | 0.30 | .00 |
| Sponemann 3 | 1205 | 66 | 0.62 | .31 | 3.00 | .47 | 1.35 | .21 | 0.85 | .32 | 1.21 | .34 | 0.44 | .48 |
| Range 2 R-1 | 1195 | 76 | — | — | 2.57 | .17 | 1.20 | .27 | — | — | — | — | 0.33 | .27 |
| Fish Lake 2 | 1176 | 86 | 0.65 | .08 | 2.90 | .00 | 1.20 | .08 | 0.65 | .08 | 0.85 | .18 | 0.30 | .33 |
| Range 2 R-3 | 1016 | 64 | — | — | 2.70 | — | 1.40 | — | — | — | — | — | 0.40 | — |
| Range (Dohack) | 1015 | 50 | — | — | 2.88 | .22 | 1.34 | .17 | — | — | — | — | 0.39 | .04 |
| Range 2 R-2 | 0883 | 73 | — | — | 2.78 | .17 | 1.35 | .06 | — | — | — | — | 0.37 | .11 |

nine-component set could exhibit precisely the opposite trend [Figure 2B]). The groups in turn could be seen as separate dart and arrow samples, the gap marking the approximate time of abrupt transition between them. In this view, a technological transition produces abrupt, discontinuous change in metric attributes that is blurred as a consequence of minor random variation in the time-ordering of sites and components. Only more data can settle the issue, and they are being sought.

Recall, though, that classified arrows occur on both sides of the gap and exhibit a consistent time-dependent size decrease. Unless the classification results are wrong—and this is less likely for arrows than for darts (Thomas 1978:471)—a single continuous trend spans the apparent gap in the data. If classification results are accurate, then the gap is just the incidental product of limited data, not the necessary consequence of an abrupt technological change.

**Table 9. Correlation of Attributes with Age B.P., Arrow Points Only**

| | By site | | By component | |
|---|---|---|---|---|
| Attribute | r | p | r | p |
| Length, stem | .76 | .08 | .83 | .01 |
| Length, maximum | .79 | .03 | .78 | .00 |
| Width, shoulder | .83 | .02 | .75 | .00 |
| Width, neck | .88 | .02 | .79 | .01 |
| Width, base | .91 | .01 | .79 | .01 |
| Thickness | .79 | .04 | .75 | .00 |

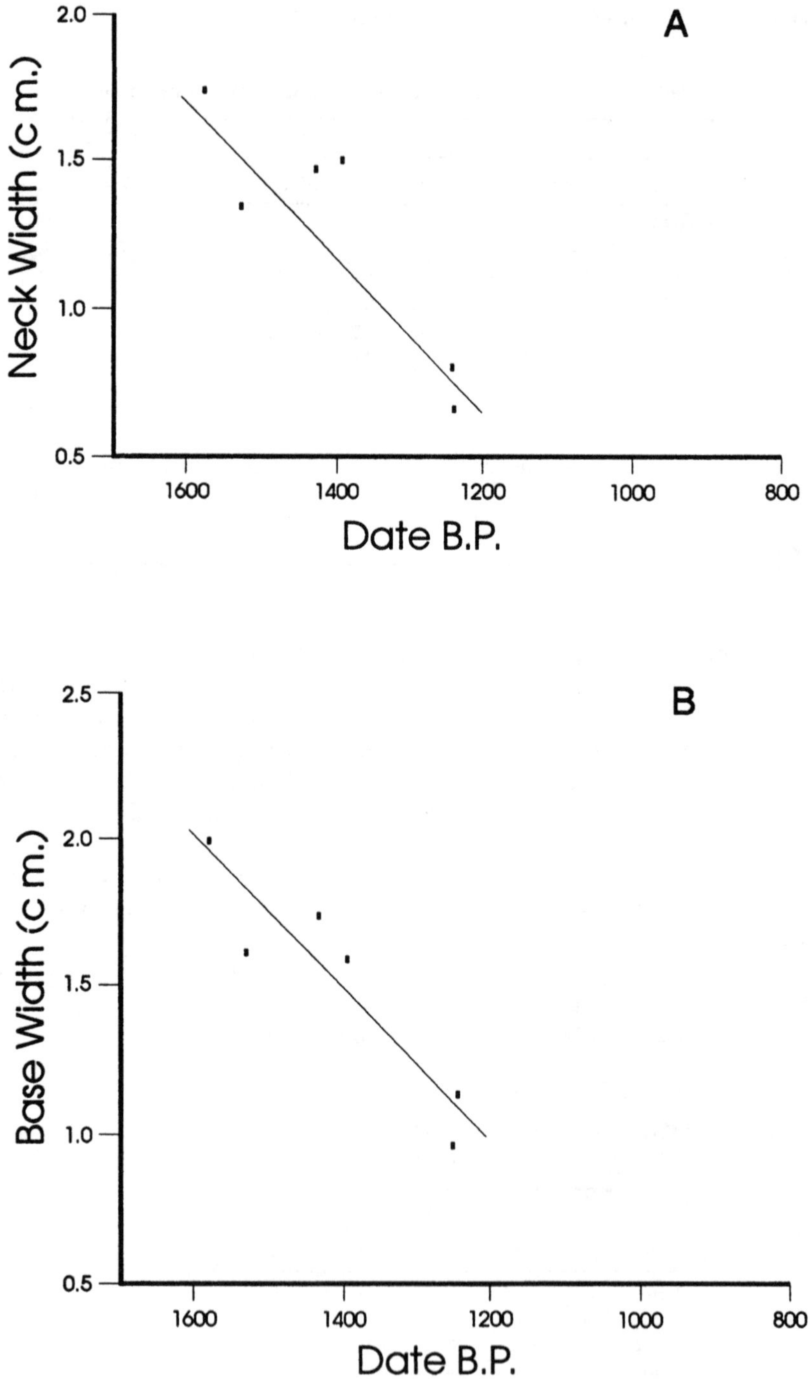


**Figure 3.** Arrow neck and base width against time-ordered sites.

It is worth noting that raw material selection does not explain the chronological trend found in American Bottom point dimensions. Systematic studies indicate relatively few differences in the mechanical or flaking properties of available cherts in the American Bottom (Kelly 1984:30–33; Moshage 1987), though some surely exist. There exists a slight tendency for Burlington chert to be favored for use over time, as Kelly (1984:Table 2.1) noted for the Merrell Tract, but no haft-element dimensions differ significantly by raw material. Maximum length of specimens does differ significantly (F=3.53, p=.02), with Burlington specimens shorter than those composed of other chert.

### Discussion

These preliminary results can be improved in several ways. First and foremost, more cases will lend greater confidence to results. They will also permit the use of multiple regression techniques. Following Plog and Hantman's (1990:445) practice, 7–10 cases are required for each independent variable. The current site-level data set contains only seven cases, making multiple regression inadvisable. Second, the chronological scope of the analysis must be broadened to minimize the possibility of reifying short-term patterns of change (Frankel 1988:41–42). At a minimum, a span of time comparable to Braun's ca. 2200–1000 B.P. interval ultimately must be studied. Radiocarbon data must also be calibrated. Third, confidence limits of the regression must be calculated. Finally, the set of steps described by Braun (1985:512–534, 1987), from smoothing and interpolation to the extraction of logistic trends, must be undertaken.

Equally intriguing time-dependent trends in the size of probable arrows are found in other areas. Seeman and Munson (1980) used discriminant functions to define three classes of triangular points in southwestern Indiana assignable to the successive Yankeetown, Angel and Caborn-Welborn Phases. There, base or maximum width declined steadily by about 9.3%, and maximum thickness by about 16.7%, over the roughly six-century span from the earlier Yankeetown to the later Caborn-Welborn Phase.

This continuous variation forms a time-dependent trend that cannot be explained by raw material or other traditionally identified factors. As argued above, such variation demands explanation, and one is proposed in the following section.

## EXPLAINING CONTINUOUS VARIATION

Optimal foraging theory (OFT) is a family of theories governing subsistence behavior under specified or assumed conditions. OFT already has earned varying receptions in anthropology, from great enthusiasm in some quarters to undisguised scorn in others. With a sincere desire to avoid polemics, this paper uses an element of OFT—the diet breadth model, one of OFT's earliest and most elegant formulations (MacArthur and Pianka 1966)—with a measure of apprehension befitting its sometimes dubious relevance to the extreme complexities of culture process (Martin 1983; Dwyer 1985; Yellen 1986). This model has achieved at least qualified success in anthropology, accounting in part for the range of subsistence classes—the diet breadth, that is—characterizing forager (Winterhalder 1981; Hawkes et al. 1982; Hill and Hawkes 1983) and horticultural (Hames and Vickers 1982) societies.

The model's form and assumptions are discussed at length in MacArthur and Pianka (1966); Hawkes et al. (1982:387–388) and Winterhalder (1981:23–25) furnish anthropological discussions. Briefly, diet breadth (DB) varies inversely with the abundance of the most important resources. That is, as resources of higher acquisition efficiency or return rates increase in abundance, resources acquired at lower efficiency are excluded from the diet. Since the return rates of maize and associated cultigens typically are high, societies which rely heavily on them should have lower DB than societies which do not. Strong ethnographic warrant exists for this view (Saffirio and Scaglion 1982; Hill and Hawkes 1983; Milton 1984; Redford and Robinson 1987; Speth and Scott 1989).

Although American Bottom Middle Woodland, Late Woodland, and Emergent Mississippian patterns of plant exploitation were similar in some ways, important trends are evident in archaeobotanical assemblages (Johannessen 1984:202–203). Compared to Middle Woodland practices, Late Woodland subsistence was characterized by a slightly expanded list of cultigens (e.g., squash became fairly common) and by considerably higher densities of plant remains overall. Starchy seeds comprise perhaps 50–60% of Middle Woodland seed assemblages, 76% of the overall Late Woodland assemblage, and about 90% of the Emergent Mississippian one (Johannessen 1984:201–202; Parker 1991:418). Riley et al. (1994:490) have reported maize from the Middle Woodland Holding site dating to between 170 B.C. and A.D. 60. On present evidence, however, there is no clear trend in maize use until the definite increase in the Emergent Mississippian phase (Johannessen 1984:201–203; Riley et al. 1994:495–496). (Maize, however, may have been harvested in Middle and Late Woodland times in its green stage, therefore being less susceptible to carbonization [Kelly et al. 1984:125].) Perhaps corresponding to the clear trend in use of native seed-bearers and possible trend in tropical cultigens are apparent reductions in nut consumption. We are accustomed to viewing the appearance of maize as relatively abrupt, producing a fundamental discontinuity with earlier cultivation practices. But the American Bottom evidence is equally consistent with a view of incremental, perhaps even continuous or cyclical, change in the range of crop plants used, and in the contribution that the most important ones made to the diet.

Kelley (1992) noted that general similarities in faunal assemblages can mask significant differences and important trends. Future work in the American Bottom will seek the sometimes subtle evidence of subsistence change that can be revealed in faunal assemblages (Klein and Cruz-Uribe 1984:97–98; Kelley 1992). Until then, available data furnish at least a partial view of subsistence trends in the region.

American Bottom Late Woodland and late prehistoric faunal assemblages are somewhat impoverished from an apparent combination of poor preservation and a limited range of site types available for study (Kelly and Cross 1984:224–231). As far as general trends are recognizable, diversity of terrestrial mammals is comparatively high in earlier assemblages from Columbia Quarry and Mund, but low in other, mostly later, ones. This observation accords with the predictions of the DB model. Moreover, the mammalian assemblage from succeeding Emergent Mississippian sites is smaller still, again consistent with expectations. Yet faunal remains are so limited and fragmentary as to greatly qualify any conclusions. On available evidence, there is little at odds with DB-model predictions and some data that may support them.

Following the MacArthur and Pianka model, DB should become narrower through time, with increasing reliance on cultigens, native or tropical. Among ways in which this change might occur would be the removal of low-ranking prey species from the diet in stepwise fashion. With rare exceptions in the midcontinent until Mississippian times, chipped stone tools were not used in plant cultivation. However, as DB declines, the faunal component of the subsistence economy—acquisition of which is directly linked to stone tools—should decline in breadth or diversity as well. That is, fewer game species—only those of highest acquisition efficiency—should continue to be taken as cultivated foods assume a greater role in the economy (Yellen 1986). Since points are used as projectiles in hunting (cf. Odell 1988), the postulated economic trends may have consequences for the American Bottom lithic assemblages.

Subject to the proviso that projectile performance is influenced by many factors, Christenson (1986a:119) suggested that range and accuracy vary inversely with projectile length and shaft diameter and that thrust varies positively with these dimensions. In general, then, smaller projectiles are more accurate but less powerful over a wider launching range. Some ethnographic data support these suggestions (e.g., Watanabe 1975:65,76). At least in part, point size and form can be conditioned by these performance criteria. This logic can be used recursively, to argue that the realized size and form of projectile points can reflect—and thus provide a means of estimating—the ballistic properties of the original projectiles of which they formed only a part.

Among point attributes, thickness (Guthrie 1983; Patterson 1985:81; Seeman 1992:42) and width (Thomas 1978; Patterson 1985:81–82; Christenson 1986a; Lorentzen 1989:14; Seeman 1992) are widely considered the most responsive to design and performance criteria. Christenson (1986b:21) argued for a close metric relationship between projectile shaft diameter and point neck width. Thomas (1978:467) noted such a relationship in ethnographic collections, and showed that width at the maximum point—usually the shoulder—and at the neck are the most important discriminators between dart and arrow points (1978:470). Patterson (1985) supported this conclusion in a study of archaeological specimens, although he made *a priori* assumptions about which specimens were dart points and which were arrows. Neck width displays the most consistent stratigraphic, and hence chronological, trends in Forbis's (1960) archaeological study. Fawcett and Kornfeld's (1980) time-dependent change in Plains point metrics is most pronounced and statistically significant with respect to neck width, yielding a regression function (1980:66, Fig. 1) that links reduction in neck width to time over an interval much longer than that under study here.

Combined with the DB model, Christenson's (1986a, 1986b) tentative conclusions about projectile size and ballistic properties can be used to predict trends in projectile size and form. As faunal DB declined with agricultural intensification, the number of acceptable individual game targets declined as well. Some ethnographic data support the proposition that subsistence change occurred in a manner consistent with the DB model (Dwyer 1974; Winterhalder 1981). That is, it occurred in a series of discrete steps involving the addition or deletion of prey species. As Winterhalder (1981:86) noted among the boreal forest Cree:

> On a hunting trip at a given point in the annual cycle the forager is prepared to pursue the set of species which have a net acquisition rate of roughly 1,500 kcal/hour or better. When the capture rate falls below this general value, dissatisfaction is expressed, and the pursuit of that species often stopped.

Similar behavior has been noted among the New Guinea Etolo (Dwyer 1974, 1985), a horticultural society not unlike late prehistoric midcontinental societies in some respects. There, market penetration brought about a decline in hunting effort. Instead of reducing their effort proportionally among all prey species, however, Etolo hunters continued to invest the same amount of effort in the pursuit of favored species and simply abandoned systematic attempts to take others (Dwyer 1974:287). That is, changes in hunting patterns were accomplished by deletion of entire species from the set of acceptable targets.

The average time and effort required to locate acceptable targets, denoted "search time" or $T_N^S$ by MacArthur and Pianka (1966:603), should increase as a direct consequence of a decline in DB. All else being equal, a reduction in the number of acceptable targets means that the time required to locate each one must rise. At least one way to offset an increase in $T_N^S$ is to reduce "pursuit time" or $T_N^P$ correspondingly. This can be accomplished by an increase in the range and accuracy of the projectiles used to bag targets. An increase in thrust might also be desired, but this imposes design conflicts with increases in range and accuracy (Christenson 1986a:119; Cundy 1989:71). Considering the relatively scant importance in ethnographic accounts ordinarily attached to thrust (e.g., Lee 1979:133–134), it seems more reasonable to expect design changes in projectiles for greater range and accuracy at the expense of thrust. Again following Christenson, this may be accomplished by reducing the length and width of projectile shafts.

Overall projectile size and form are closely, but not perfectly, related to the similar properties of their points. On limited evidence, shaft diameter varies positively with base and/or neck width of arrow points and negatively with point length (Thomas 1978:467; Hamilton 1982:27). If projectile length and shaft diameter must decline to achieve improvements in accuracy and range, then projectile point width, especially neck and base width, also should decline.

This argument assumes that stalking was a major hunting practice in aboriginal eastern North America (cf. Waselkov 1978). If animals were taken principally in snares or by using a variety of drive and surround techniques, then the bow-and-arrow would have played a secondary role, at best, in procurement. If snares were used, projectiles would not have been needed at all. If mass acquisition techniques were used, most kills would have occurred at very close quarters, obviating any concern for accuracy, range or other ballistic performance requirements. Fortunately, primary (e.g., Wenhold 1936; Smith 1966:32) and secondary sources (Swanton 1946:312–321; Reidhead 1981; McCabe and McCabe 1984) have amply documented the importance of individual stalking of large game in aboriginal eastern North America. It is unnecessary to argue that either individual stalking or mass capture techniques were favored over the other. Obviously, mass capture was practiced in many areas (Waselkov 1978), but the practices were not mutually exclusive. As Calderon notes, for instance, winter drives among the Timucua of Florida were replaced at other times in the annual cycle by individual stalking (Wenhold 1936:13).

The argument further requires that the stalking of individual prey was sufficiently time-consuming and arduous to encourage efforts to reduce its costs. This assumption, though plausible, is difficult to evaluate. At least one primary source, however, lends it credence:

> Their manner of rambling through the woods to kill deer, is a very laborious exercise, as they frequently walk twenty-five or thirty miles through rough and smooth grounds, and fasting, before they return back to camp, loaded. (Williams 1930:432)

It is reasonable to assume that work of this nature, under changing conditions requiring closer budgeting of time and effort, would motivate aboriginal hunters progressively to seek to increase the range and accuracy of their weaponry. Among the practices designed to secure these results would be the continuous reduction in base and neck width and in overall point dimensions as observed in the American Bottom archaeological record.

## DISCUSSION

The preceding argument might explain the size reduction of points, but it does not necessarily explain the origins of the transition from dart to arrow. This transition may have occurred for substantially historical reasons, as Blitz (1988) suggests. But at least in part, it could also have occurred in response to the need to sustain a trend toward declining projectile size. That is, American Bottom hunters may have approached the physical limits of atlatl and dart technology during the process, and begun to experiment with and develop the bow and arrow as an alternative delivery system. At present, of course, this is no more than a plausible suggestion that does not explain the apparently contemporaneous adoption of the bow-and-arrow across large areas of North America (Blitz 1988).

Furthermore, the argument assumes that metric attributes of points responded to steady, directional selection over a considerable time interval. It assumes, in effect, a simple cause-and-effect relationship between a constant need for increasing range and accuracy of projectiles and a constant response involving size reduction in points, especially in neck width. Yet Braun's (1987) similar, if far more sophisticated, analysis of chronological trends in Woodland vessel-wall thickness shows how a sustained trend can actually be the composite product of a set of independent, at times opposing, selective factors. The changing intensity, direction and timing of these factors can reveal a great deal about the adaptive practices and processes of prehistoric cultures. Similar analysis of point assemblages might reveal similarly complex and fascinating selective factors, a possibility to consider in future research.

Most paleoanthropologists acknowledge that study of the fossil record describes the evolutionary process experienced by hominids but does not explain it; we know *what* happened, but not necessarily *how* or *why* it happened. Similarly, the study of continuous change in material culture such as projectile points describes a process but does not explain it. As Cotterell and Kamminga (1990:6) noted, "Today there is no robust theory of evolution in material culture." The diet-breadth model presented above stands as a possible explanation of changing metric attributes of American Bottom points, but others surely will be proposed. Stimulating thought along these lines, in fact, is an important goal of this paper.

In an important study, Odell (1988) reached conclusions directly at odds with one of this study's central theses: that bow-and-arrow adoption occurred no earlier than the first millennium A.D. There exist Old World examples of the kinds of arrow points that Odell sees (Shott 1993:435), and Patterson (1993) has adduced archaeological data that may support Odell. Odell's interpretation is not addressed directly here, but the debate on this issue should stimulate productive research.

To distinguish among alternative explanations for the timing of the dart-to-arrow transition and for changing point dimensions, we require a program of experimentation that can accurately measure the performance of complete dart and arrow projectiles (Schiffer and Skibo 1987:598). Using launching devices like Knecht's (1991:124) calibrated cross-bow to control and record the energy invested in firing, the relationship between point dimensions and ballistic and other performance criteria like penetration and breakage rates can be measured. Actualistic studies—experiments designed to resemble real-world contexts of use—could gauge the importance of criteria like firing rate, ease of use in cover, and the like, and how they are related to point dimensions. An experimental program of this nature need not presume the evolutionary theory approach advocated by Dunnell (1992; see also Cotterell and Kamminga 1990:10–11), but it shares the virtues of that approach by basing analysis upon well established universal theory developed in other fields.

The possible social dimensions and implications of the dart-to-arrow transition also deserve brief comment. Freeman (1985) described a process of cooption of innovations by early adopters. He introduced the concept of preemption rent to explain how, contrary to the tenets of conventional diffusion theory, those early adopters transformed a temporary advantage into a permanent condition of privileged access. Freeman's examples involve Kenyan agricultural practices that are intensive in both labor and capital. Prehistoric hunting techniques are not ordinarily viewed in similar terms, and indeed probably were not so intensive. But if technological innovations like the bow-and-arrow were invested with a measure of symbolic significance, it is at least possible that a similar process of preemption may have occurred with their adoption. We customarily regard such innovations as equally advantageous to all members of a social group, but we might at least consider the alternative possibility that they created or exacerbated inequities within Woodland societies. For instance, if the innovation was advantageous to all, then it should have been quickly adopted. But the evidence presented here suggests a gradual adoption. Perhaps the adoption was for some material reason, but alternatively it may reflect some socially inspired resistance by preemptors to widespread adoption.

## SUMMARY

This paper has addressed the timing and nature of the dart-to-arrow transition, and the nature and explanation of continuing variation in point size and form in Late Woodland assemblages of the American Bottom region. Using Thomas' (1978) classification function to identify dart and arrow specimens, it suggests that the transition began somewhat earlier, and unfolded less abruptly, than we commonly believe. Time-dependent change in point size was documented and provisionally explained by reference to behavioral ecology theory.

It seems clear that the several point types defined for the American Bottom region effectively partition a continuum of metric variation, rather than form a time-series of normative ideals. Although the data used here are limited, they at least suggest the considerable potential of continuous variation in material culture as a source of chronometry independent of conventional techniques and as a phenomenon requiring explanation. Braun (1985, 1987) has explored these potentials at length for ceramics, in the same general region. Further work on bifaces and other artifact classes will exploit even more of it.

If theory is developed more fully to link performance requirements to point size and form on the one hand, and economic and sociopolitical properties of aboriginal cultures on the other, then points can serve as more than simple time markers. They will instead become gauges of changing economic practices and their sociopolitical context. It is an extremely long stride from here to the methodological virtuosity and close empirical control of Braun's (1985, 1987) time-series analyses of Woodland earthenware, but the detailed study of continuous change in biface size and form holds considerable promise for broader and more detailed anthropological inference from archaeological remains.

## ACKNOWLEDGMENTS

Thanks are due to George Odell of the University of Tulsa for the invitation to participate in the Second Tulsa Conference on Lithic Analysis. On present evidence, we disagree fundamentally on the subject of this paper. Its appearance here demonstrates George's admirable commitment to spirited but civil scholarly discourse, a model that more archaeologists should emulate. Thanks are due as well to other participants in the conference, whose incisive comments measurably improved the paper. Steven Shackley of the University of California-Berkeley kindly provided metric data for the late prehistoric Tolowa specimens.

## NOTE

1. Stemmed, notched, and triangular bifaces that bear recognizable haft elements are commonly termed "points" because they are assumed, not unreasonably, to have functioned as points or tips of launched projectiles of various kinds. This implicit assumption can be evaluated by classification analysis (Thomas 1978; Shott 1993). Failing such analysis, we probably should term such artifacts "bifaces," for a property that can be plainly observed rather than assumed. "Point," though, is firmly entrenched in common use and is retained here, especially since the assumption underlying this identification is explicitly tested.

## REFERENCES

Amick, D. S. 1994. Technological Organization and the Structure of Inference in Lithic Analysis: An Examination of Folsom Hunting Behavior in the American Southwst. In *The Organization of*

*North American Prehistoric Chipped Stone Tool Technologies*, edited by P. Carr, pp. 9–34. International Monographs in Prehistory, Ann Arbor, MI.

Bareis, C. J., and J. W. Porter (editors). 1984. *American Bottom Archaeology: A Summary of the FAI-270 Project Contribution to the Culture History of the Mississippi River Valley.* University of Illinois Press, Urbana.

Binford, L. R. 1962. A New Method of Calculating Dates from Kaolin Pipe Stem Samples. *Southeastern Archaeological Conference Newsletter* 9:19–21.

Blitz, J. H. 1988. Adoption of the Bow in Prehistoric North America. *North American Archaeologist* 9:123–145.

Braun, D. P. 1985. Absolute Seriation: A Time-Series Approach. In *For Concordance in Archaeological Analysis: Bridging Data Structure, Quantitative Technique, and Theory,* edited by C. Carr, pp. 509–539. Westview, Kansas City.

Braun, D. P. 1987. Coevolution of Sedentism, Pottery Technology, and Horticulture in the Central Midwest, 200 B.C.-A.D. 600. In *Emergent Horticultural Economies of the Eastern Woodlands,* edited by W. Keegan, pp. 153–181. Occasional Paper No. 7. Center for Archaeological Investigations, Southern Illinois University, Carbondale.

Christenson, A. L. 1986a. Projectile Point Size and Projectile Aerodynamics: An Exploratory Study. *Plains Anthropologist* 31:109–128.

Christenson, A. L. 1986b. Reconstructing Prehistoric Projectiles from Their Stone Points. *Journal of the Society of Archer-Antiquaries* 29:21–27.

Churchill, S. E. 1993. Weapon Technology, Prey Size Selection, and Hunting Methods in Modern Hunter-Gatherers: Implications for Hunting in the Paleolithic and Mesolithic. In *Hunting and Animal Exploitation in the Later Paleolithic and Mesolithic of Eurasia,* edited by G. Peterkin, H. Bricker and P. Mellars, pp. 11–24. Archaeological Papers of the American Anthropological Association, No. 4.

Cole, F.-C., and T. Deuel. 1937. *Rediscovering Illinois: Archaeological Explorations in and around Fulton County.* University of Chicago Press, Chicago.

Cotterell, B., and J. Kamminga. 1990. *Mechanics of Pre-Industrial Technology: An Introduction to the Mechanics of Ancient and Traditional Material Culture.* Cambridge University Press, Cambridge.

Cundy, B. J. 1989. *Formal Variation in Australian Spear and Spearthrower Technology.* BAR International Series 546, Oxford.

Deller, D. B., and C. J. Ellis. 1988. Early Paleo-Indian Complexes in Southwestern Ontario. In *Late Pleistocene and Early Holocene Paleoecology and Archaeology of the Eastern Great Lakes Region,* edited by R. Laub, N. Miller and D. Steadman. *Bulletin of the Buffalo Society of Natural Sciences* 33:251–263.

Doolittle, W. E. 1984. Agricultural Changes As an Incremental Process. *Annals of the Association of American Geographers* 74:124–137.

Dunnell, R. C. 1992. Archaeology and Evolutionary Science. In *Quandaries and Quests: Visions of Archaeology's Future,* edited by L. Wandsnider, pp. 209–224. Southern Illinois University Press, Carbondale.

Dunnell, R. C., and J. K. Feathers. 1992. Late Woodland Manifestations of the Malden Plain, Southeast Missouri. In *Stability, Transformation, and Variation: The Late Woodland Southeast,* edited by M. Nassaney and C. Cobb, pp. 21–45. Plenum, New York.

Dwyer, P. D. 1974. The Price of Protein: Five Hundred Hours of Hunting in the New Guinea Highlands. *Oceania* 44:278–293.

Dwyer, P. D. 1985. A Hunt in New Guinea: Some Difficulties for Optimal Foraging Theory. *Man* (n.s.) 20:243–253.

Fawcett, W. B., and M. Kornfeld. 1980. Projectile Point Neck-Width Variability and Chronology on the Plains. *Wyoming Contributions to Anthropology* 2:66–79.

Fischer, A. 1989. Hunting with Flint-Tipped Arrows: Results and Experiences from Practical Experiments. In *The Mesolithic in Europe,* edited by C. Bonsall, pp. 29–39. John Donald, Edinburgh.

Forbis, R. 1960. The Old Woman's Buffalo Jump, Alberta. *National Museum of Canada Bulletin* 180:56–123.

Fortier, A. C., F. A. Finney, and R. B. Lacampagne. 1983. *The Mund Site (11-S-435).* American Bottom Archaeology, FAI-270 Site Reports, Volume 5. University of Illinois, Urbana.

Fortier, A. C., T. O. Maher, and J. A. Williams. 1991. *The Sponemann Site: The Formative Emergent Mississippian Sponemann Phase Occupations (11-Ms-517).* American Bottom Archaeology, FAI-270 Site Reports, Volume 23. University of Illinois, Urbana.

Frankel, D. 1988. Characterising Change in Prehistoric Sequences: A View from Australia. *Archaeology in Oceania* 23:41–48.

Freeman, D. B. 1985. The Importance of Being First: Preemption by Early Adopters of Farming Innovations in Kenya. *Annals of the Association of American Geographers* 75:17–28.

George, R., and R. Scaglion. 1992. Seriation Changes in Monongahela Triangular Lithic Projectiles. *Man in the Northeast* 44:73–81.

Gould, R. A. 1966. *Archaeology of the Point St. George Site and Tolowa Prehistory.* University of California, Publications in Anthropology, Vol. 4.

Guthrie, R. D. 1983. Osseous Projectile Points: Biological Considerations Affecting Raw Material Selection and Design among Paleolithic and Paleoindian Peoples. In *Animals and Archaeology 1: Hunters and Their Prey,* edited by J. Clutton-Brock and C. Grigson, pp. 273–294. BAR International Series 163, Oxford.

Hall, R. L. 1980. An Interpretation of the Two-Climax Model of Illinois Prehistory. In *Early Native Americans: Prehistoric Demography, Economy, and Technology,* edited by D. Browman, pp. 401–467. Mouton, The Hague.

Hames, R. B., and W. T. Vickers. 1982. Optimal Diet Breadth Theory as a Model to Explain Variability in Amazonian Hunting. *American Ethnologist* 9:358–378.

Hamilton, T. M. 1982. *Native American Bows,* second edition. Special Publication No. 5, Missouri Archaeological Society.

Hawkes, K., K. Hill, and J. O'Connell. 1982. Why Hunters Gather: Optimal Foraging and the Aché of Eastern Paraguay. *American Ethnologist* 9:379–398.

Hill, K., and K. Hawkes. 1983. Neotropical Hunting among the Aché of Eastern Paraguay. In *Adaptive Responses of Native Amazonians,* edited by R. Hames and W. Vickers, pp. 139–188. Academic Press, New York.

Johannessen, S. 1984. Paleoethnobotany. In *American Bottom Archaeology: A Summary of the FAI-270 Project Contribution to the Culture History of the Mississippi River Valley,* edited by C. Bareis and J. Porter, pp. 197–214. University of Illinois Press, Urbana.

Kelley, D. B. 1992. Coles Creek Period Faunal Exploitation in the Ouachita River Valley of Southern Arkansas. *Midcontinental Journal of Archaeology* 17:227–264.

Kelly, J. E. 1984. Late Bluff Chert Utilization on the Merrell Tract, Cahokia. In *Prehistoric Chert Exploitation: Studies from the Midcontinent,* edited by B. M. Butler and E. May, pp. 23–44. Center for Archaeological Investigations, Occasional Paper No. 2. Southern Illinois University, Carbondale.

Kelly, J. E., F. A. Finney, D. L. McElrath, and S. J. Ozuk. 1984. The Late Woodland Period. In *American Bottom Archaeology: A Summary of the FAI-270 Project Contribution to the Culture History of the*

*Mississippi River Valley*, edited by C. Bareis and J. Porter, pp. 104–127. University of Illinois Press, Urbana.

Kelly, J. E., A. C. Fortier, S. J. Ozuk, and J. A. Williams. 1987. *The Range Site: Archaic through Late Woodland Occupations*. American Bottom Archaeology, FAI-270 Site Reports, Volume 16. University of Illinois Press, Urbana.

Kelly, J. E., S. J. Ozuk, and J. A. Williams. 1990. *The Range Site 2: The Emergent Mississippian Dohack and Range Phase Occupations*. American Bottom Archaeology, FAI-270 Site Reports, Volume 20. University of Illinois Press, Urbana.

Kelly, L. S., and P. G. Cross. 1984. Zooarchaeology. In *American Bottom Archaeology: A Summary of the FAI-270 Project Contribution to the Culture History of the Mississippi River Valley*, edited by C. Bareis and J. Porter, pp. 215–232. University of Illinois Press, Urbana.

Klein, R., and K. Cruz-Uribe. 1984. *The Analysis of Animal Bones from Archeological Sites*. University of Chicago Press, Chicago.

Knecht, H. 1991. The Role of Innovation in Changing Early Upper Paleolithic Organic Projectile Points. *Techniques et Culture* 17–18:115–144.

Lee, R. B. 1979. *The !Kung San: Men, Women and Work in a Foraging Society*. Cambridge University Press, Cambridge.

Lorentzen, L. H. 1989. Form and Function of the Chodistaas and Grasshopper Springs Projectile Points. Unpublished ms. on file, Department of Anthropology, University of Arizona, Tucson.

Lynott, M. J. 1991. Identification of Attribute Variability in Emergent Mississippian and Mississippian Arrow Points from Southeastern Missouri. *Midcontinental Journal of Archaeology* 16:189–211.

MacArthur, R. H., and E. R. Pianka. 1966. On Optimal Use of a Patchy Environment. *American Naturalist* 100:603–609.

Martin, J. F. 1983. Optimal Foraging Theory: A Review of Some Models and Their Applications. *American Anthropologist* 85:612–629.

McCabe, R. E., and T. R. McCabe. 1984. Of Slings and Arrows: An Historical Retrospective. In *White-Tailed Deer: Ecology and Management*, edited by L. Hall, pp. 19–72. Stackpole, Harrisburg.

Milner, G. R. 1984. *The Julien Site (11-S-63)*. American Bottom Archaeology, FAI-270 Site Reports, Volume 7. University of Illinois, Urbana.

Milton, K. 1984. Protein and Carbohydrate Resources of the Maku Indians of Northwestern Amazonia. *American Anthropologist* 86:7–27.

Moshage, M. 1987. Appendix 1. Description of Material Types. In *The George Reeves Site*, edited by D. McElrath and F. Finney, pp. 393–399. American Bottom Archaeology, FAI-270 Site Reports, Vol. 15. University of Illinois Press, Urbana.

Odell, G. H. 1988. Addressing Prehistoric Hunting Practices through Stone Tool Analysis. *American Anthropologist* 90:335–356.

Parker, K. E. 1991. Sponemann Phase Archaeobotany. In *The Sponemann Site: The Formative Emergent Mississippian Sponemann Phase Occupations*, edited by A. Fortier, T. Maher and J. Williams, pp. 377–419. American Bottom Archaeology, FAI-270 Site Reports, Vol. 23. University of Illinois Press, Urbana.

Patterson, L. W. 1985. Distinguishing between Arrow and Spear Points on the Upper Texas Coast. *Lithic Technology* 14:81–89.

Patterson, L. W. 1993. Current Data on Early Use of the Bow and Arrow in Southern North America. *Ohio Archaeologist* 43:21–24.

Plog, S., and J. L. Hantman. 1990. Chronology Construction and the Study of Prehistoric Culture Change. *Journal of Field Archaeology* 14:439–456.

Redford, K. H., and J. G. Robinson. 1987. The Game of Choice: Patterns of Indian and Colonist Hunting in the Neotropics. *American Anthropologist* 89:650–667.

Reidhead, V. 1981. *Optimization and Food Procurement at the Prehistoric Leonard Haag Site, Southeastern Indiana: A Linear Programming Approach.* Indiana Historical Society, Prehistory Research Series No. 6(1).

Riley, T., G. Walz, C. Bareis, A. Fortier, and K. Parker. 1994. Accelerator Mass Spectrometry (AMS) Dates Confirm Early *Zea Mays* in the Mississippi River Valley. *American Antiquity* 59:490–498.

Saffirio, G., and R. Scaglion. 1982. Hunting Efficiency in Acculturated and Unacculturated Yanomama Villages. *Journal of Anthropological Research* 38:315–327.

Sahal, D. 1981. *Patterns of Technological Innovation.* Addison-Wesley, Reading, MA.

Schambach, F. F. 1982. An Outline of Fourche Maline Culture in Southwestern Arkansas. In *Arkansas Archaeology in Review,* edited by N. Trubowitz and M. Jeter, pp. 132–197. Arkansas Archeological Survey, Research Series, No. 15, Fayetteville.

Schiffer, M. B. 1979. A Preliminary Consideration of Behavioral Change. In *Transformations: Mathematical Approaches to Culture Change,* edited by C. Renfrew and K. Cooke, pp. 353–368. Academic Press, New York.

Schiffer, M. B., and J. Skibo. 1987. Theory and Experiment in the Study of Technological Change. *Current Anthropology* 28:595–622.

Seeman, M. F. 1992. The Bow and Arrow, the Intrusive Mound Complex, and a Late Woodland Jack's Reef Horizon in the Mid-Ohio Valley. In *Cultural Variability in Context: Woodland Settlements of the Mid-Ohio Valley,* edited by M. Seeman, pp. 41–51. MCJA Special Paper No. 7. Kent State University Press, Kent, Ohio.

Seeman, M. F., and C. A. Munson. 1980. Determining the Cultural Affiliation of Terminal Late Woodland-Mississippian Hunting Stations: A Lower Ohio Valley Example. *North American Archaeologist* 2:53–65.

Shott, M. J. 1990. Stone Tools and Economics: Great Lakes Paleoindian Examples. In *Early Paleoindian Economies of Eastern North America,* edited by K. Tankersley and B. Isaac, pp. 3–43. Research in Economic Anthropology, Supplement 5. JAI Press, Greenwich, CT.

Shott, M. J. 1992. Radiocarbon Dating as a Probabilistic Technique: The Childers Site and Late Woodland Occupation in the Ohio Valley. *American Antiquity* 57:202–230.

Shott, M. J. 1993. Spears, Darts, and Arrows: Late Woodland Hunting Techniques in the Upper Ohio Valley. *American Antiquity* 58:425–443.

Smith, J. 1966. *The Generall Historie of Virginia, New-England, and the Summer Isles.* UMI Press, Ann Arbor.

Spencer, L. 1974. Replicative Experiments in the Manufacture and Use of a Great Basin Atlatl. In *Great Basin Atlatl Studies,* edited by T. Hester, M. Mildner and L. Spencer, pp. 37–60. Ballena Press, Ramona, CA.

Speth, J. D., and S. Scott. 1989. Horticulture and Large Mammal Hunting: The Role of Resource Depletion and the Constraints of Time and Labor. In *Farmers as Hunters: The Implications of Sedentism,* edited by S. Kent, pp. 71–79. Cambridge University Press, Cambridge.

Swanton, J. R. 1946. *The Indians of the Southeastern United States.* Bureau of American Ethnology Bulletin No. 137. Smithsonian Institution, Washington, D.C.

Thomas, D. H. 1978. Arrowheads and Atlatl Darts: How the Stones Got the Shaft. *American Antiquity* 43:461–472.

Waselkov, G. A. 1978. Evolution of Deer Hunting in the Eastern Woodlands. *Midcontinental Journal of Archaeology* 3:15–34.

Watanabe, H. 1975. *Bow and Arrow Census in a West Papuan Lowland Community: A New Field for Functional-Ecological Study.* Occasional Papers in Anthropology, No. 5. University of Otago, New Zealand.

Wenhold, L. L. 1936. A 17th Century Letter of Gabriel Diaz Vara Calderon, Bishop of Cuba, Describing the Indians and Indian Missions of Florida. *Smithsonian Institution, Miscellaneous Collections* 95(16). Washington, D.C.

Williams, S. C. 1930. *Adair's History of the American Indians.* Promontory, New York.

Winterhalder, B. 1981. Optimal Foraging Strategies and Hunter-Gatherer Research in Anthropology: Theory and Models. In *Hunter-Gatherer Foraging Strategies: Ethnographic and Archaeological Analyses,* edited by B. Winterhalder and E. Smith, pp. 13–35. University of Chicago Press, Chicago.

Yellen, J. 1986. Optimization and Risk in Human Foraging Strategies. *Journal of Human Evolution* 15:733–750.

Yerkes, R. 1992. Review of "Early Paleoindian Economies of Eastern North America," edited by K. Tankersley and B. Isaac. *North American Archaeologist* 13:67–71.

# Part V

# Technique and Methodology

The distinction between theory and method has never been clearly delineated. "Theory," as used here, follows Bettinger (1987:128) in subsuming both major theories such as evolution and "Middle-Range Theory" (Binford 1977; Raab and Goodyear 1984). The latter includes, but is not restricted to, considerations of site formation processes. If "theory" concerns principles underlying certain phenomena, "method" concerns ways to investigate those principles.

It is difficult to conceive of any objectively oriented study of the world, or any part of the world, that failed to contain both method and theory, unless the object of the "study" is simply to observe. If any interactive manipulation is intended, then the researcher must begin with a concept of how the underlying principles work (i.e., theory), no matter how rudimentary or wrong that idea might be. Without such a concept it is impossible to proceed, because to do so would entail snatching randomly at hundreds of competing directions; and without any sort of explanatory model, a "right" answer might not even be recognized. By the same token, manipulative research requires method, because method governs the process used to provide the solution to the problem.

During the past few years the point has frequently been made that archaeologists (Moore and Keene 1983; Saitta 1983), and especially lithic analysts (Cross 1983; Torrence 1989), have emphasized particularistic concerns and methodological advances over generalist behavioral approaches. An entire book has been devoted to the "Law of the Hammer," which

> suggests that, although the methods we use are often appropriate to the task, too frequently a given method is used simply because it is the tool currently in hand. In these cases the pounding has produced more than a little noise, yet little anthropological understanding of prehistory has been constructed. (Moore and Keene 1983:4–5)

The problem with methodological dominance is that method may determine the nature of the theory that is developed, thereby perpetuating the status quo or "normal" science (Saitta 1983:302). While acknowledging this danger, it is nevertheless difficult to deny that the types of problems that can be addressed in archaeological research are, to some degree, dependent on technical advances. Landsat photography, for example, has spurred a consideration of distributional relationships that cannot fail to have had some effect on our perception of the theoretical principles governing them (Ebert 1988). The point is that theories and techniques are interdependent. Neither can exist without the other, and advances in one often affect the other.

The close association between theory and method can be amply demonstrated in lithic analysis, for our perceptions of prehistoric stone tool use, discard, manufacture and procurement have changed not only with alterations in behavioral theory, but also with changes in type of data and method of data accumulation. If lithic analysis is to advance beyond the "Law of the Hammer," it will have to demonstrate more clearly than ever before the anthropological relevance of its methods.

Two recently developed techniques, tool use-wear and piece refitting, have made major impacts on the ways lithic data are perceived. We are fortunate in this volume to be able to provide examples of both of these analytical modes.

Use-wear analysis, developed and popularized during the 1950s through 1970s by Sergei Semenov (1964), has experienced a tumultuous recent history, though the critical introspection that characterized its early days has mellowed somewhat. At this moment the most popular use-wear technique employs a relatively high-magnification, metallurgical microscope with incident lighting (Keeley 1980).

In this volume Marvin Kay examines tools from experimental and archaeological (Clovis) context for the purpose of reconstructing minute details of the hafting and use of specific tools. In the process he proposes a technical improvement to the older high-power technique. His system employs a microscope with high-magnification capabilities, but possesses reflected light Nomarski optics that allow for three-dimensional views of tool surfaces. Kay's technique is largely untested, but it deserves to be investigated as an alternative to high-power techniques already in existence.

The other technique highlighted here involves the refitting of pieces to one another, an analytical process of venerable antiquity among pottery analysts but only recently experiencing renewed interest among lithic analysts (Van Noten 1978; Hofman and Enloe 1992). Toby Morrow's refitting analysis of artifacts from the Early Archaic Twin Ditch site in Illinois has provided a considerable amount of useful information. For example, it has established a minimal disturbance level for the site; the location of two knapping episodes, each of short duration; and an interpretation of tool curation and use behavior.

These are informative parameters and they persuasively demonstrate the effectiveness of the refitting technique.

## REFERENCES

Bettinger, R. L. 1987. Archaeological Approaches to Hunter-Gatherers. *Annual Review of Anthropology* 16:121–142.

Binford, L. R. 1977. General Introduction. In *For Theory Building in Archaeology*, edited by L. Binford, pp. 1–10. Academic Press, New York.

Cross, J. R. 1983. "Twigs, Branches, Trees, and Forests: Problems of Scale in Lithic Analysis. In *Archaeological Hammers and Theories*, edited by J. Moore and A. Keene, pp. 87–106. Academic Press, New York.

Ebert, J. I. 1988. Remote Sensing in Archaeological Projection and Prediction. In *Quantifying the Present and Predicting the Past*, edited by W. Judge and L. Sebastian, pp. 429–492. Bureau of Land Management, U. S. Department of the Interior, Denver.

Hofman, J. L., and J. G. Enloe (editors). 1992. *Piecing Together the Past: Applications of Refitting Studies in Archaeology*. BAR International Series 578, Oxford.

Keeley, L. H. 1980. *Experimental Determination of Stone Tool Uses: A Microwear Analysis*. University of Chicago Press, Chicago.

Moore, J. A., and A. S. Keene. 1983. Archaeology and the Law of the Hammer. In *Archaeological Hammers and Theories*, edited by J. Moore and A. Keene, pp. 3–13. Academic Press, New York.

Raab, L. M., and A. C. Goodyear. 1984. Middle-Range Theory in Archaeology: A Critical Review of Origins and Applications. *American Antiquity* 49:255–268.

Saitta, D. J. 1983. The Poverty of Philosophy in Archaeology. In *Archaeological Hammers and Theories*, edited by J. Moore and A. Keene, pp. 299–304. Academic Press, New York.

Semenov, S. A. 1964. *Prehistoric Technology*. Translated by M. W. Thompson. Adams and Dart, Bath, England.

Torrence, R. 1989. Tools as Optimal Solutions. In *Time, Energy and Stone Tools*, edited by R. Torrence, pp. 1–6. Cambridge University Press, Cambridge.

Van Noten, F. L. (editor). 1978. *Les Chasseurs de Meer*. Dissertationes Archaeologicae Gandenses 18. De Tempel, Brugge.

*Chapter* **11**

# *Microwear Analysis of Some Clovis and Experimental Chipped Stone Tools*

Marvin Kay

## ABSTRACT

Stone, bone or ivory implements used to kill and butcher a menagerie of now largely extinct animals are found at late Pleistocene terrestrial big game kills. Tool function of big game kill artifacts is often assumed but not confirmed by reference to artifact form and archaeological context. Experimentally produced microwear traces, however, do provide an empirical basis to judge the likely use of the archaeologically derived artifacts. Microwear on experimental tools and Clovis points from Colby unequivocally show consistent patterns of tool use as projectile points and butchering tools; evidence of site-specific haft binding technique, of tool maintenance, and use-life histories. A further evaluation of the "Keeley method" of identifying micropolishes, how they form and rates of formation, indicates that conventional incident light microscopy is severely limited in its application for microwear studies.

## INTRODUCTION

Inspired by S. A. Semenov's (1964) pioneering studies and later expositions by Lawrence H. Keeley (1974, 1980; Newcomer and Keeley 1979), many recent

**MARVIN KAY** • Department of Anthropology, University of Arkansas, Fayetteville, Arkansas 72701.

*Stone Tools: Theoretical Insights into Human Prehistory*, edited by George H. Odell. Plenum Press, New York, 1996.

microwear analysts of chipped stone artifacts have dealt mainly with unifacially flaked or unflaked tool working edges in attempts to define tool function and material worked, or "contact material." These researchers usually employ similar incident light microscopes (and, less commonly, scanning electron microscopes [SEM]) at magnifications generally greater than 100 diameters but in differing ways and often toward differing ends. Semenov's primary impact has been in Russia and the former republics of the Soviet Union, where "traceological" studies of microscopic polishes, striations and edge damage are used to delineate tool function, or use. Empirically grounded in thousands of experimental tool use replications carried out under naturalistic field conditions by either Semenov or his successor, G. F Korobkova, and her staff, the traceological approach has placed priority on the analysis of striations on tool edges and surfaces as the key to understanding tool use and, to a lesser degree, materials worked (Phillips 1988; Korobkova, personal communication, 1991). Echoing comments Semenov and other traceologists have made, Cotterell and Kamminga (1990:159) succinctly identified the heuristic potential of abrasive microwear evidence: "The importance of use-scratching . . . is that it provides evidence of how a tool was oriented during use as well as about the nature of the material worked."

The "Keeley method" (Newcomer et al. 1986, 1988; Rees et al. 1991), now popular in western Europe and North America, contrasts significantly to the traceological approach in its focus upon identification of materials worked by stone tools from the analysis of micropolishes. Micropolishes, in the words of one analyst (Moss 1987:475), are "a visual phenomenon, a visible alteration to the flint surface" impervious to chemical and mechanical cleaning (but see Moss 1987:475). The polishes are generally not regarded as having been caused by abrasion, but rather are additive—to quote Keeley (1980:166–167), "frictional heat" and the "melting" of contact materials impart distinctive optical qualities. Among the more celebrated examples of the additive polishes, according to Keeley (1980:60–61), is corn gloss (but see Ahler 1979:312) due to the spreading of plant opal. Evidence commonly used to interpret micropolishes includes "the surface texture, brightness and degree of development of the polish, and the extent and location of the polished area," according to Bamforth (1988:12). Of these attributes, polish brightness is regarded as the prime signature of different contact materials and the measure to distinguish among different polishes, even on surfaces of different textures (Keeley 1980:23–24, 62–63; see also the comments by Bamforth 1988:13). Thus Keeley concludes that "hide polishes are relatively dull and rough[,] corn glosses . . . are bright and smooth" (1980:62); wood polish is "very bright [and] very smooth in texture" (1980:35); and meat cutting polish "varies in brightness but . . . [is] relatively dull . . . [and] can be distinguished by its different surface texture and by its pronounced greasy luster" (1980:53). Building in part on Vaughan's (1985) recognition of degrees of polish development and the conclusion that there are overlaps among some polishes, as noted initially by Keeley (1980:56, 61), Bamforth (1988:20–23) calculated identification error rates of "20–30%, but [varying] considerably depending on how long a tool was used."

Keeley (1980:176) included rounding and edge damage along with polish in defining tool function and material worked, and others inspired by his research have developed explicit and reasonable criteria for recording wear trace data. Hurcombe (1988:4) provided a useful, although not exhaustive, list: "for polish—brightness, texture, distinctiveness, invasiveness, location both on the tool and in respect to the edge, plus edge roundness; for striations—depth, width, orientation with respect to the edge and number within a defined area; for attrition—its extent from the edge and the presence of edge damage or crushing."

In comparing aspects of the traceological approach and the Keeley method, I summarize in detail in this paper a microwear analysis of a very limited number of experiments in unifacial and bifacial chipped stone tool use. For further illustration, I include an equally small number of Clovis artifacts from the Colby archaeological site in Wyoming. These artifacts and experimental replicates of Clovis points and likely butchering tools are representative of microwear evidence I have analyzed for other Clovis big game kills, and have larger ramifications beyond the purposes of this paper that are not fully developed here. All of the materials were supplied at various times since 1985 by George C. Frison, who has published descriptions of his experiments and of the Colby archaeological site (Frison 1986a,b, 1989). The experimental microwear examinations were conducted between 1985 and 1988; the Colby specimens have been unearthed since 1992. In the case of the experimental replicates, Frison submitted them without further comment about the specifics of their use or the manner in which they were either held or, in the case of the points, bound to a foreshaft. I have described aspects of the experimental microwear examinations at the Society for American Archaeology meetings in 1986 (New Orleans, Louisiana) and 1988 (Phoenix, Arizona), and at the 1991 Second Soviet-American Symposium on Upper Paleolithic-Paleoindian Adaptations in Denver, Colorado.

The microwear patterns are independent of stone materials and are observable at magnifications greater than 100 diameters. The overall methodology originates from, complements and provides, in my estimation, more specific information about tool use than is available through either macroscopic or low-power magnification (<100X) microwear studies alone. As an advocate of the latter approach (Odell 1990; see also for references) noted, the low-power magnification examinations are likely to be coarse-grained evaluations of worked material hardness. The analyses described here take substantially longer to perform and also require more expensive technical equipment than needed for either macroscopic or low-power microscopic tool use evaluations. The added expense in time and equipment is, in my opinion, justified by the results.

In developing information of general theoretical interest about both the design criteria for specific tool classes and their intended use, this analysis further documents patterns of haft binding wear traces, evidence of tool maintenance, and use-life histories. The latter have ramifications for evaluations of manufacturing, curation and intrinsic (if not functional) values of stone tools (Shott 1989:22–27), whereas the former provides a clue to group composition and individual differences in hafting technique that may have been important in tool usage.

## MATERIALS USED

Based on North American late Pleistocene archaeological evidence concerning terrestrial big game kill, Frison (1986b, 1989) conducted chipped stone tool experiments in "killing" and butchering mortally wounded African elephants (*Loxodonta africana*) in Zimbabwe. These modern-day pachyderms differ in evolutionary histories and possibly predator-avoidance mechanisms from their extinct North American counterparts. But, as proboscidians, similarities far outweigh differences insofar as their likely effects on chipped stone tools go. The "killing" experiments used bifacially flaked, Clovis projectile point replicates, hafted onto foreshafts attached to spears and used to penetrate elephant carcasses by both thrusting and throwing facilitated with an atlatl. The Clovis point replicates are mainly fine-grained cherts of several lithologies. Carcass butchering was done with either a fine-grained, porcelanite ovate biface or a large, Spanish Diggings (Hartsville Uplift, Wyoming) coarse-grained quartzite flake, backed by steep unifacial retouch on one edge. The biface, however, was lost in the butchering experiments.

The experimental results are valuable in two respects. First, insight is afforded into procedures and effort needed to kill and butcher extinct proboscidians, and the likely artifact by-products from these operations. Second, documented stone projectile points and a butchering tool are available for an independent evaluation of tool function—namely, microwear analysis that is the subject of this paper.

The discovery and subsequent excavation of the Colby site, a mammoth (*Mammuthus columbi*) kill and butchery site with evidence of meat piles, or "insurance caches," produced four reasonably fine-grained chert, bifacially flaked, chipped stone points. As Frison (1986a:91–96) noted, three are tear-drop in shape and differ in outline from "classic" Clovis points, but share technological attributes of manufacture—not restricted to but including, of course, basal fluting. In outline and other respects, the fourth does resemble the typical Clovis point. One of the tear-drop points has an obliquely fractured tip that was not recovered in the excavation. It seems likely to be a by-product of impact fracturing as initially described from archaeological materials by Witthoft (1968:32; see also Frison 1971:81, 1974:95–99; Frison et al. 1976:42–44) and subsequently experimentally replicated by Ahler (1971) and others (Barton and Bergman 1982; Bergman and Newcomer 1983; Odell and Cowan 1986). However, its form is sufficiently ambiguous in appearance to be able to be confused with a bending or snap fracture perhaps caused by specimen manufacture or resharpening; Frison (1986a:94) does, indeed, document the possibility that this specimen was reworked. The other three Colby specimens have relatively sharp tips, but not ones that are typically needle-like in form or sharpness. Fragile, easily damaged needle-sharp tips, however, are an essential design criterion for penetration with bifacially fashioned projectile points (Frison et al. 1976), and are associated specifically with Clovis points (Graham et al. 1981). Their absence from the Colby specimens could be explained as a function of either projectile point use breakage or intended usage as something other than a projectile point. In either case, macroscopic and low-powered microscopic evidence is insufficient to certify which is the more likely. Nor would these approaches deal effectively with otherwise seemingly undamaged tools (to include, in this study, the two experimental Clovis point replicates) or ones whose breakage could be confused with prior errors of manufacture.

In sum, the sample of experimental and archaeologically derived tools would be analytically ambiguous were they subjected only to either conventional macroscopic and/or low-power microscopic analysis. They provide an instructive test of the higher magnification approaches, and do so for artifacts that normally have been excluded—namely, bifacially flaked tools whose formalized shapes are functionally subdividable. Diagnostic microwear evidence exists on all specimens.

## METHODOLOGY

The original intent of the microscopic examination methods was to insure comparability of results with Keeley's published micropolish descriptions and photomicrographs. The microwear approach used here (Figure 1) allows for sequential macroscopic and microscopic observations and is patterned after Keeley, but eliminates the two-dimensional image problem he faced (Keeley 1980:12–14). Although lower magnifications are employed, most helpful is the intermediate magnification range of 100 to 400 diameters using a reflected-light, differential-interference microscope with polarized light Nomarski optics (Hoffman and Gross 1970). The optical qualities of this microscope are superior to those of most incident-light microscopes presently used for microwear studies, but—to my knowledge—have received only minimal attention by microwear analysts in western Europe and America (see Dumont 1982 for a quantified application using a similar microscopy system). Its Nomarski optics capability is ideally suited to lithic microwear analysis, because its color divisions of polarized light allow for three-dimensional views of tool surfaces. Image resolution, or clarity, increases with magnification (in contrast to incident-light microscopes but similar to SEM), while depth of field decreases. Unlike SEM, metallic coating of a tool surface is unnecessary. Striations and micropolishes are readily observed and easy adjustment of the Nomarski optics allows for the best display of wear traces. Micropolishes are optically bright and characteristically undifferentiated, based

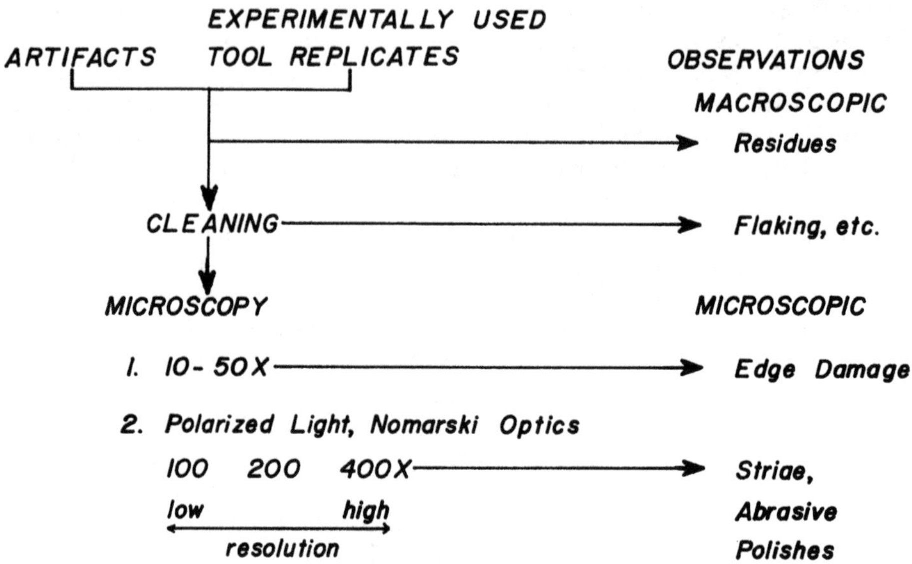


Figure 1. Methodological flowchart of microwear analysis.

on observations of micropolishes from tool use replications on other contact materials not further described here. For most archaeological specimens, it is also unnecessary to carry out the extensive and potentially destructive chemical cleaning Keeley (1980:11) advocates to insure that the microscopic inspection of a tool surface or edge is unimpeded by organic or inorganic residues.

The experimental tool replicates (Figures 2, 3) were, however, chemically cleaned in accordance with Keeley's general approach. The major difference was in substituting KaOH for NaOH to remove organic residues and to reduce the likelihood of chemically patinating the tool surfaces, a problem with NaOH (Keeley 1980:11) but not KaOH. The archaeological specimens were not subjected to any chemical cleaning beyond the occasional use of methyl alcohol to remove oily films from tool surfaces. These came from the plastocene clay used to mount the points on a slide plate or from previous handling prior to my analysis. Latex surgical gloves were worn while handling both the chemically cleaned and uncleaned specimens, except when lateral edges were felt for edge irregularities and basal grinding.

Parenthetically, one of the two experimental replicates sent by Frison in May, 1985, after he first removed it from its foreshaft but without further cleaning, had a girdle of dried blood in the haft element area (Figure 3). Once the major blood residues had been removed from this specimen, blood residues were found in step fractures as microthin coatings, or films, that apparently collect by capillary

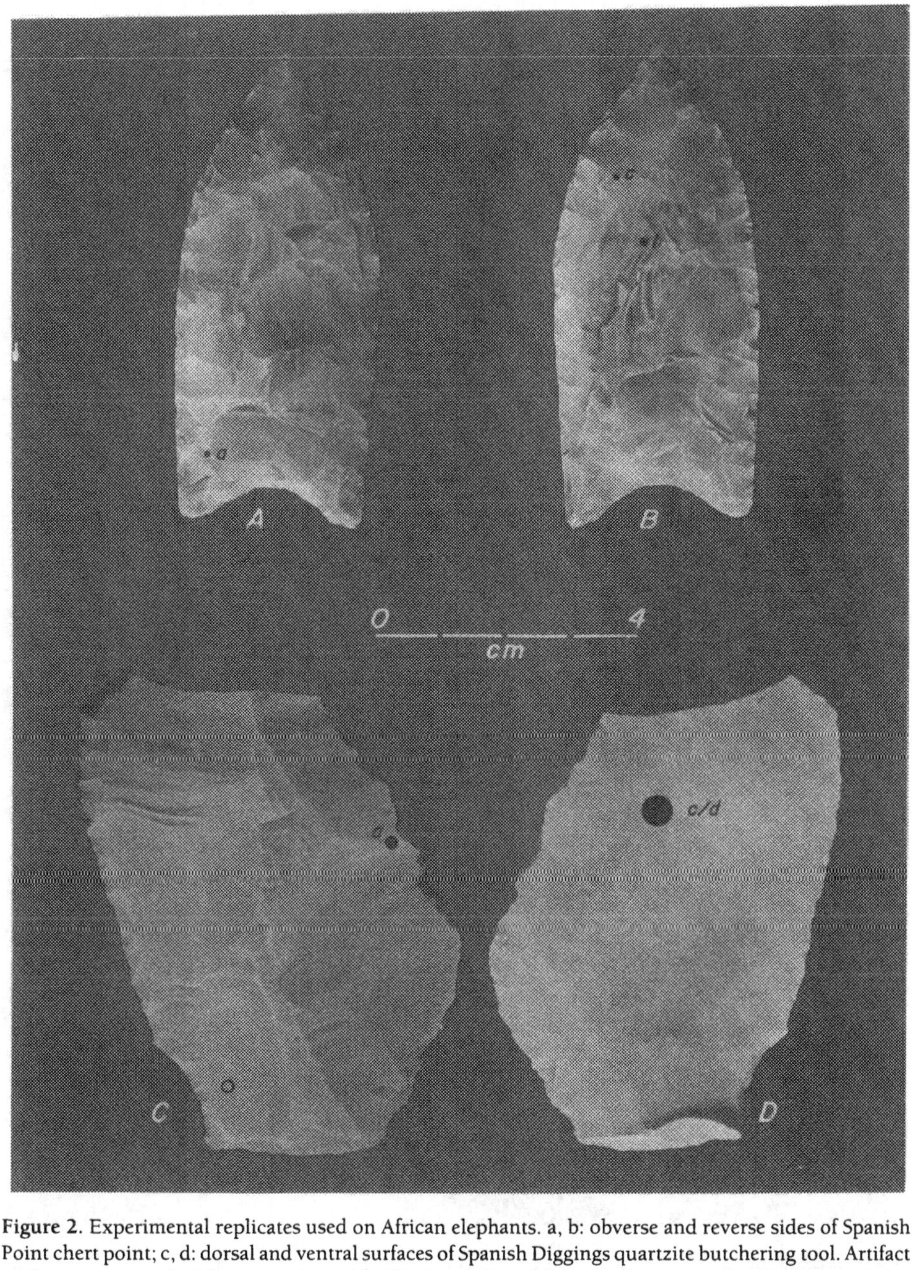


**Figure 2.** Experimental replicates used on African elephants. a, b: obverse and reverse sides of Spanish Point chert point; c, d: dorsal and ventral surfaces of Spanish Diggings quartzite butchering tool. Artifact surfaces are coated with ammonium chloride "smoke" and letters adjacent to closed circles refer to photomicrographs.

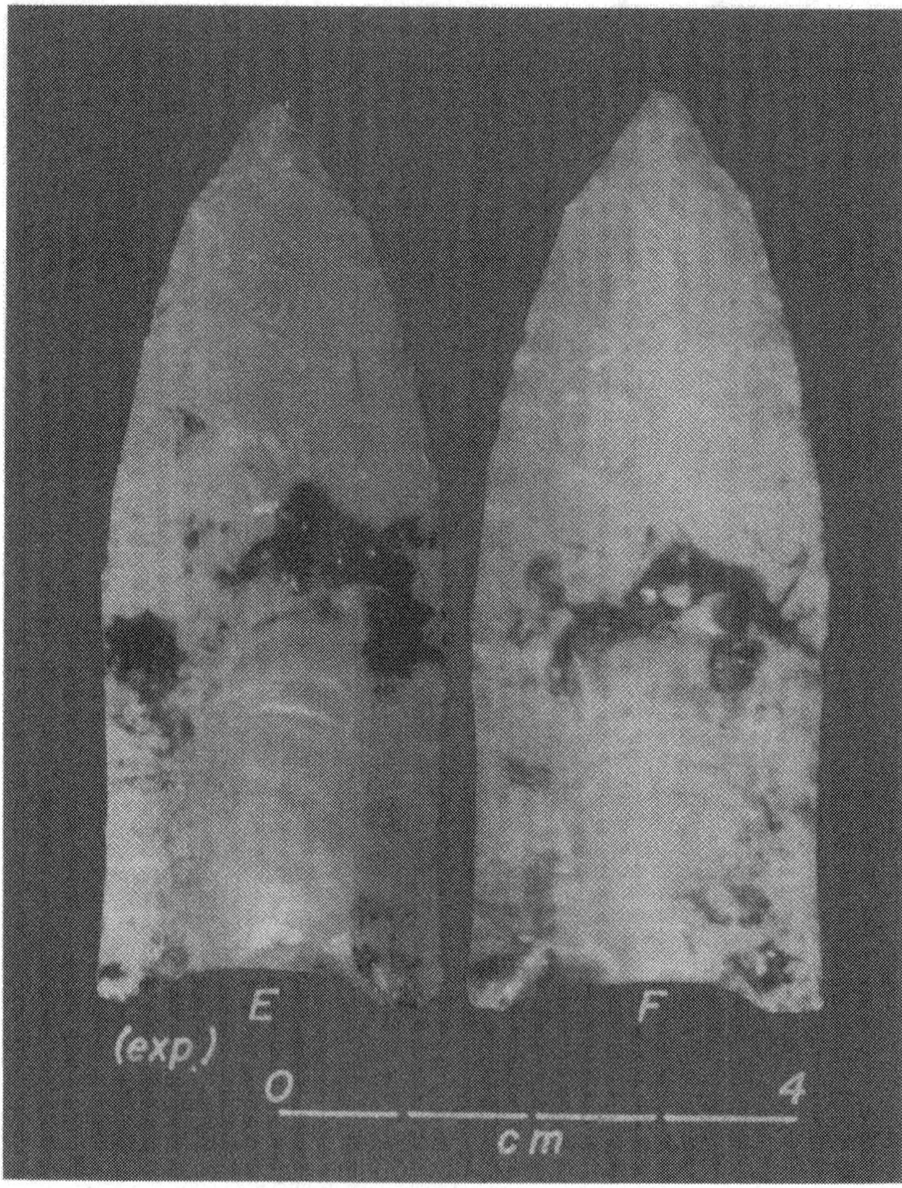


Figure 3. Experimental Clovis point replicate of Edwards chert prior to removal of elephant blood residues.

action. Potential blood residues undoubtedly occur in analogous locations on archaeological specimens and should be evaluated (see Hyland et al. 1990).

At magnifications greater than 100 diameters, wear traces can be distinguished from edge damage and overall flaking characteristics. Indeed, at these magnifications the concern to reliably separate edge damage from intentional bifacial retouch (Frison 1979:264–267) becomes irrelevant to the technofunctional analysis of tool wear. Bifacially flaked stone tools can be evaluated for microwear traces in the same manner as unifacially flaked or unretouched stone tools, because the higher magnification wear traces cross-cut or extend beyond the grosser details of microflaking. In either case, oriented micropolishes and striae can be located, photomicrographed, and measured as needed.

To evaluate wear traces the full range of magnifications was employed, going from 100 to 200 and then 400 diameters and using both bright and dark field illumination. Tool surface scans of both surfaces and the edges of the tools were accomplished at magnifications of 100 or even 200 diameters. Further examinations at 400 diameters were done whenever possible to better characterize wear traces. Photomicrographs were routinely taken with a 4X5 camera back and Polaroid Professional 55 positive/negative film. A green filter and dark field illumination were used in many cases to improve, or enhance, the contrast of the photomicrographs. The orientation and location of the photomicrograph were recorded relative to that of the artifact. Photomicrograph size was determined by photographing a micrometer disk scale at the appropriate magnifications. Wear traces were further described, even when not photomicrographed. The location of wear traces discovered at intermediate-range magnifications relative to edge damage was also noted in the comparisons between macroscopic and lower-powered magnifications. In most cases there was a direct correlation between edge damage and wear traces near or at a tool edge.

Minor but important additions to the overall methodology have also been made. Subsequent to the examination of the experimental tool replicates, the orientation data for photomicrographs have been systematized. Initially, I noted (and continue to note) the photomicrograph orientation on a contact print with respect to which end of the photomicrograph faced the proximal or distal end of the artifact. I now routinely position a photomicrographed artifact still-mounted on its microscope slide plate onto either an enlarged photograph or outline drawing of the artifact and then inscribe the edge of the slide plate onto the illustration; by so doing, it is possible to get a reasonable match with the actual photomicrograph orientation. I also, on occasion, take overlapping photomicrographs to further show the arrangement of wear traces within a photomosaic. Because slight changes in the dip of the object can dramatically change the microscopic image, it is also useful and instructive to position an artifact or tool replicate on a slide plate and then examine it again for wear traces. For archival and other analytical purposes, I would recommend usage of silicon casts.

In the present study the most crucial information concerns: (1) the placement, orientation, and cross-cutting relationships among use-wear striations, plus their width, depth, and number; (2) for polishes, the degree of development, texture, area, and placement on a tool surface or edge; and (3) for edge damage, its relationship to other use-wear traces.

## EXPERIMENTAL TOOL MICROWEAR ANALYSIS RESULTS

### Butchering Use Wear

Although coarse-grained, the quartzite flake tool (Figure 2) exhibits unambiguous and functionally diagnostic use-wear traces. For purposes of understanding tool function, most informative are the clearly striated micropolishes (Figure 4) that originate at the cutting edge and could be oriented relative to tool morphology. Note that this micropolish is on the dorsal face and is intermediate in development between a merely abrasively dulled surface and ones where surface

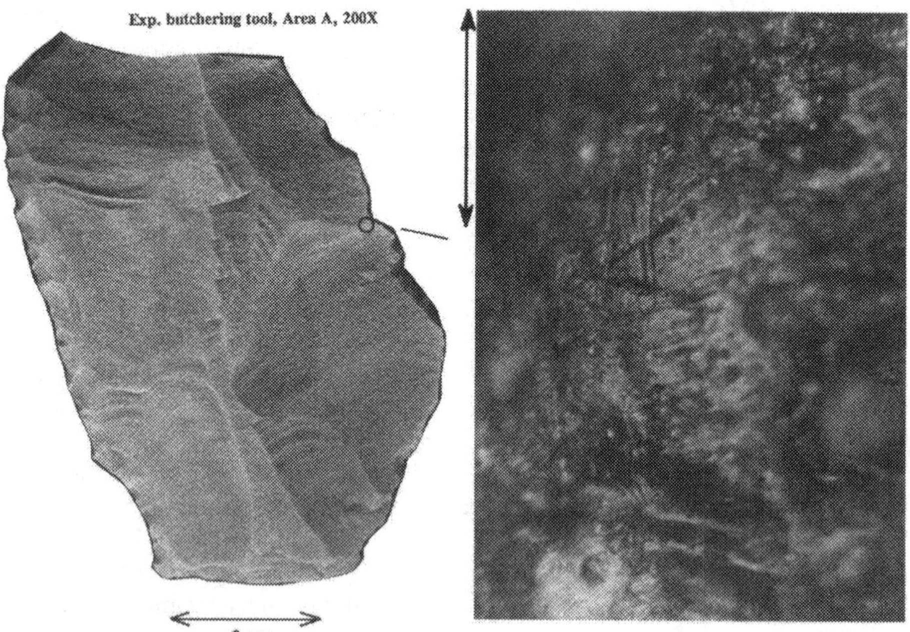


Figure 4. Spanish Diggings quartzite butchering tool: intermediately developed polish and striations near edge at dorsal surface area A (circled on specimen), 200X. (Note: the orientation of all photomicrographs [Figures 4–15] is normal to the longitudinal axis of the artifact; the arrowed scale bar on the left side of the photomicrograph is the same for each photomicrograph and is 0.1 mm.)

microtopography is obliterated (Figure 5a). Intermediate and most extensively developed polishes, however, occur in the approximate center of the ventral flake surface as discontinuous patches in areas of elevated microtopography, and indicate an asymmetry in tool manipulation.

The most developed abrasive polishes are uniformly flat, nearly featureless (note the striation, however, in Figure 5a), and highly reflective due to abrasive planing, which is all the more remarkable given the quartzite's coarseness. Because these polishes occur exclusively on the ventral surface of this experimentally produced tool replicate and in conjunction with other clearly striated, intermediately developed polishes (Figure 5b), this extensive abrasive planing is use-related in an unambiguous manner. It should not be confused with the "bright spots" produced in the rubbing of rocks in a moist soil medium, or with some other natural abrasive process that might have affected the exterior (distal) surface of the tool prior to its manufacture.

Keeley (1980:28, pls. 8,10) referred to similar "abrasive beveling" caused by hard hammer retouch as being "unlike any type of microwear." Vaughan's (1985:185–187) descriptions and photomicrographs (1985:134) of artifacts from Cassegros (but not of tool replicates) allow no room for doubt that he, too, witnessed the same thing. He described for elevated microtopography "glossy polish spots" of nonuse origin that occur as both flat and featureless, and ripply polishes. He further observed, "In comparing the special polish spots to other *known* microwear polishes, it is clear that the highly smoothed surface texture, the ripply and very flat topographies, the restricted size of the spots, and their random distribution over the flints do not resemble the features of *any* micropolish produced in the use and nonuse tests reported . . . or by any other researcher (sic)" (emphasis added). He added that it is likely to form by "a process of dissolution and reformation of silica gel" in postdepositional movement of rocks within a moist soil medium. Nevertheless, it clearly occurs as a by-product of the experimental elephant butchering, but clearly not in a random fashion, nor by the dissolution and redeposition of a silica gel. As we shall see later, it also occurs in a functional, or use-related way on one of the points from the Colby site.

The intermediate polishes are visually similar to the impact and haft binding polishes of the projectile points, described and illustrated later. They would be indistinguishable were it not for the origin and orientation of striae and polishes. In theory, the extensively developed polishes could occur on chipped stone projectile points also, but are less likely for reasons of intensity of use. In essence, a projectile point is a one- or two-stroke tool. It goes in and may come out. Microwear develops at least in part as microparticles dislodge from the tip and form an abrasive film that drags along the blade surface with further penetration. Carcass butchering is a prolonged and repetitive process of multiple cutting strokes in one or several directions. The time invested in butchering, as opposed to piercing, a carcass is necessarily greater. And so is the likelihood of more extensive damage to the cutting edge. Cutting edge damage, causing detachment

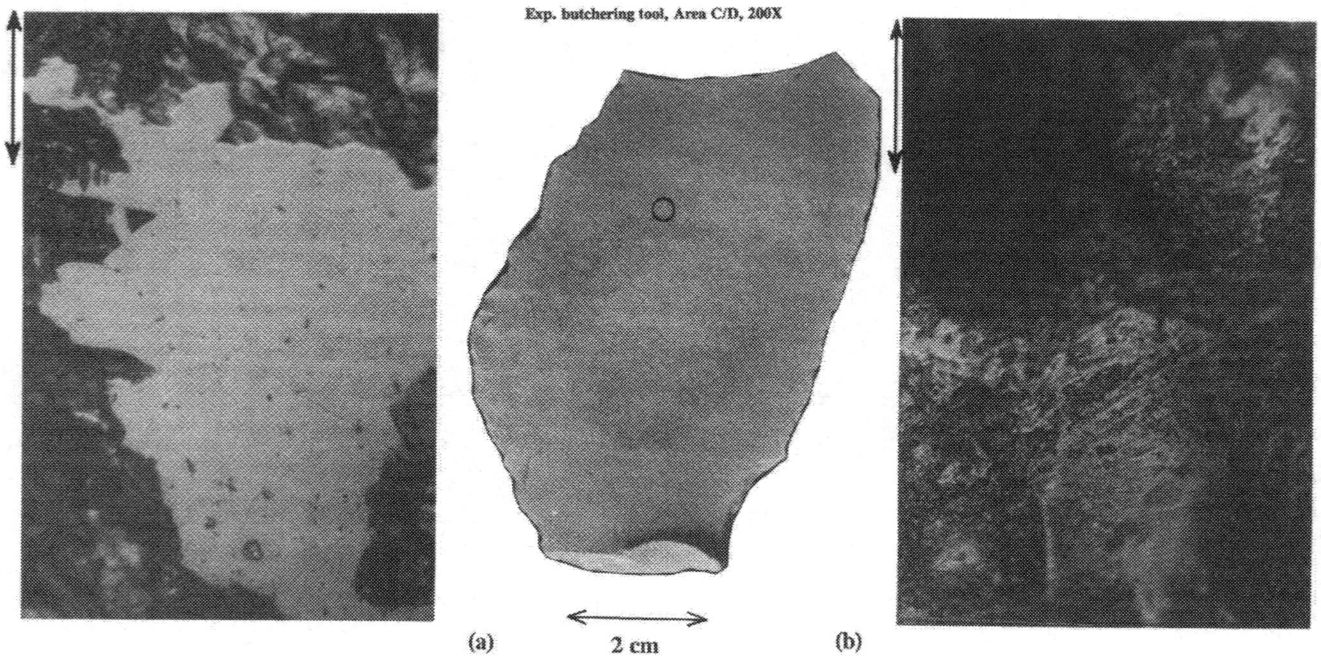


**Figure 5.** Spanish Diggings quartzite butchering tool interior at ventral surface area C/D′ (circled on specimen), 200X: (a) extensively developed polish and striations (EDP); (b) intermediately developed polish and striations (IDP).

of small flakes and microparticles, would have been one, nearly constant, source of an enveloping "cloud" of carcass-imbedded abrasives. With specific reference to the elephant butchering experiments using the quartzite tool, it is clear that, irrespective of lithology, the main factors controlling polish development are repetition of use and retention of a sharp, if not stable, cutting edge. The unprepared or ill-prepared edge of even an extremely coarse-grained cutting tool will, with sufficient use and edge failure, cause development of extensive micropolishes.

Macroscopic damage and microwear evidence point to the tool's unretouched right edge as the cutting edge, completely consistent with Frison's (1989:780) account of its use. The adjacent ventral surface bore the brunt of carcass contact, extending almost two-thirds across its width, and constitutes the tool's "leading" surface. Were the implement held in the right hand, for illustration, its dorsal surface and striking platform would rest in the palm with fingers and thumb grasping its surfaces adjacent to the laterally backed left edge; the ventral surface and unretouched right edge would be either facing or in an "up" position.

Butchering proceeded through a combination of sawing and slicing strokes, and produced progressive cutting edge attrition. The result is a characteristic pattern of intersecting striae, most convincingly scored on the dorsal surface adjacent to the edge (Figure 4) but also observed repeatedly on the ventral surface (Figure 5b). Sawing strokes in a back-and-forth rocking motion produced, at a tangent to the cutting edge, intersecting striae with a measured angle difference of about 70 degrees. These contrast with slicing stroke striae that parallel the cutting edge and indicate an overall direction of cutting normal to the tool's longitudinal axis.

## Projectile Point Use-Wear

Two Clovis replicates (Figures 2, 3) are made of fine-grained cherts, as are most of the Clovis points examined. They were made by Bruce Bradley from light gray, Edwards Plateau chert from Texas and a nearly black, Spanish Point chert from Wyoming. The former was used in spear thrusting experiments to pierce seven dead African elephants; the latter, in atlatl spear-throwing at downed elephants, which proved quite successful in deeply penetrating their carcasses.

Microwear evidence on the projectile points is unambiguous and functionally diagnostic. Their identification follows the functional differences of tool morphology readily apparent on chipped stone points (Binford 1963:196–199). In the case of the Clovis point replicates (and the points from Colby also), the haft element is assumed to correspond with the laterally ground, proximal portion of the point. The blade element corresponds to the area distal to the haft element, which has unground lateral edges and ends at the tip.

Wear traces of impact and carcass penetration begin near the tip and extend toward the haft along the blade. Polishes occur only on flake ridges or high spots, or sloping surfaces distal to high spots or flake ridges; that is, on the leading aspect of the tool facing its point of penetration. Polishes often occur with similarly-oriented striae (Figure 6). Impact striae also occur in non-polished areas and often connect polish patches. The smooth, finely striated polishes have a "melted" appearance, or texture. Their greatest development is somewhat back from the tip, perhaps as much as one-quarter to one-third the total blade length from the tip, and may be indicative of tip resharpening after use. The polishes are spotty and generally discontinuous. They, along with the impact striae, are not oriented or associated with blade edges and surfaces adjacent to the edge. Rather, their orientation and nearly symmetrical bifacial placement are consistent with one

**Figure 6.** Impact polish and striations at area C (circled on specimen) on Spanish Point chert Clovis replicate, 400X.

direction only, the direction of tip impact and penetration. This microwear evidence is indicative of but one function: that of a projectile point.

Two other microwear types (initial haft preparation or grinding, and haft binding to a foreshaft) are apparent on the projectile points. These wear traces vary with the direction of lateral edge grinding and hafting approaches. Lateral edge grinding is local to the edge and often is overlain by other microwear evidence. Haft binding wear traces are observed at the lateral edge and, as illustrated in Figure 7, on interior flake arrises on the faces of the haft element adjacent to the channel flake scars of the basal flutes. Their orientation and placement are consistent with functional requirements to bind the tool to its haft and with subsequent mechanical forces exerted on the haft element during tool use.

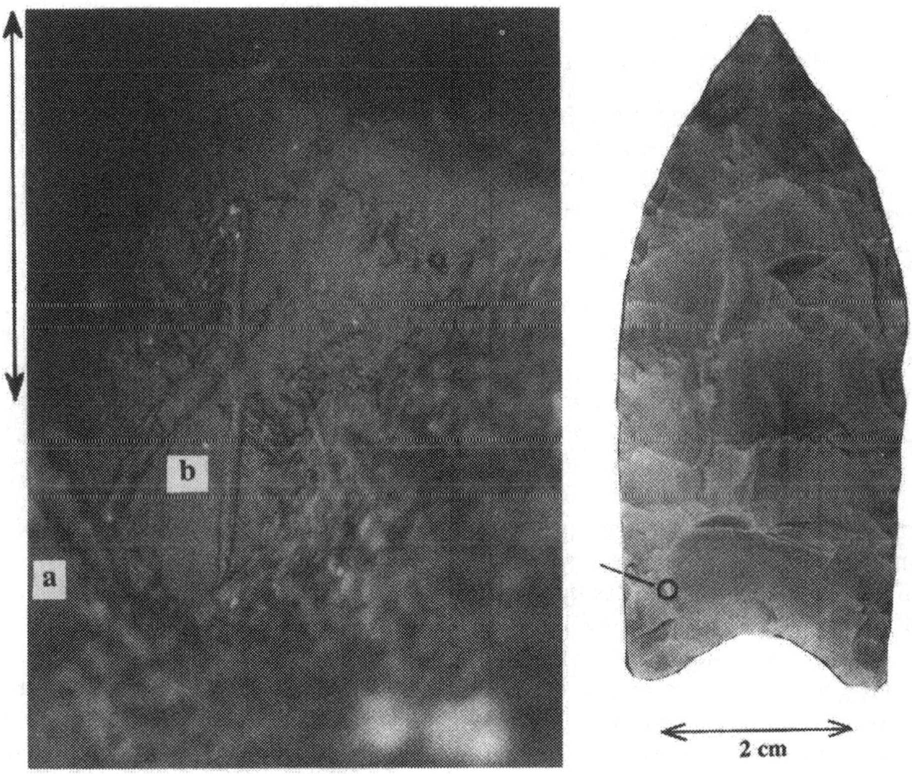


**Exp. point, Area A, 400X**

**Figure 7.** Haft binding polish and striations at area A (circled on specimen) on Spanish Point chert Clovis replicate, 400X: (a) haft binding striae; (b) cross-cutting impact striae.

Mechanical forces of tool use that affect, and are affected by, haft binding approaches have their counterpart at the macroscopic level in haft element breakage observed in experimentation or in the refitting of conjoinable fragments of archaeologically obtained tool elements. At the microscopic level their effects are noted, if only partially, by striae whose origin and orientation are consistent with blade element usage and that override, or cross-cut, haft binding wear traces, as shown in Figure 7. Haft binding polishes, however, are generally smooth in texture and similar to impact polishes in this respect. Their associated oblique or transverse striae (Figure 7) record the effects of haft binding or lateral movement of the projectile within the foreshaft during use. In these respects, the differences between the experimental replicates and some, but not all, of the points from the Colby site are substantial. In most instances, the Clovis points were bound to foreshafts more efficiently than the replicates, resulting in less lateral movement and more effective use as projectiles.

## Clovis Point Microwear Synopsis

All four specimens from the Colby site exhibit diagnostic microwear patterns that are best described by two words: fantastic and astonishing! These are fantastic in being extremely detailed and complex striated micropolishes; astonishing, in that the points display distinctive use-lives, unambiguous evidence of recycling, or variation in hafting technique.

Three of the four points from the Colby site are stylistically different from "classic" Clovis points. All are fashioned from cherts available in the Big Horn Mountains to the east or northeast of the site, and they share a "typical Clovis percussion-flaking technology" (Frison 1986a:91). Two of the points are of an identical maroon Phosphoria Formation chert, but were found in two different bone piles (Frison 1986a:91; Todd and Frison 1986:46), whereas the third, probably of Madison Formation chert, was found by the discoverer of the site, Donald Colby.

The two maroon points exhibit almost identical microwear patterns, indicative of use exclusively as projectile points (Figure 8). The unidirectionality of impact striae and almost razor sharpness of their blade elements are indicative of single usage only. It appears that both were lost, or at least were not recovered, after use. This inference is consistent with the archaeological context "at the bottom of bone pile 2" (Frison 1986a:94) of the smaller point. The larger maroon point's tip was broken on impact. Given the apparent accessibility of this point within bone pile 1, it is surprising to me that it was not further reworked to correct the tip breakage. Probably these two specimens most faithfully document the intended tear-drop shape of these points, which have deeply bifurcated bases. The larger of the two, and the other deeply bifurcated base specimen found by Colby, appear to have blood residues at the ends of hinge and step fractures.

Before moving on, I wish to further direct attention to Figure 8b, which shows the impact striae on the larger point from bone pile 1. Abrasive microparticles

**Figure 8.** Impact striae and abrasive microparticles viewed at 400X, area A (solid circle on specimen), Colby specimen C3408, the maroon Phosphoria Formation point from bone pile 1: (a) normal exposure; (b) overexposed and highlighting abrasive particles. Tick marks on either side of specimen denote extent of lateral edge grinding. Artifact was not chemically cleaned.

remain on the surface of the striated area and within the striae, and have been there ever since the tool was broken in use. Needless to say, the presence of the abrasive microparticles came as a surprise. Upon further examination of other photomicrographs of both archaeological and experimental tools, I find that these are not at all uncommon, whether or not the specimens have been subjected to intensive chemical cleaning. The fact that chemical cleaning does not destroy the particles indicates that they are nonorganic, whereas their context within the striae demonstrates that they would have to be as hard or harder than the tool surface they scar.

The point originally found by Colby and the "classic" Clovis point of blue Phosphoria Formation chert from the first bone pile have contrasting blade element wear traces that express a more complex use history. The microwear evidence is supportive of Frison's (1986a:92, 94) macroscopic evaluations of the blade elements of these two points, but is more specific. Based on cross-cutting relationships that express an order among the striation patterns (Figure 9), the two were repeatedly used as projectiles; however, initial projectile usage was at an oblique angle to the present longitudinal axes of the two points. Presumably, after damage to their tips, both were reworked into approximately their present form, and were reused as projectiles. They then were employed as heavy-duty butchering tools. Although reasonably sharp, their blade edges exhibit substantial edge damage and the final series of intersecting transverse striations are clearly oriented to, and originate at, the blade edges. The photomosaic (Figure 10) shows the full arrangement of patchy, elevated microwear traces in the center of the "classic" Clovis specimen from bone pile 1, followed by some of the microwear details from this photomosaic of final projectile point usage and subsequent heavy-duty butchering, shown in Figures 11–13. Although I did not similarly document the experimental flake butchering tool, it, too, would have shown a similar distribution of intermediate (Figure 5b) and extensively (Figure 5a) developed abrasive polishes. Also note in Figure 11 the detail of abrasive planing polish consistent with Frison's elephant butchering experiments, the cross-cutting sequences of striae that show two episodes of projectile point usage followed by at least one episode of heavy-duty butchering use (Figures 12, 13), and the lodged abrasive particle and its track that occurred during carcass butchering (Figure 13). The typological identification of the "classic" Clovis point is as much an artifact of use-life, recycling, and maintenance as it is a measure of its original intended form.

Colby site point hafting technique is reasonably clear and consistent for all but the smaller of the two maroon points found at the base of bone pile 2. Haft wear traces on this point are largely overridden by secondary impact striations (Figure 14), as exhibited by the experimental point replicates (Figure 7). The other three points exhibit a similar hafting technique of horizontal wrapping, best illustrated on the blue Phosporia, classic Clovis point (Figure 15). This is most similar to the haft binding microwear produced when Frison (1989:772) bound the experimental point replicates to foreshafts.

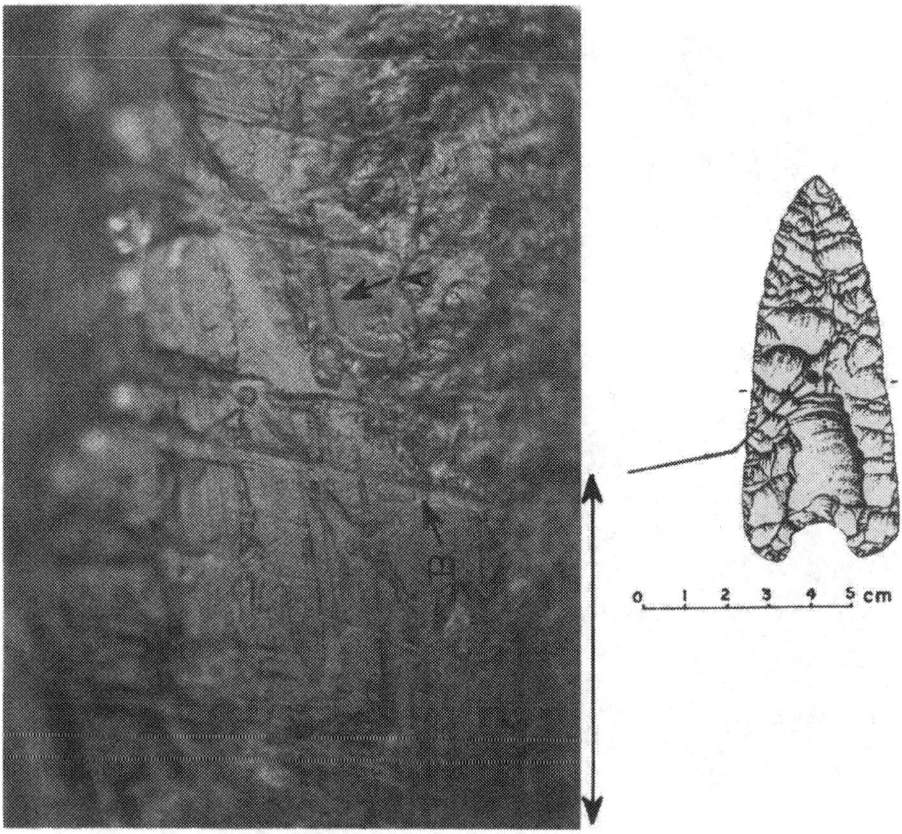


**Original Colby point, Area B, 400X**

Figure 9. Area B (solid circle on specimen) blade element micropolish with impact (a) and cross-cutting butchering (b) use-wear striations, 400X, on the original point found by Donald Colby at the Colby Site. Tick marks on either side of specimen denote extent of lateral edge grinding. (Note: Figures 9–14 photomicrographs were taken with a green filter; photomicrographs in Figures 8–15 are of uncleaned tools.)

# IMPLICATIONS

Microwear described in this paper models expectable, instructively contrastive wear traces produced in prehistory by killing or butchering big game. Further light is also shed on four general issues of microwear analysis: (1) the identification of microwear traces, (2) their potential functional significance, (3) their formation, and (4) their potential variation among stone lithologies.

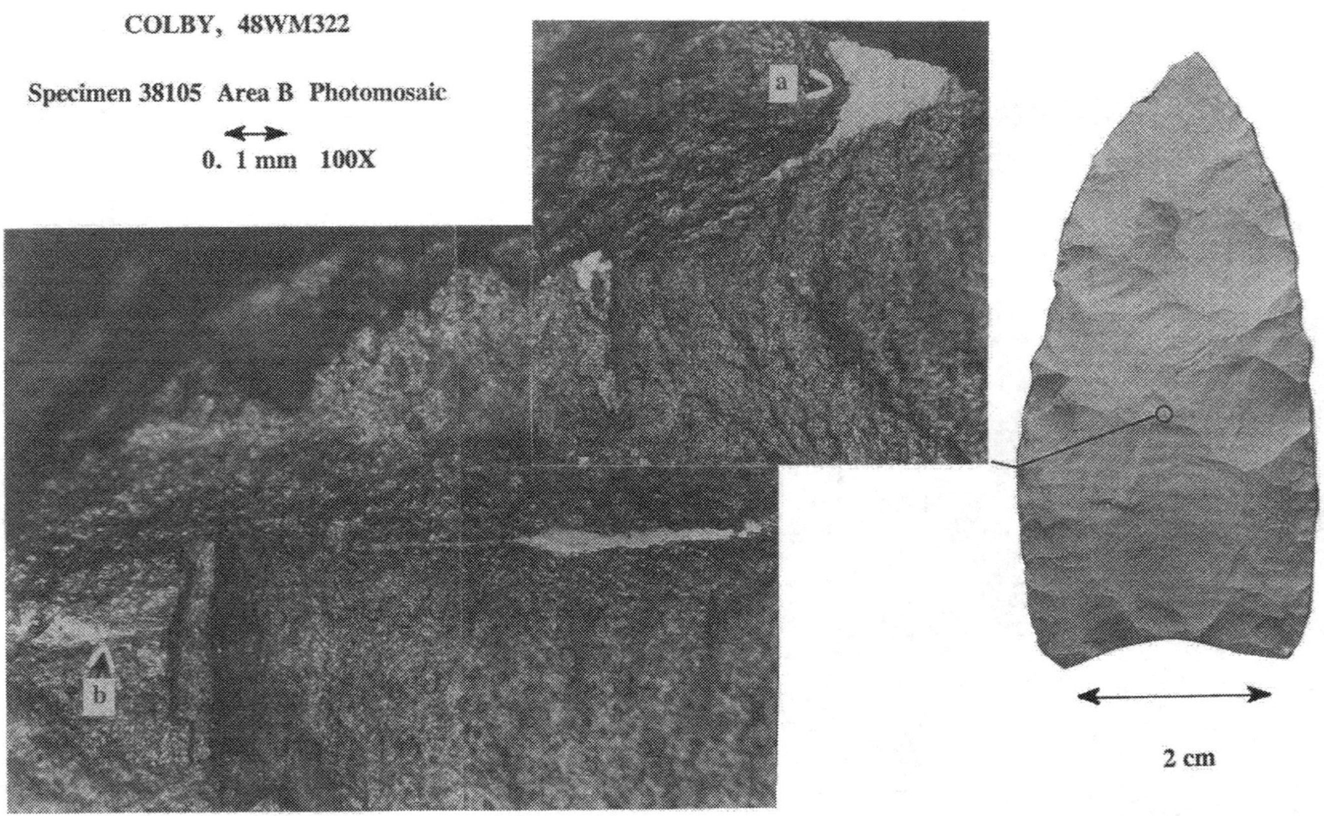


**Figure 10.** Photomosaic of microwear at 100X, area B (circled on specimen), Colby specimen 38105 (the blue Phosphoria point), showing a sequence of tool use beginning as a projectile point and ending as a heavy-duty butchering tool. Arrows at: (a) refers to detail in Figure 11; (b) refers to details in Figures 12 and 13.

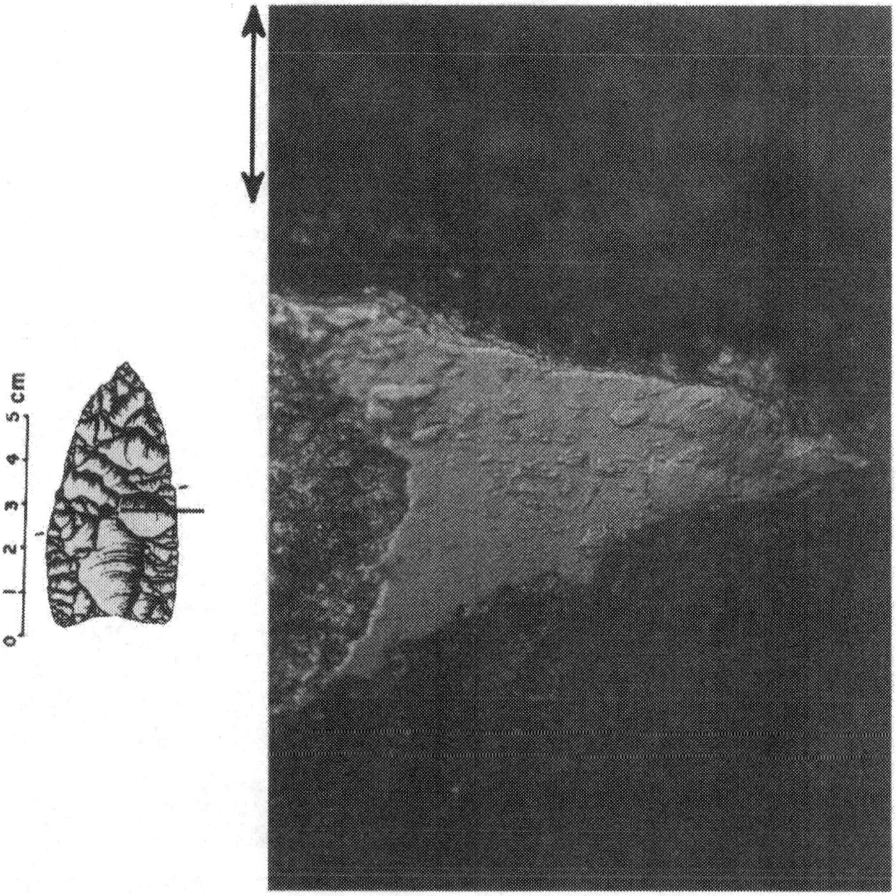


**Colby 38105, Area B detail. 200X**

Figure 11. Abrasive planing of microtopography, area B (solid circle on specimen), Colby specimen 38105 at 200X, as noted in Figure 10; compare with Figure 5a. Tick marks on either side of specimen denote extent of lateral edge grinding.

Abrasive microwear polishes produced in the killing and dismemberment of big game have similar optical and physical qualities. Regardless of stone lithology or its relative grain coarseness, microwear polishes develop progressively (consistent with Vaughan's [1985:28–29] polish stages, but not actually resembling his "generic weak" and "smooth-pitted" polishes). They may culminate in a highly reflective, flat and nearly featureless form, or an extensively developed polish (EDP). Microwear polish formation may be analogous to lapidary wheel

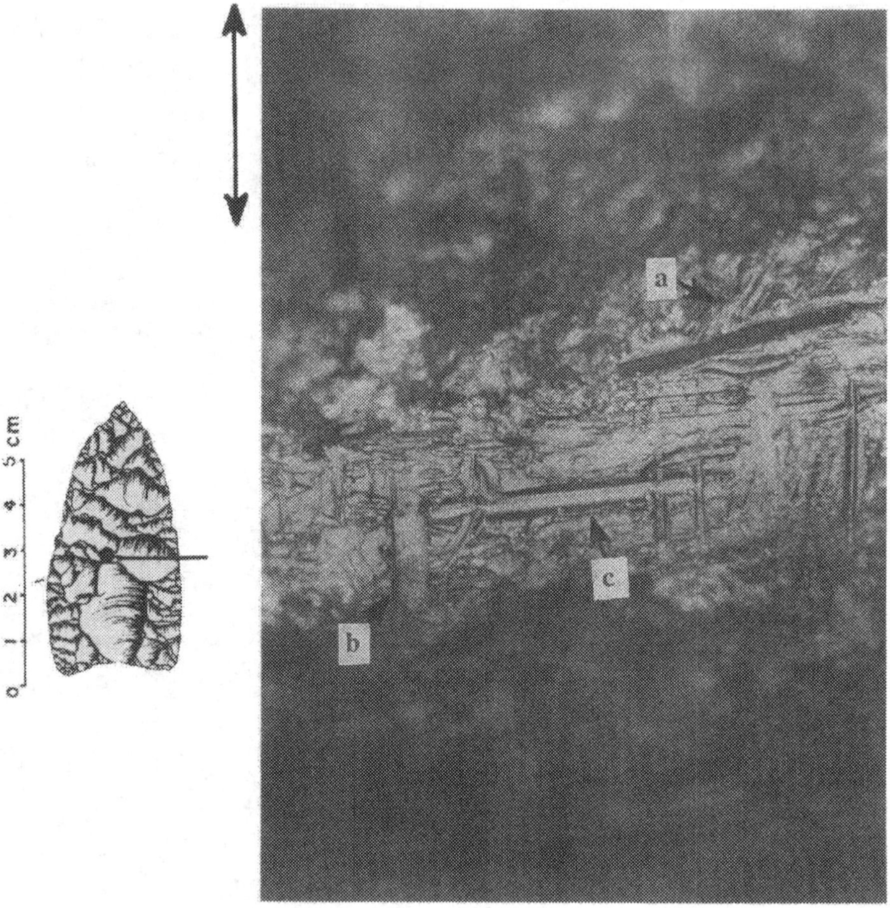


**Colby 38105, Area B detail, 200X**

Figure 12. Complex patterns of intersecting striae, area B (solid circle on specimen), Colby specimen 38105 at 200X, as noted in Figure 10. Arrows at: (a) first use as projectile point; (b) second use as projectile point; (c) subsequent use as butchering tool. Tick marks on either side of specimen denote extent of lateral edge grinding.

polishing in which the smaller the abrasive particle, the finer the polish. Inspection of butchering microwear polishes bears out this relationship; compared to intermediately developed polishes (IDP), EDP striations are smaller and more numerous, and only a few vestigal troughs of the larger IDP striations remain.

Abrasive microwear polishes thus may be seen to be the end product of general, rather than functionally specific, processes. Microwear polishes form as

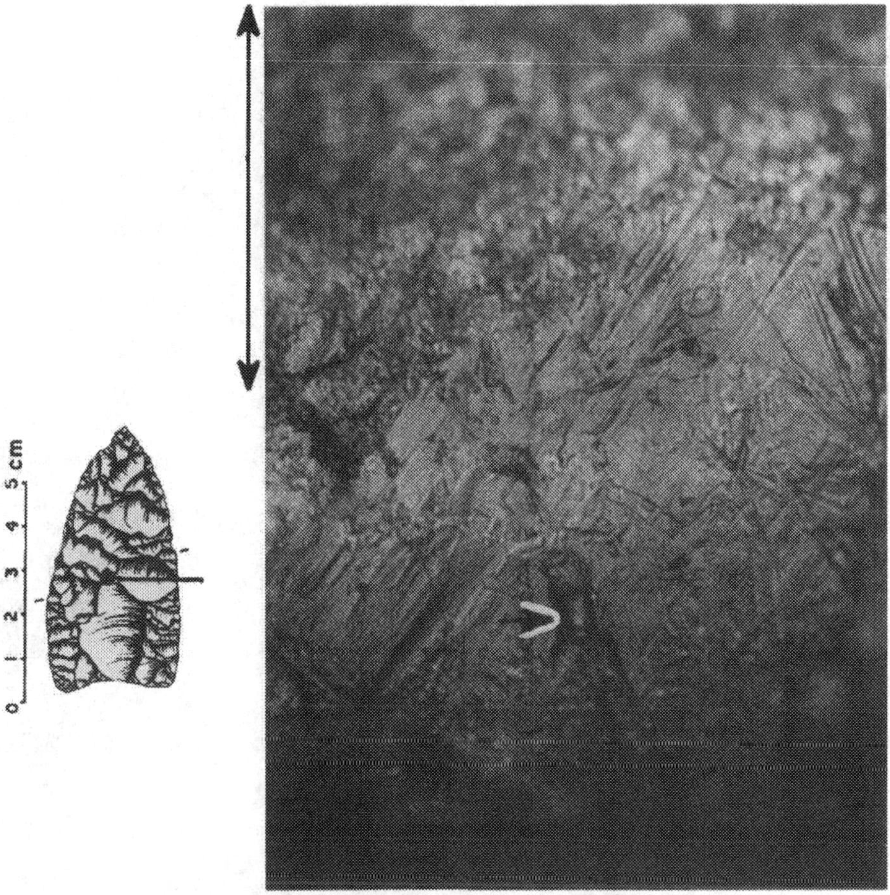


**Colby 38105, Area B detail, 400X**

**Figure 13.** Complex patterns of intersecting striae, area B (solid circle on specimen), Colby specimen 38105 at 400X, as noted in Figure 10. Note (at arrrow) the abrasive particle lodged in the surface and its track. Tick marks on either side of specimen denote extent of lateral edge grinding.

abrasive microparticles dislodge from tool edges or surfaces, or are introduced from the material worked, and they develop in direct proportion to frictional forces at the time of use. They should be regarded as most informative of frictional forces, and should not be considered as qualitative indices of tool function or material worked. Degree of abrasive polish development, micropolish location and associated striae orientation, however, are indicative of tool use, prehension or hafting mode, and material worked, to the extent a worked material is understood to have specific frictional coefficients that would result in a discrete microwear trace.

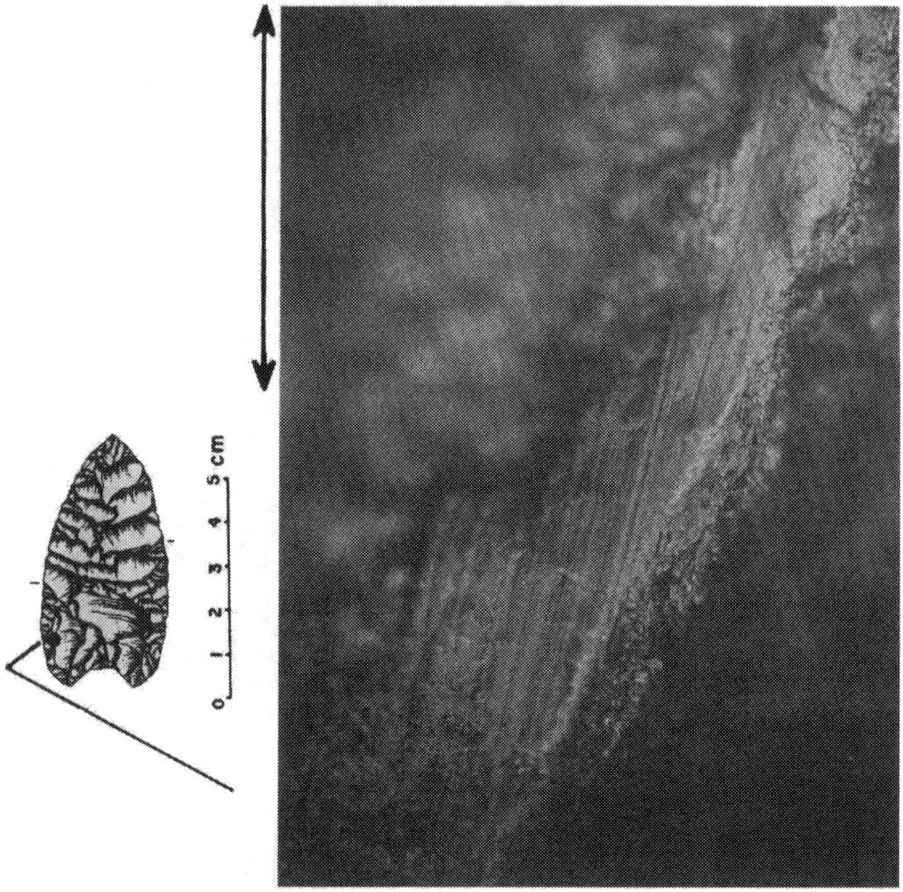


**Colby 38107, Area B, 400X**

**Figure 14.** Haft location, secondary impact(?) striae at 400X, Area B (solid circle on specimen), Colby specimen 38107, the maroon Phosphoria Formation point from bone pile 2. Tick marks on either side of specimen denote extent of lateral edge grinding.

The killing or butchering of big game is likely to result in a tool's contact with several soft and hard tissues such as hair, hide, muscle, fat, internal organs, tendons, sinew, bone or other osseous material. All of these would have individual frictional coefficients whose arithmetic mean might approximate the intrinsic resistance, or drag, a stone tool would encounter in a stroke. Regardless of its actual value, the cumulative resistance per stroke should be greater for a butchering tool than for a projectile point of similar size and shape. The manner of employment results in a different overall tool orientation relative to a carcass. A

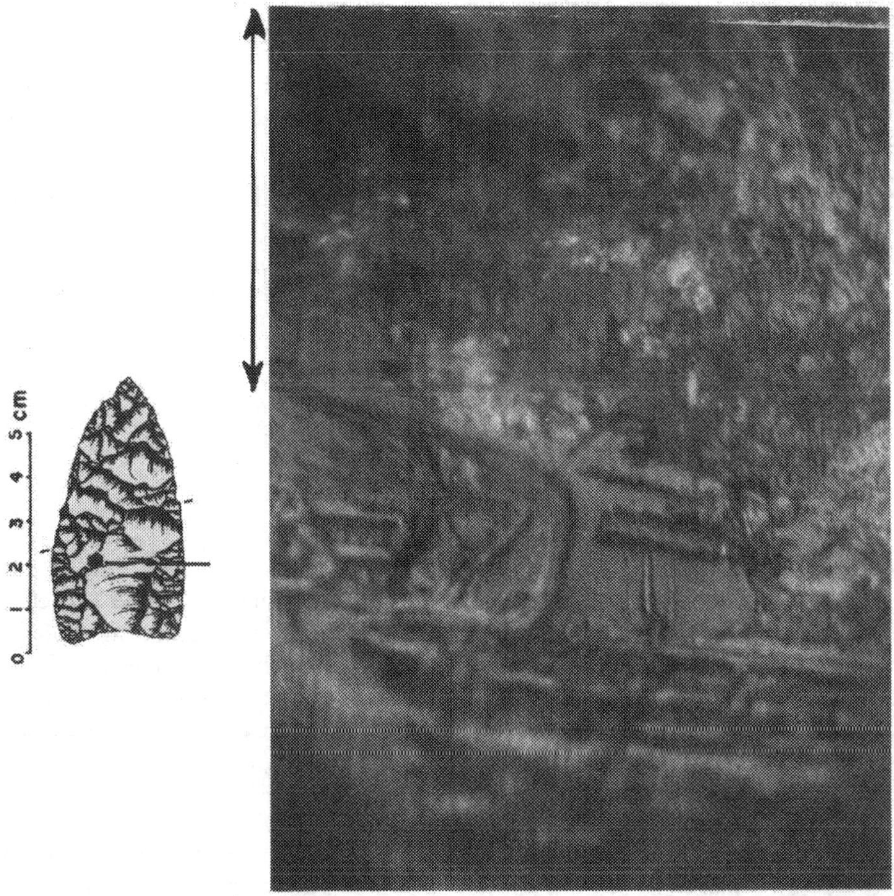


**Colby 38105, Area A, 400X**

**Figure 15.** Area A (solid circle on specimen) haft element micropolish and striations from haft binding, 400X, blue Phosphoria chert Colby point, specimen 38105, bone pile 1. Tick marks on either side of specimen denote extent of lateral edge grinding.

butchering tool or knife will have most of the length of an edge in contact with a carcass, thus increasing drag relative to that of a projectile whose penetration originates at the tip. A Clovis point's needle-like tip and razor sharp, straight edges are engineered to minimize drag on penetration, and its form is functionally constrained. Similar formal constraints need not exist for a butchering tool, and aspects of shape may not be fully indicative of tool use. Thus the recycling of projectile points from the Colby site as butchering tools is not merely a theoretical possibility, but a reality. Point recycling, in this instance, has little to do with engineering specifications for projectiles, but rather with the usability of a sharp edge or edges.

Microwear analysis affords a reliable and accurate means to determine tool function independent of tool form. Regardless of tool form, we may conclude that a used chipped stone butchering tool will display diagnostic microwear signatures. In their order of functional specificity, these include:

1. intersecting polish striations originating at, and oriented transversely to, the cutting edge,
2. polish striations parallel and adjacent to the cutting edge, and
3. nearly continuous incipient abrasive polish development on surfaces beginning at or above the cutting edge.

In contrast, projectile point microwear should originate at the tip and exhibit polishes and striations oriented with the point's longitudinal axis.

## CONCLUSIONS

Hafted bifacial tools are sophisticated chipped stone artifacts. In the case of Clovis projectile points, form does follow function. This relationship is made clear by microwear analysis, which affords an unambiguous evaluation of functionally diagnostic wear traces found on the haft or blade element. For purposes of tool use classification, the blade element wear traces are the more informative. For neither the Clovis replicate nor the Clovis points was sustained or repeated usage needed for wear traces to form. In contrast to prevailing expectations (Vaughan 1985:28–29; Hurcombe 1988:3; Bamforth et al. 1990:414), wear trace formation occurs simultaneously with use and is an indelible indicator of usage.

Haft wear traces serve other interpretative needs. In the finishing stages of manufacture, final basal grinding of Clovis points is common and results in wear traces that may be easily distinguished from evidence of subsequent haft binding (Keeley 1982:799–800). An individual tool-by-tool comparison may show substantial differences in haft wear traces. The heuristic potential of this knowledge is considerable and relates directly to the delineation of Clovis group composition at individual sites. For other similarly homogeneous and culturally diagnostic hafted bifacial implements, there should be an equal interpretative value. Relative to lithologic differences, the abrasive wear traces identified in this study do not differ significantly. That is, their form and places of occurrence are sufficient for correct identification and interpretation.

The projectile point microwear evidence from the Colby site highlights patterns of tool usage, maintenance, and recycling that is complementary to, but substantially more elaborate than, Frison's experimental study. The initial and clearly intended function was as highly effective projectile points for killing mammoths. In this respect, it is important to recognize that implement form is functionally constrained. The needle-sharp tips and symmetrical, straight, razor sharp blade edges of these points were deliberately designed to reduce drag on penetration into an animal carcass, and to cause a lethal wound. Although these design criteria are reconstructable from the experimental and archaeological evidence, they illustrate attributes that were dynamically adjusted over the life of a tool and that have variable archaeological visibility. The needle-sharp tip is essential for initial penetration. As a design feature, however, it is inherently fragile and unstable, and possesses low archaeological visibility. It is relatively easy to retip a projectile point that has sustained minor impact damage, the major result being a reduction in total length without necessarily altering its further usability. Retention of an artifact as a projectile point would continue as long as it was possible to accommodate the competing demands of maintaining an aerodynamic form essential for initial penetration and drag reduction, while addressing technological considerations of hafting and conserving overall object size for subsequent maintenance.

Two points from the Colby site were used exclusively as armatures. Their microwear evidence indicates a single use episode only. They were either lost, or impact damaged and discarded. When the other two points are considered, it seems likely that these specimens were lost in the Colby mammoth kill because, had they been retrieved, it was well within the group's competence to rework them. The smaller of the two may have been substantially reworked prior to its documented use because, as shown in Figure 14, there is a remnant of either a fractured area or the original flake striking platform on the left lateral edge just proximal to the tip. The flake scars originating from this edge override the ones from the right edge, and there is less extensive haft element lateral grinding on this edge.

The other two points were used several times as projectiles. As judged by the measurable difference of about 30 degrees in the orientation of the initial impact striae relative to the present longitudinal axis of one of these points (Figure 13), it was apparently damaged sufficiently that extensive modification was required to retip it before it was reused as a projectile point. This probably explains the difference in the extent of haft element grinding on this point. Its left lateral edge has been extensively reworked in a way consistent with a switch in the longitudinal axis following initial projectile point usage, which resulted in a reworking (and removal) of the formerly ground distal haft element portion of the left lateral edge. Ultimately, however, both artifacts were employed at the Colby site as mammoth butchering knives; the final microwear traces are indistinguishable in form from that produced in Frison's elephant butchering experiments. This final employment as cutting tools capitalized on the presence of servicable blade edges in an ad hoc fashion and was incidental to the intended function of these artifacts as projectile points.

Taken as a group, the combined contextual, macroscopic and microscopic evidence for the artifacts from the Colby site is unequivocal in illuminating the fact that these were, or had the potential to be, extensively curated tools. From a technological perspective they share in the constellation of attributes common to and representative of the Clovis culture. But when use-life is factored out of the forms, from a stylistic perspective these projectile points are sufficiently different from Clovis to warrant a separate taxonomic status as Colby points.

Lastly, I am mindful of the criticism of small experimental samples as being, if only in theory, unlikely to express a suitable range of variation in microwear evidence. However, if carefully controlled, even a small number of experiments may be extremely valuable, and an economical way to systematic assessment. This, indeed, is the case here, but I would not quibble with those who favor more experimental replications. The concordant results of the experimental and archaeological microwear evidence are unequivocal and compelling; they may also be compared directly with the descriptions and photomicrographs of Keeley and other advocates of contact material polishes. Further replicative studies undoubtedly would add to the interpretations offered here. They would not, however, be required to critically assess and evaluate the evidence presented. This evidence stands or falls on its own merits.

## ACKNOWLEDGMENTS

Jerome C. Rose was instrumental in recommending the reflected-light, differential-interference microscope with polarized light Nomarski optics. George C. Frison kindly provided the Colby artifacts. Photographic enlargements for Figures 2 and 3 were done by the Illinois State Museum. The automated photomicrographic system and the microscope were purchased through grants awarded by the University of Arkansas Office of Academic Affairs and J. William Fulbright College of Arts and Sciences. An earlier version of this paper was critically read and commented upon by one anonymous reviewer, and by Douglas B. Bamforth, David J. Meltzer, and George H. Odell. This paper is a revision of one presented at the 1993 lithics conference organized by George H. Odell and held at the University of Tulsa. Comments by Odell and the other participants, by C. Vance Haynes, and by an additional anonymous reviewer, were helpful in making revisions; any errors of fact or intent, however, are my own.

## REFERENCES

Ahler, S. A. 1971. *Projectile Point Form and Function at Rodgers Shelter, Missouri.* Missouri Archaeological Society, Research Series, Number 8.

Ahler, S. A. 1979. Functional Analysis of Nonobsidian Chipped Stone Artifacts: Terms, Variables, and Quantification. In *Lithic Use-Wear Analysis,* edited by B. Hayden, pp. 301–328. Academic Press, New York.

Bamforth, D. B. 1988. Investigating Microwear Polishes with Blind Tests: The Institute Results in Context. *Journal of Archaeological Science* 15:11–23.

Bamforth, D. B., G. R. Burns, and C. Woodman. 1990. Ambiguous Use Traces and Blind Test Results: New Data. *Journal of Archaeological Science* 17:413–430.

Barton, R. N. E., and C. A. Bergman. 1982. Hunters at Hengistbury: Some Evidence from Experimental Archaeology. *World Archaeology* 14:237–248.

Bergman, C. A., and M. H. Newcomer. 1983. Flint Arrowheaed Breakage: Examples from Ksar Akil, Lebanon. *Journal of Field Archaeology* 10:238–243.

Binford, L. R. 1963. A Proposed Attribute List for the Description and Classification of Projectile Points. In *Miscellaneous Studies in Typology and Classification*, edited by A. M. White, L. R. Binford, and M. L. Papworth, pp. 193-221. University of Michigan, Museum of Anthropology, Anthropological Papers No. 19.

Cotterell, B., and J. Kamminga. 1990. *Mechanics of Pre-industrial Technology*. Cambridge University Press, Cambridge.

Dumont, J. 1982. The Quantification of Microwear Traces: A New Use for Interferometry. *World Archaeology* 14:206–217.

Frison, G. C. 1971. The Buffalo Pound in Northwestern Plains Prehistory. *American Antiquity* 36:71–91.

Frison, G. C. 1974. *The Casper Site*. Academic Press, New York.

Frison, G. C. 1979. Observations on the Use of Stone Tools: Dulling of Working Edges of Some Chipped Stone Tools in Bison Butchering. In *Lithic Use-Wear Analysis*, edited by B. Hayden, pp. 259–268. Academic Press, New York.

Frison, G. C. 1986a. Human Artifacts, Mammoth Procurement, and Pleistocene Extinctions as Viewed from the Colby Site. In *The Colby Mammoth Site*, edited by G. C. Frison and L. C. Todd, pp. 91–114. University of New Mexico Press, Albuquerque.

Frison, G. C. 1986b. Mammoth Hunting and Butchering from a Perspective of African Elephant Culling. In *The Colby Mammoth Site*, edited by G. C. Frison and L. C. Todd, pp. 115–134. University of New Mexico Press, Albuquerque.

Frison, G. C. 1989. Experimental Use of Clovis Weaponry and Tools on African Elephants. *American Antiquity* 54:766-784.

Frison, G. C., M. Wilson, and D. J. Wilson. 1976. Fossil Bison and Artifacts from an Early Altithermal Period Arroyo Trap in Wyoming. *American Antiquity* 41:28–57.

Graham, R. W., C. V. Haynes, D. C. Johnson, and M. Kay. 1981. Kimmswick: A Clovis-Mastodon Association in Eastern Missouri. *Science* 213:1115–1117.

Hoffman, R., and L. Gross. 1970. Reflected-Light Differential-Interference Microscopy: Principles, Use and Image Interpretation. *Journal of Microscopy* 91:149–172.

Hurcombe, L. 1988. Some Criticisms and Suggestions in Response to Newcomer et al. (1986). *Journal of Archaeological Science* 15:1–10.

Hyland, D. C., J. M. Tersak, J. M. Adovasio, and M. I. Siegel. 1990. Identification of the Species of Origin of Residual Blood on Lithic Material. *American Antiquity* 55:104–112.

Keeley, L. H. 1974. Technique and Methodology in Microwear Studies: A Critical Review. *World Archaeology* 5:323–336.

Keeley, L. H. 1980. *Experimental Determination of Stone Tool Uses: A Microwear Analysis*. University of Chicago Press, Chicago.

Keeley, L. H. 1982. Hafting and Retooling: Effects on the Archaeological Record. *American Antiquity* 47:798–809.

Moss, E. H. 1987. A Review of "Investigating Microwear Polishes with Blind Tests." *Journal of Archaeological Science* 14:473–481.

Newcomer, M. H., R. Grace, and R. Unger-Hamilton. 1986. Investigating Microwear Polishes with Blind Tests. *Journal of Archaeological Science* 13:203–217.

Newcomer, M. H., R. Grace, and R. Unger-Hamilton. 1988. Microwear Methodology: A Reply to Moss, Hurcombe and Bamforth. *Journal of Archaeological Science* 15:25–33.

Newcomer, M. H., and L. H. Keeley. 1979. Testing a Method of Microwear Analysis with Experimental Flint Tools. In *Lithic Use-Wear Analysis*, edited by B. Hayden, pp. 195–205. Academic Press, New York.

Odell, G. H. 1990. Brer Rabbit Seeks True Knowledge. In *The Interpretive Possibilities of Microwear Studies*, edited by B. Graslund, H. Knutsson, K. Knutsson, and J. Taffinder, pp. 125–134. Societas Archaeological Upsaliensis. AUN 14, Uppsala.

Odell, G. H., and F. Cowan. 1986. Experiments with Spears and Arrows on Animal Targets. *Journal of Field Archaeology* 13:195–212.

Phillips, P. 1988. Traceology (Microwear) Studies in the USSR. *World Archaeology* 19:349–356.

Rees, D., G. C. Wilkinson, R. Grace, and C. R. Orton. 1991. An Investigation of the Fractal Properties of Flint Microwear Images. *Journal of Archaeological Science* 18:629–640.

Semenov, S. A. 1964. *Prehistoric Technology.* Cory, Adams and MacKay, London.

Shott, M. J. 1989. On Tool-Class Use Lives and the Formation of Archaeological Assemblages. *American Antiquity* 54:9–30.

Todd, L. C., and G. C. Frison. 1986. Taphonomic Study of the Colby Site Mammoth Bones. In *The Colby Site*, edited by G. C. Frison and L. C. Todd, pp. 27–90. University of New Mexico Press, Albuquerque.

Vaughan, P. C. 1985. *Use-Wear Analysis of Flaked Stone Tools.* University of Arizona Press, Tucson.

Witthoft, J. 1968. Flint Arrowpoints from the Eskimo of Northwestern Alaska. *Expedition* 10:1–37.

*Chapter* 12

# *Lithic Refitting and Archaeological Site Formation Processes*

## A Case Study from the Twin Ditch Site, Greene County, Illinois

TOBY M. MORROW

## ABSTRACT

The lithic assemblage recovered from Horizon 2 of the Twin Ditch site is the focus of an ongoing refitting study and provides the data employed in the analysis presented here. The abundance of lithic tools and manufacturing debris recovered from this relatively pristine Early Archaic base camp has allowed for the reassembly of numerous broken stone tools and for the reconstruction of several chipped stone tool manufacturing episodes through refitting. These refitted artifacts provide a wealth of spatial and technological data that can be applied to numerous archaeological research questions. More specifically, the refitted tools and debitage from Horizon 2 provide details concerning archaeological site disturbance processes, modes of refuse accumulation, and the occupational history of the site. The results of the present analysis indicate that the Horizon 2 cultural deposits have been subjected to minimal post-occupational disturbance, that lithic debris

**TOBY M. MORROW** • Office of the State Archaeologist, University of Iowa, Iowa City, Iowa 52242.

*Stone Tools: Theoretical Insights into Human Prehistory*, edited by George H. Odell. Plenum Press, New York, 1996.

accumulations on the site represent both primary knapping loci and secondary refuse deposits, that at least two distinct occupational episodes are represented on the site, and that the individual occupations of the site were probably of relatively short duration.

## INTRODUCTION

This paper will explore various applications of lithic refitting to the investigation of archaeological site formation processes in general and to the study of a prehistoric hunter-gatherer habitation site in particular. An intensive program of refitting the chipped stone tools and manufacturing debris recovered from Horizon 2 of the Twin Ditch site, a buried Early Archaic habitation in the Lower Illinois River Valley, is currently under way. Data derived from this study will be used to address three key issues: (1) archaeological site disturbance processes, (2) intrasite spatial patterning and modes of refuse accumulation, and (3) the occupational history of the site. In particular, the potential for refitting to investigate archaeological site occupation span is explored.

Site occupation span is an important, though oftentimes elusive, variable in archaeological interpretations. This issue has particular relevance to studies of prehistoric hunter-gatherers. While opinions vary as to whether Paleoindian and Early Archaic period populations in the midcontinental United States practiced a system of residentially mobile foraging or logistically organized collecting (after Binford 1980), or some combination of the two, there are persuasive indications that these earlier hunter-gatherers were generally less sedentary than those of the subsequent Middle and Late Archaic periods (Morse 1971, 1975; Schiffer 1975a, 1975b; Brown and Vierra 1983; Brown 1985). Therefore, site occupation span could provide a valuable scale for comparing past hunter-gatherer adaptations and for monitoring changes in mobility.

## THE TWIN DITCH SITE

The Twin Ditch site is located on the floodplain of the Illinois River in western Greene County, Illinois. The site was discovered during a 1983 cultural resource survey of the Spankey and Eldred Levee and Drainage District conducted for the U.S. Army Corps of Engineers, St. Louis District (Hassen and Hajic 1984). At the time of the site's initial discovery, concentrations of Archaic period stone tools and flaking debris were noted in the spoil piles dredged from two roughly parallel drainage ditches.

Excavations at the southern portion of the Twin Ditch site, known as Locality 1, were conducted by the Center for American Archaeology's Education Program from 1987 to 1990. These excavations revealed the presence of two distinct cultural strata, designated Horizons 1 and 2 (Morrow 1987). Horizon 1 is a mixed, multicomponent late Early Archaic-to-Middle Archaic (ca. 9000 to 6000 B.P.) deposit lying 30 to 70 centimeters below the present floodplain surface. The Horizon 1 deposits were contained within a variegated clay, silt, and sand matrix that showed extensive bioturbation in the form of numerous crayfish krotovina. Horizon 2, the deeper and earlier of the two cultural strata, was little affected by this form of bioturbation. This older and better preserved occupation surface is the subject of the following discussion and provided the refitting data that is the focus of the present study.

Horizon 2 and Horizon 1 are separated by a 20-to-70 cm thick layer of sterile sand. The lower Horizon 2 is a remarkably well preserved mid-Early Archaic habitation site resting on and within a silty clay loam paleosol that represents a land surface exposed by Illinois River down-cutting at around 9750 B.P. (Hajic 1987). The Horizon 2 paleosol exhibits a gently undulating surface with at least 40 cm of vertical relief that approximates the topography of the site at the time of occupation. Eight radiocarbon dates are available from Horizon 2, ranging from 9510±100 B.P. to 8740±70 B.P., with five of the dates clustering between 9400 and 9100 B.P. Diagnostic artifacts recovered from the Horizon 2 deposits include several Thebes knives and St. Charles points, along with one Dalton point and one Holland point (a shouldered Dalton variant).

A total of 194 m$^2$ of the Horizon 2 deposits has been excavated, including a contiguous block of 165 m$^2$ (Figure 1). These excavations cover an estimated

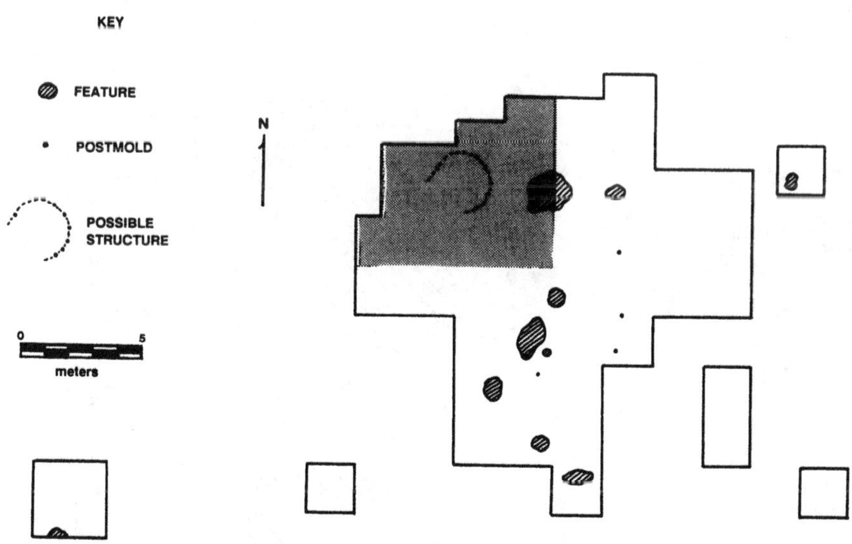


**Figure 1.** Plan map of Twin Ditch, Horizon 2 excavations. Shaded area indicates the portion of the block excavation selected for the debitage refitting study.

15–20% of the known Horizon 2 site area. Tools and worked stone items were piece-plotted and other debris were collected within 2-by-2 meter excavation units during the 1987 season and by 1-by-1 meter quadrants during the 1988 through 1990 seasons. The removal of the overlying sand layer followed the natural contours of the Horizon 2 paleosol. From this position, excavation units were dug by 10 centimeter arbitrary levels and, in some cases, by five-centimeter levels. Flotation samples were collected from each excavation unit and from all identified feature contexts. All other sediment was screened through 1/4 inch hardware cloth.

A series of hearth features and adjacent ash and charcoal concentrations was encountered in the excavation block. These hearth areas were routinely encircled by moderate-to-dense concentrations of stone tools and tool fragments, chipped stone manufacturing debris, and faunal remains. In areas excavated 2 or more meters away from these hearth features, tool and debris densities were generally sparse (Figure 2). During the 1989 and 1990 excavation seasons, a possible house floor measuring 2.0-to-2.5 meters in diameter was uncovered in the northwestern portion of the main excavation block. There are subtle indications, in the form of superimposed hearth layers and vertically separated concentrations of lithic debris in the northwestern portion of the block excavation, that two or more occupational episodes are represented in the Horizon 2 deposits.

Floral remains have been recovered from Horizon 2, but potential plant foods are not well represented and only wood charcoal is common. Because of the waterlogged condition and clayey soil matrix of Horizon 2, the absence of non-wood charcoal may merely reflect the difficulties involved in floral recovery.

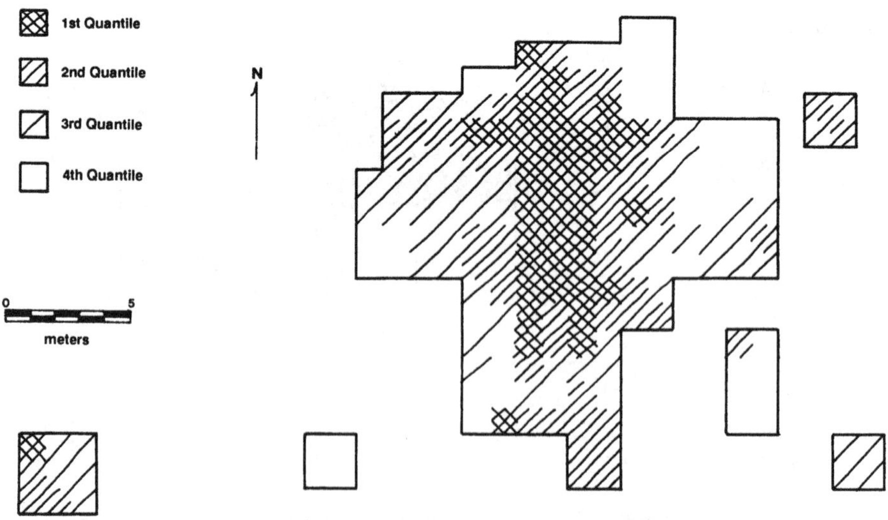


Figure 2. Density of chipped stone debris in Horizon 2.

The faunal assemblage is dominated by white-tailed deer and various species of fish. Bird bone is common, and small mammal elements are also represented.

The Horizon 2 materials are interpreted as representing a seasonal, perhaps seasonally reoccupied, base camp. The abundance of fish remains (and the *apparent* absence of nut charcoal), along with the site's location, suggest a late spring and/or summer season of occupation(s).

## HORIZON 2 LITHIC TECHNOLOGY

In excess of 32,000 pieces of chipped stone manufacturing debris and more than 700 chipped stone tools, tool fragments, and production rejects have been recovered from Horizon 2. Burlington chert, abundant in the uplands and secondary stream gravels in this portion of the Illinois River valley (Meyers 1970), is the most common raw material in the collection, at around 95 percent. The remaining raw materials in the assemblage are predominantly other west-central Illinois cherts derived from the Keokuk, St. Louis, and Ste. Genevieve Formations.

The chipped stone assemblage recovered from Horizon 2 is dominated by large bifacial tools and the products and by-products associated with their manufacture. There are few indications of a core/flake reduction strategy on the site; only one block core and one core fragment were recovered. The majority of the formal (e.g., end scrapers, side scrapers, and gravers) and informal (e.g., retouched and utilized flakes) flake tools in the assemblage were not made on flake blanks derived from the reduction of block cores. Rather, they exhibit striking platform and dorsal flake scar pattern attributes that are fully consistent with flakes derived from the primary and secondary trimming, thinning and shaping of large bifacial tools. In other words, large bifacial blanks served as cores for many of the flake tools recovered from the site, in addition to providing preforms for the manufacture of refined bifacial tools (Morrow 1988).

Three morphologically distinct classes of bifacial tools are abundantly represented in the assemblage: adzes, Thebes knives, and St. Charles points (Figure 3). Adzes (n = 40) are the single most common formal tool category in the Horizon 2 assemblage. Adzes from Horizon 2 are similar to those described as Dalton adzes from sites in northeastern Arkansas (Morse and Goodyear 1974). These tools are typically thick and comparatively narrow with a distinctly plano-convex cross-section, convex bit edges, and heavy haft grinding down the full length of both lateral edges. The majority of the adzes recovered show macrofractures, edge crushing, and/or light use-polish suggestive of a wood chopping function. Since the finished adzes are generally equal in thickness to many of the stage 2 bifaces (after Callahan 1979) in the assemblage, their manufacture generally involved little thinning. Rather, most of the flaking was directed towards shaping the tool, that is, forming the tool's hump-backed cross-section, straightening the lateral edges, and tapering the bit end.

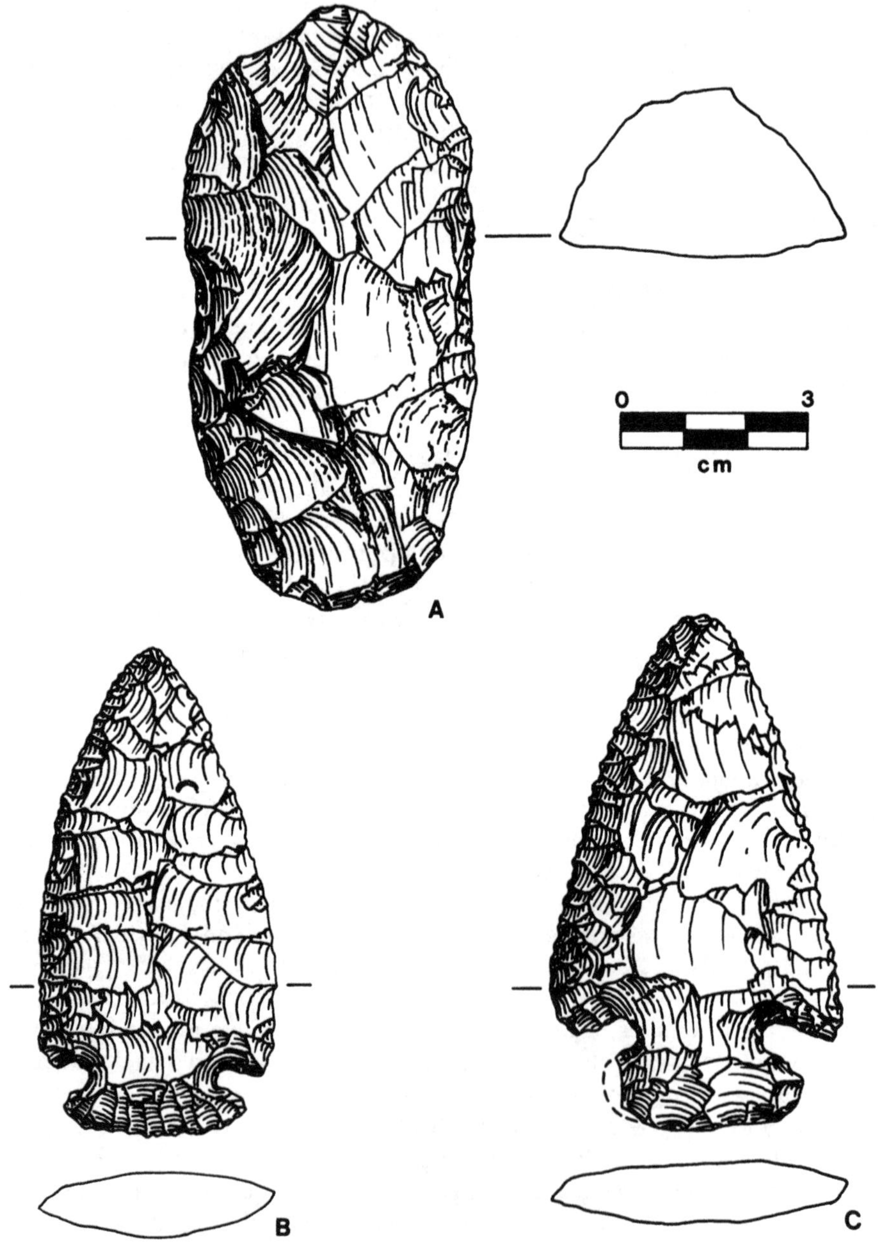


**Figure 3.** Characteristic bifacial tools from Horizon 2: a) adze; b) St. Charles projectile point; and c) Thebes knife.

Thebes knives (n = 22) are characterized by broad, relatively flat blades, deep diagonal side or corner notches, and heavy haft grinding. The Thebes knives recovered from the site show no evidence of impact fractures and they are interpreted as general purpose cutting, sawing, and prying tools. Blade edge resharpening was accomplished by unifacial alternate beveling, occurring exclusively on the left margin of the tool when the tip is oriented upward. Thebes knives were manufactured through a more complex reduction sequence than the adzes were. End thinning and subsequent lateral thinning reduced the irregular stage 2 blanks into regularly shaped, thinned (stage 3) blanks. Secondary thinning was accomplished through the removal of randomly oriented, transverse thinning flakes resulting in broad, flattened blades with a few widely spaced flake scars.

St. Charles points (n = 7) are markedly narrower than Thebes knives and exhibit distinctly bi-convex or lenticular cross-sections. Like the Thebes form, they are characterized by deep, diagonal notches and heavily ground haft elements. Breaks exhibited by the St. Charles points are consistent with their primary use as projectile points. Lateral edge resharpening in the form of unifacial alternate beveling, again along the left blade margin, suggests secondary use as hafted cutting tools. St. Charles points were manufactured through the same general sequence as were Thebes knives; however, the final percussion flake removals are more closely spaced, generally horizontally oriented, and terminate at or near the biface midline. Only one of the seven St. Charles specimens recovered from Horizon 2 is unbroken, and three have been recycled into hafted drills.

Dalton points (n = 2) are not well represented in the Horizon 2 assemblage. These generally narrower lanceolate points clearly seem out of place in an assemblage otherwise dominated by broader, deeply notched point/knife forms. They may represent artifacts deposited during a separate and earlier occupation of the site, or items scavenged from an earlier site and brought in by the site occupants. Alternatively, they may demonstrate a "transition" from lanceolate to notched point/knife forms. Future research on the Horizon 2 assemblage will hopefully clarify this situation. The two Dalton points recovered exhibit different overall morphologies and flaking patterns. The narrower, more "classic" Dalton specimen exhibits random percussion thinning, and was resharpened by steep, bifacial pressure flaking. The broader, lightly shouldered Holland variety point possesses blade proportions and flaking reminiscent of the St. Charles points, and was resharpened through unifacial alternate bevelling along the left side.

## REFITTED BIFACES AND FLAKE TOOLS

Refits are abundant in both the tools and flaking debris recovered from Horizon 2. Prior to the present study, the refitting effort had concentrated on the fragmented bifacial artifacts and broken flake tools from the entire Horizon 2 excavations. Of the 552 tool fragments identified, 287 (52 percent) have been

Figure 4. Spatial distribution of refitted bifaces and flake tools.

successfully conjoined (Figure 4). The high frequency of refits and the often short distances separating refitted pieces suggest that the Horizon 2 deposits have suffered minimal post-occupational disturbance. This proposition can be further evaluated through the analysis of the distributional patterns exhibited by refitted artifacts.

## SITE DISTURBANCE PROCESSES

A major factor influencing the distribution and relative frequency of refittable artifacts within a site is the level of site disturbance. Various post-occupational processes can mix, rearrange, and even remove artifacts from the contexts of an archaeological site. Sealed beneath a blanket of sterile sand that was probably deposited some time around 9000 B.P., the Horizon 2 paleosol appears to have been largely insulated from both cultural and natural disturbance processes since that time. The more obvious sources of post-occupational disturbance that the site could have been subjected to prior to being buried (e.g., bioturbation and erosion) appear to have had a minimal impact. One method of assessing these types of disturbance is through the examination of refitting data in both the vertical and horizontal dimensions.

Figure 5. Vertical profile of refitted bifaces and flake tools. Vertical exaggeration 2:5:1.

The Horizon 2 paleosol surface did exhibit some traces of erosion, but its effects appear to have been minimal and localized. Narrow, linear rills incised through the paleosol surface were encountered in several locations within the Horizon 2 excavations. These rills rarely exceeded 10-to-15 cm in depth or width and were regularly spaced at 2-to-3 m intervals, all trending in a north-south direction. Filled with sand, these erosional rills appear to post-date the occupation of the Horizon 2 surface. The erosion of these rills may have displaced an indeterminate (but probably small) number of artifacts. Fortunately, the effects of rill erosion appear to be isolated to the rills themselves. The horizontal distribution of refit linkages (see Figure 4) shows no preferred orientation along north-south lines, which would be expected had the entire surface been subjected to sheet erosion and deflation.

The relative lack of extensive sheet erosion and deflation on the Horizon 2 paleosol is also demonstrated in the vertical dimension. Throughout the majority of the Horizon 2 excavations, the main vertical concentrations of cultural debris were encountered 5-to-10 centimeters below the contact between the overlying sterile sand and the paleosol surface. The debris followed the undulating contours of the paleosol surface in a blanket-like fashion. The vertical distribution of refitted artifact linkages also conforms generally to the topography of the paleosol (Figure 5).

A minority of refits, however, transcend this generally consistent pattern. When piece size is compared to the overall vertical separation of refitted fragments, it becomes clear that the smaller artifact fragments exhibit the greatest degree of vertical disturbance (Figure 6). A maximum of 17 cm of vertical separation is demonstrated by the refitted bifaces and bifacial tools from Horizon 2; however, fully half of the refitted pieces were located within 3 cm of each other based on relative depth below the paleosol surface, and 75% were located within 6 cm of each other. The greater potential of smaller size fragments to have been vertically displaced is consistent with the findings of several studies of site disturbance processes that have investigated the effects of human trampling (e.g., Stockton 1973; Villa and Courtin 1983; Gifford-Gonzalez et al. 1985). Intermittent soil wetting and drying, periodic frost action, and a low level of bioturbation

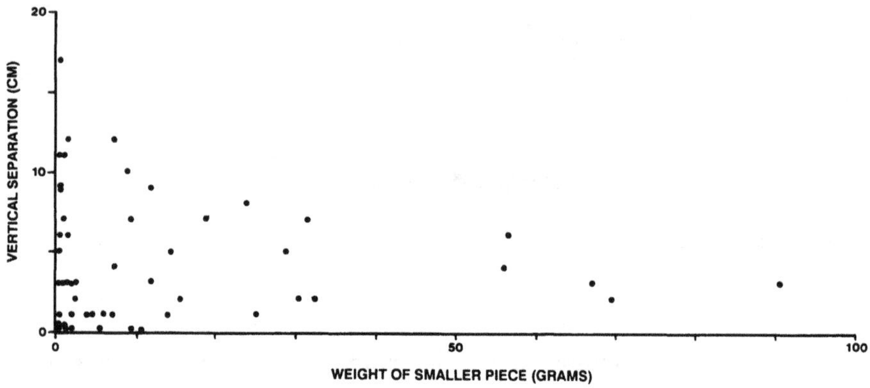


Figure 6. Scatterplot showing the relationship between the vertical distance separating two refitted tool fragments and the weight of the smaller refitted piece.

may also have contributed to the vertical displacement of artifacts on the site (see Wood and Johnson 1978).

## LITHIC DEBRIS SCATTERS AND ARCHAEOLOGICAL SITE FORMATION PROCESSES

Different categories of cultural refuse (i.e., stone, bone, and wood) are subject to differential processes of primary deposition, secondary disposal, disturbance, and decay. Most organic trash, e.g., wood scraps and bone, can be eliminated by burning (Schiffer 1987:70). Carnivores may scavenge and disperse bone (Kent 1981). Chipped stone debris, an inorganic substance, is susceptible to several disturbance and dispersal processes, but it is largely unaffected by others. In this section, I discuss the deposition of chipped stone debris scatters and the cultural disturbance processes most likely to disperse them.

Chipped stone tool manufacture is a reductive process that results in the accumulation of flaking debris. Different reduction strategies result in characteristically distinct patterns of products and by-products, and this ultimately affects their deployment and discard. A blade industry, for example, yields a relatively high tool-to-debris ratio (Sheets and Muto 1972). The manufacture of a single bifacial tool, on the other hand, can produce hundreds of waste flakes (Newcomer 1971; Henry et al. 1976; Patterson and Sollberger 1978). The following discussion is framed around the characteristics of a bifacial tool industry such as that represented at Twin Ditch.

For the sake of the present discussion, bifacial tools are defined as chipped stone artifacts that have been flaked on both faces to the extent that all or most of the original natural spall or flake blank surface has been removed. Bifacial tools are typically made through complex and protracted reduction sequences that can be subdivided into arbitrary stages of manufacture (Callahan 1979). Because of the complexity of the manufacturing process, production failures are not infrequent, and broken or otherwise unsatisfactory bifaces were often discarded before completion of the intended tool (Johnson 1979).

The manufacture of a single bifacial tool results in the deposition of an accumulation, literally a small pile, of chipped stone debitage. These initially discrete concentrations of debris are susceptible to several cultural processes that act to disperse them laterally across the surface of a site and work them downward into the soil matrix.

Though often referred to as "waste flakes," the by-products of biface manufacture are not necessarily useless trash (Kelly 1988). A pile of "waste flakes" is also a pile of potential tools. Selected on the basis of size, general morphology, edge angles, and sharpness, many such flakes were transformed through retouch into various tools or were used without subsequent modification for a variety of tasks. The selection of flakes for tools could bring about the wide dispersal of flakes resulting from a single knapping episode, particularly when they were employed in activities conducted some distance away from the primary knapping locus. Within the debris resulting from a typical biface reduction sequence, however, a minority is likely to have been selected for use. Flake tool selection would have favored larger flakes and those with specific edge attributes; the majority of smaller and less regular flakes would probably have remained unused.

Trampling has been experimentally studied for its effects on the dispersion of material. Vertical displacement by trampling is limited by the penetrability, or relative looseness, of the soil matrix (Schiffer 1987:126). In firm substrates (silts and clays), small items are more likely to be worked down into the subsurface. Gifford-Gonzalez et al. (1985) observed vertical displacements of 1-to-4 cm in their experiments. Villa and Courtin (1983) reported vertical displacement of up to 8 cm and horizontal movement of up to 85 cm. Larger items are more susceptible to lateral displacement on firm substrates (Stevenson 1991:271–272).

Once chipped stone debris accumulates in a particular area of a site, it may interfere with other activities requiring a less cluttered surface. The intentional movement or removal of debris takes materials from their primary locations and puts them into secondary contexts (Schiffer 1972, 1987). Stevenson (1991:273–276) distinguished between expedient refuse clearing (brushing or tossing debris aside) and systematic refuse disposal (picking up and dumping debris elsewhere). Regardless of the method employed, secondary displacement tends to be biased toward larger and more obtrusive items. Small pieces of debris are likely to be missed by such cleanup activities, and many remain at the primary locus (McKellar 1983).

The three processes described above—selection and utilization of flakes for tools, trampling, and secondary disposal—can be considered as ongoing, concurrent factors influencing the distribution of lithic debris from the time they are deposited to the end of a site's occupation. Their effect would be cumulative; e.g., flake scatters laid down early in the occupation of a site are subject to more trampling, sweeping, dumping, etc., than the debris deposited from later reduction episodes (e.g., Winter 1990). Stevenson (1985, 1991) argued that the debris laid down in the final days of a site's occupation are generally not subject to secondary displacement and remain as clustered primary debris (see Fisher and Strickland [1991:223] for an ethnoarchaeological example). This holds important implications for the interpretation of spatial patterning on archaeological sites that have suffered little post-occupational disturbance: the relative dispersal of debris concentrations should approximate their sequence of deposition.

Further, lithic debris concentrations laid down during separate and distinct periods of site occupancy would be subjected to different patterns of disturbance and dispersal. The specific locations of facilities, such as hearths and domestic structures, probably varied from one occupation to another. The positions of these facilities form the structural framework of the habitation and ultimately affect where various activities take place and where refuse is allowed to accumulate. Therefore, variation in the distribution of refuse generated by a single, temporally discrete behavioral episode (like the debitage produced during the manufacture of a single bifacial tool) should accurately reflect its relationship to contemporary features on the site. This being the case, the distribution and condition of refuse generated around an active hearth should be markedly distinct from the materials deposited over the same area before and after the feature was in use.

The distribution of debitage resulting from a tool manufacturing episode that preceded the placement of a hearth in the same area should show no discernible relationship to the position or limits of the feature. Further, flakes and tool fragments that were present at or slightly below the ground surface before a hearth was established over them would exhibit reddening, smoking, and/or thermal fracturing as a result of heat exposure.

Lithic debris laid down as a result of activities that were conducted around a hearth at the time the feature was in active use would mostly be distributed in a crescentic arc around the feature. Since chert has a tendency to explode violently when exposed to the direct heat of a wood fire, the site occupants would have made some effort to keep the chipped stone debris they generated out of an actively burning campfire. An occasional flake or tool fragment might have been accidentally dropped or scuffed into a fire, but the degree of burning should be much less than that observed on flakes that preceded the use of a hearth. For the most part, secondary displacement of lithic debris through scuffage, sweeping, and dumping should exhibit a marked pattern of dispersal around and away from the contemporarily used hearth.

Chipped stone debris and tool fragments deposited over a hearth area after the feature was abandoned should, of course, exhibit no evidence of heat exposure, unless the burning occurred elsewhere. In addition, post-hearth debris would be freely deposited over and through the location of the former hearth and should not exhibit the concentric patterning of debris accumulations that were contemporary to the active use of the feature.

In order to apply these principles, it is necessary to isolate individual behavioral events and their material consequences. The unit of analysis to be employed in this study is the individual knapping episode as reconstructed through the refitting of debitage recovered from Horizon 2 of the Twin Ditch site. These groups of refitted flakes represent assemblages of cultural debris that were produced and deposited within a very brief interval of time, most likely a matter of minutes or, at most, a few hours. Following their initial production, these flakes were subjected to various forms of dispersal which would have varied within the spatial and sequential contexts of their deposition. As such, these refitted individual knapping episodes provide the opportunity to identify and assess the specific cultural processes that affected their distribution across the site. These observations can in turn be compared to those derived from other individual knapping episodes. The patterns that emerge from this analysis can then be used to interpret some of the sequence of events that lead to the formation of the cultural deposits excavated from the site.

## REFITTING AND SITE OCCUPATION SPAN

Schiffer (1975c) has stressed the inter-relationship between the length of a site's occupation and the assemblage of tools and debris deposited there. Simply put, Schiffer reasoned that sites occupied for greater lengths of time would contain a wider diversity of tools and debris. Conversely, sites that result from brief episodes of habitation would contain more restricted assemblages of artifacts, dominated by the types of tools that were rapidly worn out and discarded and by the most common types of debris.

Portable artifacts that were still in a usable state at the time a site was abandoned were likely to be taken away by the site occupants to the new destination. Schiffer (1975c, 1987:90) characterized artifacts brought into a newly occupied site from the previous occupation as being the "founding curate set." The assemblage of artifacts remaining at the previously occupied site constitutes the "donor curate set." Logically, as the span of time between a site's initial occupation and its abandonment decreases, the opportunity for objects manufactured there to have become depleted and discarded there also decreases.

It is in this latter regard that lithic refitting can provide a potent means of assessing some aspects of site occupation span. Artifacts manufactured at one site and then transported to a subsequently occupied site will become separated from their resulting manufacturing debris. In this situation, the manufacturing debitage remaining at the initial site, when refitted, will enclose a "ghost" of the artifact taken away. Conversely, tools brought into a habitation site from a previous occupation will not be accompanied by their resulting manufacturing debris, and in that sense, will be "orphans." Refitting provides one means of identifying "ghosts" and "orphans" within a lithic assemblage by demonstrating a linkage, or lack thereof, between the tools and manufacturing debris recovered. The combined proportion of "ghosts" and "orphans" within an assemblage would vary with the relative frequency with which finished artifacts could be reunited with manufacturing debitage recovered from the same site. Further, as occupation span increases, the relative representation of "ghosts" and "orphans" within a site assemblage should accordingly decrease.

Refitting can provide one component of an equation for calculating a more informed perspective on site occupation span. Artifact use-life is another critical variable. While artifact use-life has been addressed in some articles (e.g., Odell 1980; Shott 1989), it remains a poorly understood phenomenon that is rarely applied in archaeological analyses. Perhaps experimental studies such as that of Schultz (1992) will provide some of the insights necessary to understand this concept.

Site occupation span can also be registered in other site characteristics, such as refuse disposal patterns. Murray's (1980) survey of ethnographic data pertaining to discard locations revealed strong correlations between discard behavior and the duration of settlement. Populations who inhabited the same site for a period of less than one full season were likely to leave discarded materials in their primary locations, particularly in unenclosed or open areas within the habitation site. Regardless of occupation span, Murray found that the interiors of domestic structures were kept generally free of debris and were periodically swept or cleaned. Since refitting is well suited to identifying primary and secondary refuse deposits, information derived from reassembling lithic debitage should provide valuable insights into discard behavior.

Lithic refitting can be employed to investigate the topic of site occupation span in two substantive ways: (1) by identifying the frequency of "ghosts" and "orphans" in an assemblage of lithic tools and quantifying the relative contributions of tools made outside of the site and of tools removed when the site was abandoned; and (2) by elucidating aspects of discard behavior by distinguishing and correlating primary and secondary refuse deposits. The conventional picture of Early Archaic settlement in the American Midwest is one of generalized, short-term habitation sites (Brown and Vierra 1983; Brown 1985). In this situation we anticipate that refitting lithic debitage from Horizon 2 of the Twin Ditch site would demonstrate (1) a high frequency of "ghosts" and "orphans" and (2) only limited accumulations of refuse in secondary contexts.

## DEBITAGE REFITTING

Since the chipped stone debitage assemblage recovered from Horizon 2 is rather large and because this study is intended to be an exploratory one, only a portion of the Horizon 2 macroblock serves as the focus for the present analysis. The area chosen encompasses the northwestern quarter of the excavation block, squares 12 through 15, 34 through 38, and 54 through 57 (see Figure 1). This 44.5 $m^2$ area contains one-half of a large hearth feature (Feature 6), the total area of the suspected house floor and, in places, large concentrations of chipped stone flaking debris (see Figure 1). The paleotopography of the Horizon 2 paleosol in this area generally exhibits a southward-trending, gradual slope.

Approximately 7100 pieces of chipped stone flaking debris were recovered from the screened portions of the excavation in this area. An estimated 500 additional flakes more than 1/4 inch in diameter were recovered in the flotation samples, along with thousands of microflakes.

## METHODS

Lithic refitting has a reputation of being enormously time-consuming. Taking on a refitting project when faced with an assemblage of thousands of pieces of lithic debris would seem to be a daunting task, indeed. There are, however, several procedures that can greatly increase the efficiency of the refitting work. Refitting is a cumulative process, with each new conjoined piece reducing the remaining pool of non-refitted fragments. Some refits are easy to find, while others are more elusive. Time and frustration are both reduced by taking the simple refits out of the assemblage first and then turning attention to the more difficult cases. I employed the following hierarchical scheme for refitting the Twin Ditch debitage.

Since small flakes are very difficult to refit within a complex biface reduction sequence, I first size-sorted the flaking debris into flakes greater than and less than 2.5 cm in maximum dimension, and focused on reassembling the larger flakes. While this might seem to be over-restrictive, the production of large bifaces like those represented in the Horizon 2 assemblage does result in the creation of many large flakes. Of the total 7100 pieces of chipped-stone debitage in the sampled quarter of the block area, 1575, or roughly 22 percent, were pieces larger than 2.5 cm, so a substantial number of flakes was employed in the refitting program.

Pre-selecting for larger pieces when refitting provides several advantages and was well-suited to the goals of the present analysis. First, since the surface area of a flake is directly related to the ease with which the piece can be confidently conjoined to another flake, refitting larger pieces can be accomplished in a much more timely manner. Second, since the goal of this analysis was the reconstruction of tool manufacturing events, the reassembly of larger pieces allowed the effort to be focused upon those flakes that were most likely to be related to tool manufacture without the potential hindrance of numerous small flakes related to the resharpening and repair of finished tools. And third, since larger pieces are less prone to vertical displacement, the resulting distributional data reflect a truer characterization of the vertical relationships among various knapping episodes.

From here, all of the larger size-grade flaking debris was sorted into unique or similar raw material groups. This procedure is perhaps the single most critical step toward successful debitage refitting. Because color can be altered by patination and exposure to heat, color alone can potentially be a misleading and erroneous guide to sorting chert. Other characteristics such as mottling patterns, fossil content, translucency, and overall texture were considered to be more reliable gauges of raw material similarity than color.

The majority of the Horizon 2 assemblage is made up of locally available Burlington chert. Fortunately, this material often exhibits considerable macroscopic variation from one piece to another, allowing one to isolate generally similar pieces. At the same time, however, there may be considerable macroscopic variation *within* a single piece of Burlington chert, and "transitional" zones represented within single flakes were most helpful for reuniting what initially appeared to be completely different pieces of material. The remainder of the chipped stone assemblage is made up of various other regional chert varieties that are quite distinct from the Burlington. For the most part, each of the individual pieces of Keokuk, St. Louis, and Ste. Genevieve chert in the Horizon 2 assemblage fell into visually distinctive groups that were subsequently refitted.

Many flakes were broken from bending and snapping during removal from the parent core or biface and from trampling; others were fractured as a result of exposure to fire. Once the flakes had been sorted into similar raw material groups, the third step in debitage refitting involved separating all of the broken flakes into proximal, medial, and distal portions. Then, by matching break types, thicknesses, and profiles, the fragmented flakes were reassembled into complete flakes. This approach routinely yielded an impressive success rate. In general, the procedure enhances the success of the following refitting steps, because complete flakes with larger surface areas are easier than broken pieces to conjoin with other flakes, and they have a greater capacity to intersect the paths of any additional flake removals that preceded or followed them. In addition, complete flakes allow one to line up the striking platforms with the former margins of a core or biface once reconstruction begins.

It is generally easier to fit flakes back onto a core or biface than it is to fit flakes onto each other. Thus, the fourth phase of this refitting program involved matching flake groups to the bifaces and cores recovered from the site and reconstructing the reduction sequence for each case (backwards, of course). Several of the bifacial artifacts present in the assemblage, particularly those broken in manufacture, provided a framework for this reassembly.

Following these four steps reduced the size of the non-refitted flake pool, making the fifth and final phase of this scheme somewhat less difficult. Flake-to-flake refits in the absence of the core or biface from which they were derived are among the most challenging of refits. To accomplish this, it is especially helpful to sort the flakes according to their approximate position in the reduction sequence (i.e., thick primary flakes, bifacial thinning flakes), general morphology (flat or ventrally curved), and the presence (or absence) and type of cortex. Primary flakes tend to mesh laterally with other primary flakes, with thinning flakes tending to fit inside, or to the ventral surfaces of, the primary flakes.

## RESULTS AND INTERPRETATIONS

A total of 349 of the 1575 flakes larger than 2.5 cm have been refitted to date. This number comprises nearly 22 percent of the larger-sized flakes recovered from the northwestern quarter of the Horizon 2 excavation block. This proportion of refitted pieces would place the Horizon 2 sample among those Old World sites with the highest refitting rates (see Cziesla 1990:24–25). Fifty-eight of these refitted pieces belong to simple two-, three-, and four-piece refits involving the conjoining of flake fragments and/or flakes.

The remaining 291 flakes belong to larger groups of flakes that have been reassembled, or have been refitted to a biface recovered from the site. For the purposes of the present analysis, these are termed "refit cases." These 25 refit cases range from 1 to 38 refitted flakes (Table 1). Most of these appear to represent discrete, individual knapping episodes, although six of the refit cases are relatively incomplete and may actually represent only three or four individual tool manufacturing events. In addition, four of the refit cases involve the reassembly of one or two flakes to a biface.

Because debitage was not piece-plotted during the Horizon 2 excavations, the respective proveniences of the refitted flakes are, for the most part, known only to the nearest one-meter square and to a 5- or 10-cm level. These data do not retain the refined spatial clarity desirable for a refitting study. Nonetheless, robust vertical and horizontal patterning is well expressed in the refitted debitage.

Once the debitage refitting conducted for this study was under way, one of the things that was immediately striking was the tight clustering of the majority of the refitted groups of flakes. Large numbers of flakes derived from the reduction of a single core or biface were usually found within immediately adjacent excavation units and within single or immediately adjacent levels. These data provide further support for the observation that the Horizon 2 deposits were in a fairly pristine and undisturbed condition.

Refit case 1, by far the most complete reduction episode yet reassembled from Horizon 2, provides a remarkably detailed spatial, technological, and behavioral history of the manufacture, recycling, and discard of a bifacial adze (Figure 7). This adze was manufactured from a thick blank of St. Louis chert that had apparently been partially roughed out at another location. Thirty-eight flakes related to the manufacture of this tool have been refitted. The distribution of these flakes reveals two general clusters, one located to the west of Feature 6, the other located to the northwest of this feature (Figure 8). A large proportion

Table 1. Summary Data for the 25 Refit Cases Identified in the Northwest Quarter of the Horizon 2 Macroblock

| Case | Total flakes | Area $m^{2a}$ | Group[b] | Comments |
|---|---|---|---|---|
| 1 | 38 | 6 | Upper | Flakes refit to two fragments of a bifacial adze; one utilized flake |
| 2 | 12 | 5 | Upper | Primary and secondary flakes from an indeterminate biface |
| 3 | 7 | 2 | Upper | Flakes refit to a block core |
| 4 | 8 | 2 | Upper | Flakes refit to two fragments of a stage 2 biface; may be related to cases 15 and 16 |
| 5 | 8 | 4 | Upper | Primary and secondary flakes from an indeterminate biface |
| 6 | 11 | 2 | Upper | Primary flakes from an adze-like biface |
| 7 | 2 | — | — | Two flakes refitted to four fragments of a stage 4 biface |
| 8 | 20 | 2 | Upper | Primary flakes from a large indeterminate biface; one utilized flake |
| 9 | 22 | 4 | Upper | Flakes refit to two fragments of a stage 3 biface; one utilized flake |
| 10 | 14 | 3 | Upper | Primary flakes from an indeterminate biface |
| 11 | 13 | 5 | Lower | Secondary flakes from a refined biface |
| 12 | 13 | 7 | Lower | Secondary flakes from a refined biface |
| 13 | 6 | 5 | Upper | Primary and secondary flakes from an adze-like biface |
| 14 | 24 | 3 | Upper | Primary and secondary debris from an indeterminate biface |
| 15 | 5 | 3 | Upper | Primary flakes from an indeterminate biface; may be related to cases 4 and 16 |
| 16 | 18 | 4 | Upper | Primary flakes from an indeterminate biface; may be related to cases 4 and 15 |
| 17 | 5 | 4 | Lower | Primary flakes from an indeterminate biface |
| 18 | 25 | 4 | Upper | Primary and secondary flakes from an indeterminate biface; may be related to case 19 |
| 19 | 13 | 3 | Upper | Primary flakes from an indeterminate biface; may be related to case 18 |
| 20 | 9 | 4 | Upper | Primary and secondary flakes from an adze-like biface |
| 21 | 5 | 1 | Lower | Chert cobble broken along flawlines |
| 22 | 9 | 8 | Lower | Flakes refit to six fragments of a stage 4 biface; biface fragments are burned |
| 23 | 1 | — | — | One flake refitted to a broken stage 4 biface |
| 24 | 1 | — | — | One resharpening flake (made into an end scraper) refitted to a resharpened adze |
| 25 | 2 | — | — | Two flakes refitted to two fragments of a narrow stage 2 biface (adze blank?) |

[a]Total area from which refitted flakes were recovered; not calculated for refit cases 7, 23, 24, and 25.
[b]Based on mean depth within the Horizon 2 paleosol; not indicated for refit cases 7, 23, 24, and 25, though all four appear to belong to the upper group.

of the western flake cluster contains smaller flakes (less than 3.5 cm), and probably represents the location where the tool was made. The northwestern flake cluster contains predominantly larger flakes, including those removed both early and late in the manufacturing sequence. This most likely represents a secondary refuse deposit. One of the secondary flakes resulting from this

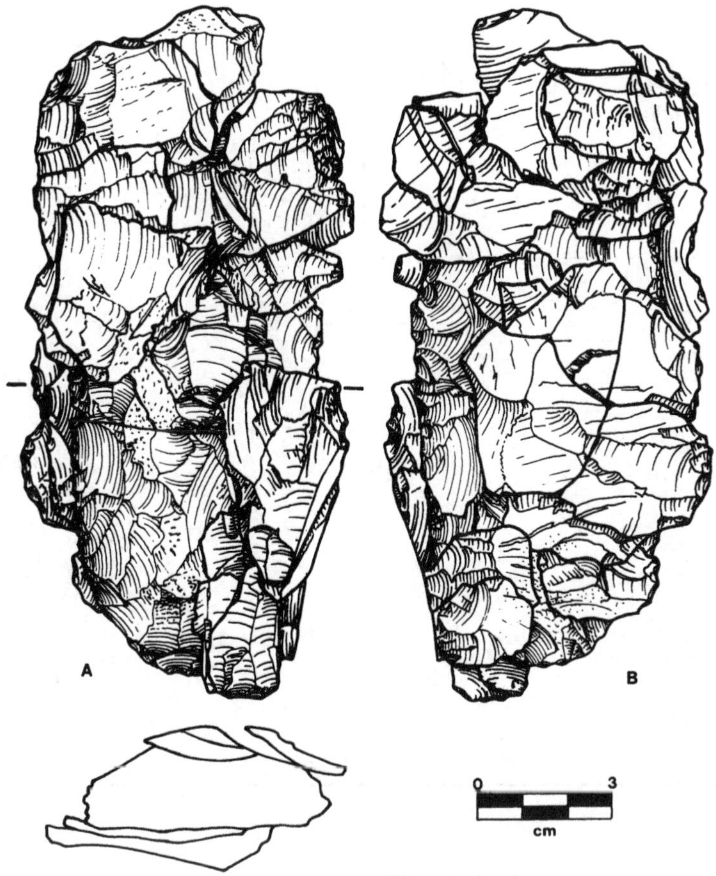


**Figure 7.** Refit case 1: a) dorsal face, b) ventral face.

reduction episode was employed as an unmodified flake tool that was discarded southwest of Feature 6.

The two halves of the adze represented in refit case 1 were recovered within 20 cm of each other near the northern margins of Feature 6. The lateral margins of this adze had been ground for hafting, and remnant portions of the bit surfaces exhibit use-polish. This adze was broken by a perpendicular bending fracture about halfway down the length of the tool, apparently as a result of heavy chopping use. The adze was close to its pristine size when it was broken and had probably been subjected to little, if any, resharpening. Following the breakage of the tool, the distal half of the adze was casually reflaked by percussion around most of its periphery. Two of the flakes resulting from this reshaping have been refitted, and they, too, were distributed around Feature 6.

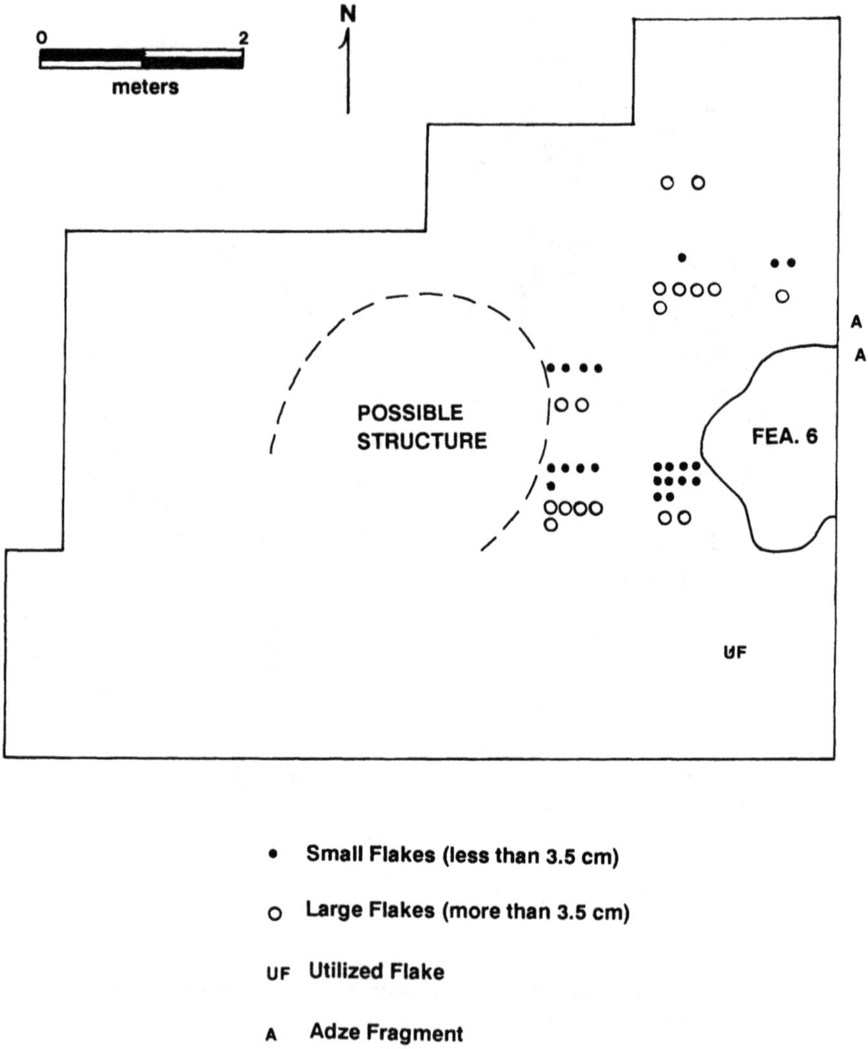


Figure 8. Distribution of refitted pieces belonging to refit case 1.

All of the refitted pieces belonging to refit case 1, including the utilized flake and the two fragments of the broken adze, were distributed in a clear arc that encircles most of the large hearth referred to as Feature 6. This pattern of dispersal indicates that the manufacture and discard of the adze were synchronous with the use of the hearth. None of these refitted pieces exhibit any evidence of exposure to heat, though this is not necessarily inconsistent with their being contemporary with the hearth.

The 25 refit cases relate to the reduction, manufacture, and/or maintenance of a variety of artifacts. Ten of these cases involve refitting to the parent cobble, core, or biface from which they were derived. In all of these, the associated cobble, core or biface, or portion thereof, was recovered in the same general area as its refitted flaking debris, or was within 1 to 2 meters of it. Refit case 21 is a fist-sized chert cobble fractured along internal flawlines during attempted reduction. Refit case 3 consists of seven flakes refitted to one of the two block cores represented in the Horizon 2 assemblage. All of the remaining 8 cases relate directly to bifaces, with 6 of them (cases 4, 7, 9, 22, 23, and 25) refitted directly to bifaces that were broken during manufacture. Refit case 1, described above, is the only example thus far identified in which manufacturing flakes have been conjoined to a finished bifacial tool that was recovered from the site. Finally, refit case 24 consists of an end scraper made on a percussion resharpening flake that refits to a resharpened adze.

The other 15 refit cases are represented only by flakes and have not been refitted to a biface or core. All of these appear to relate directly to the manufacture of bifaces and, for some of these cases, the form of the resulting biface is discernable from the reassembled flakes. At least 3 of these refit cases (6, 13, and 20) were derived from the manufacture of bifacial adzes; their respective reassembled flakes clearly show the impression of the thick, relatively narrow, humpbacked morphology that characterizes these tools. Reassembled flakes from two of these refit cases (11 and 12) reveal a broader and flatter biface cross-section that relates to the manufacture of a more refined bifacial tool, probably a Thebes knife and/or a St. Charles point. The remaining 10 refit cases are too incomplete to determine the morphology of the resulting biface, or are represented only by primary flaking debris that do not clearly distinguish the intended form of the biface being manufactured.

It is interesting that so few of the reassembled flakes have so far been refitted to finished bifacial tools that were recovered from the site. Perhaps the finished tools were discarded in other, unexcavated portions of the site. Just as plausible, however, these finished tools may have been carried to another location when the site was abandoned. In either case, several bifacial reduction episodes are represented that are not related to any of the bifaces that were recovered from the site. Conversely, there appear to be a large number of finished bifacial tools in the recovered Horizon 2 assemblage that are not accompanied by the flaking debris that resulted from their manufacture.

The apparent preponderance of "ghosts" (manufacturing debris not related to any of the bifaces recovered from the site) and "orphans" (bifaces not accompanied by their respective manufacturing debris) in the assemblage leads me to suspect that many of the recovered tools were made elsewhere and brought to the site, and that many of the tools made here were taken to other locations when the site was abandoned. This scenario would imply that the duration of the occupation(s) of Horizon 2 generally did not exceed the average use-life of the formal bifacial tools that characterize the assemblage.

Three utilized flakes have been refitted into groups of reassembled biface manufacturing flakes (cases 1, 8, and 9). While perhaps only a coincidence, all three of these flake tools were recovered from the same excavation unit located to the southwest of Feature 6. The manufacturing flakes to which they refit were all recovered to the west and north of this feature.

Excluding refit cases 7, 23, 24, and 25, which are represented by only one or two flakes refitted to a biface, the remaining 21 refit cases exhibit considerable variation in the area over which their respective reassembled flakes were distributed. These range from a minimum of 1 square meter to a maximum of 8 square meters (see Table 1). The by-products resulting from some of these reduction episodes have been subjected to more dispersal than others. When mean points for the horizontal and vertical distribution of these 21 refit cases are plotted, several clusters of flake concentrations become evident (Figures 9 and 10). These reflect the overall distribution of debitage noted in this portion of the excavation block (compare with Figure 2).

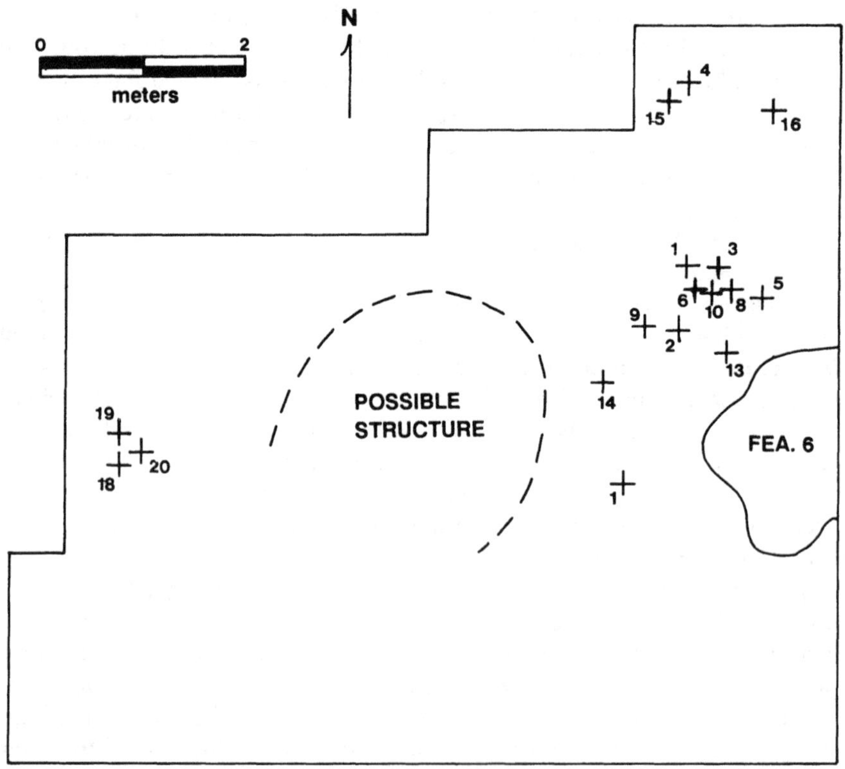


Figure 9. Mean horizontal grid positions for the 16 refit cases assigned to the upper occupation.

Vertically, these 21 refit cases fall into two distinct groups. Sixteen were concentrated within the upper 5 cm of the Horizon 2 paleosol, of which several include flakes that were recovered slightly above the paleosol surface. Flakes recovered from just above the sand/paleosol contact in the area immediately west of Feature 6 routinely refitted with flakes that were recovered beneath this contact but within the upper 5 cm of the paleosol in the area just north of the feature. Apparently, a thin veneer of fluvial or eolian sand had been deposited over the lowlying portions of the Horizon 2 paleosol before or during the latest occupation of this surface. These refit cases appear to be contemporaneous and related to the same episode of site occupation.

The 16 refit cases belonging to this upper group are distributed in one of three areas: (1) in a large cluster of flaking debris located in a crescentic arc to the west and northwest of Feature 6; (2) in a discrete concentration of flaking debris located 2 meters north of Feature 6; and (3) in a discrete concentration of flaking debris located 5-to-6 meters west of Feature 6 near the western limits of the block

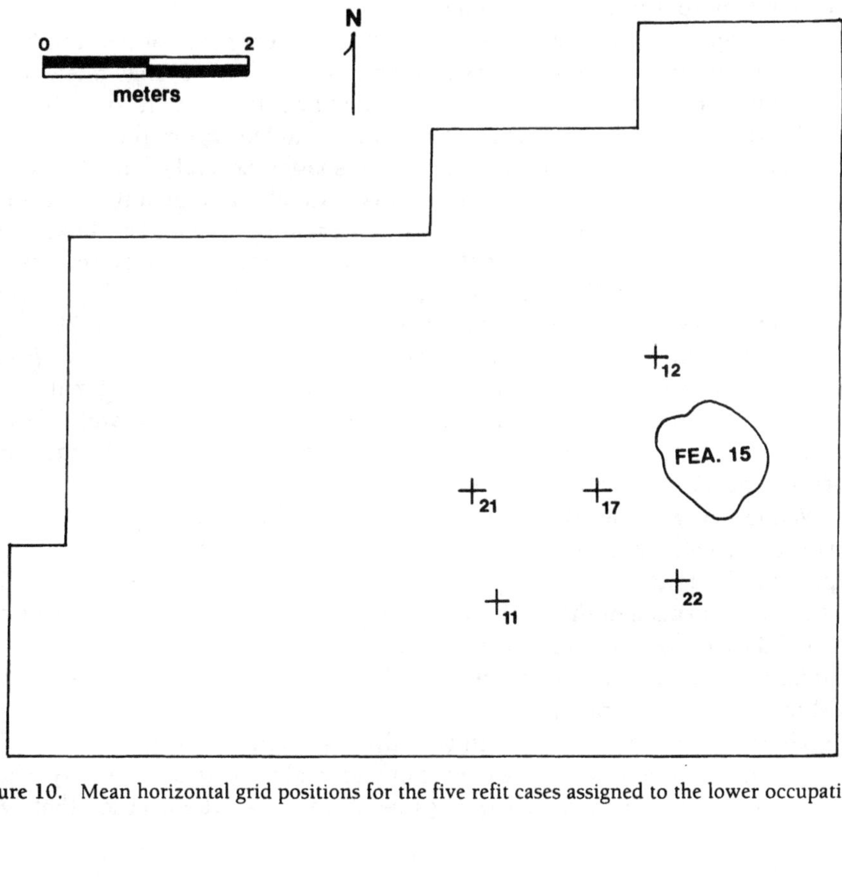


Figure 10. Mean horizontal grid positions for the five refit cases assigned to the lower occupation.

excavation. It is noteworthy that the reassembled flakes belonging to these refit cases do not intrude into the suspected house floor area. This suggests that some type of barrier (like a hut wall) kept these flakes from being scuffed into this portion of the site and/or that flakes larger than 2.5 cm were routinely removed from this area as a result of cleanup activities.

At least three of the refit cases located within the arc adjacent to Feature 6 (cases 6, 8, and 10) appear to represent secondary deposits. These consist predominantly of larger flakes (greater than 3.5 cm), and the reduction episodes that have been reconstructed appear to be incomplete. Undoubtedly, several other flakes that belong to these cases exist at the site; based on observed raw material similarities, some of them probably occur in portions of the excavation block that were not included in the present study. The distributions of flakes reassembled from each of these refit cases are very concentrated, being limited to a 2-to-3 square meter area. Further, the centers of distribution for each are located just northwest of Feature 6 in the same area that contained several of the larger flakes belonging to refit case 1. Refit cases 9 and 14, as well as the southern concentration of flakes from refit case 1, contain a broader range of flake sizes than the other groups, and these may represent primary knapping loci.

The three refit cases centered in the area 2 meters north of Feature 6 near the northern limits of the block excavation (cases 4, 15, and 16) may relate to one or two reduction episodes. Two of these include flakes that were recovered 2-to-3 meters to the south and southwest. The debris included within these three refit cases is dominated by large primary flakes and shatter. Several of the flakes were recovered within a very discrete concentration, literally a single pile, measuring about 40 by 50 cm. They are interpreted as representing a secondary flake dump associated with reduction activities that took place in the vicinity of Feature 6.

The flakes from refit cases 18, 19, and 20 were predominantly recovered from another very discrete concentration of flakes that measured about 70 by 100 cm. This concentration was located immediately west of the area of the suspected house floor and represents at least two distinct reduction episodes. Small flakes were present in this concentration but were not overly abundant, and refit case 20 is represented by only a few, predominantly large flakes. These refit cases may represent a secondary deposit of flaking debris.

While many of the flake concentrations identified in the upper portion of the Horizon 2 paleosol appear to be in secondary contexts, the distances between knapping loci and the locations of eventual secondary disposal are not great. For those reduction episodes that include related primary and secondary flake deposits, the flakes are within 1 to 3 meters of each other. It would appear, then, that secondary refuse disposal on the site was largely limited to scuffing, sweeping, and short-distance dumping.

There are no clear vertical trends in the distribution of lithic debris within the upper group of refit cases. The degree of dispersal appears to be closely related to the location and contexts of the reassembled flakes. Refit cases that were centered in the area immediately west of Feature 6 appear to be somewhat more dispersed than those located in other portions of the northwest quarter of the block. This area may have been subjected to more foot traffic or periodic refuse cleanup that resulted in a greater overall degree of dispersal. Most of the other refit cases in the upper group are tightly clustered, and these gravitate toward three areas that were probably secondary dumps. These dump locations lie along the periphery of the upper Horizon 2 occupation and, by virtue of their position within the site, were probably affected little by subsequent cultural disturbances.

Refit cases 11, 12, 17, 21, and 22 occur considerably deeper in the Horizon 2 paleosol profile, roughly 10-to-15 cm below the sand/paleosol contact. Together, these five refit cases represent a stratigraphically distinct zone of debris that relate to an earlier occupation of Horizon 2. With the exception of refit case 21, which consists of 5 fragments of a shattered chert cobble that were found in close proximity to each other, these lower refit cases are very dispersed, i.e., over areas ranging from 4 square meters to as much as 8 square meters. The distribution of flakes in these cases bears no relationship to either the suspected house floor or Feature 6. The reassembled flakes were freely distributed through these areas rather than around them, and they generally lay 5-to-10 cm beneath the house floor and the hearth.

Refit case 22 is the only example from the northwest quarter of the block excavation that exhibits evidence of burning. The 6 biface fragments associated with this refit case were recovered beneath the southern edge of Feature 6, and all exhibit reddening, potlidding, and/or thermal fractures. Though not included in Table 1, there are several additional flakes related to this refit case that were recovered from excavation units immediately east of the arbitrary boundaries of the present study area. A large proportion of the flakes recovered from beneath the *eastern* half of Feature 6 was also burned.

The nature of the lower occupation of Horizon 2 is currently unclear. Debris appears to be much less dense in the lower levels, and comparatively few large lithic items were recovered. An irregular scatter of charcoal, designated Feature 15, was identified approximately 5-to-10 cm beneath the western portion of Feature 6. This represents the only feature thus far associated with the lower occupation. With further refitting and analysis, it may be possible to trace the lower occupation across the remainder of the block and clarify its relationship with the upper occupation.

## CONCLUSIONS

The refitting analysis presented here has been an exploratory study, and has only started to tap the potential for this kind of research. Nonetheless, the data derived from the refitting of biface and tool fragments and the reassembly of chipped stone debitage from a quarter of the Horizon 2 excavation block have provided a wealth of information.

First, the horizontal and vertical relationships between refitted biface and tool fragments demonstrate that the cultural deposits contained in Horizon 2 have been subjected to minimal post-occupational disturbance. Smaller lithic items were identified as being more susceptible to vertical displacement, probably as a result of human trampling in concert with low levels of natural disturbances. These observations were supported by the results of refitting debitage from the sample area of the block excavation.

Second, the contents of reassembled groups of flakes from the northwest quarter of the excavation block have been used to differentiate primary knapping loci from secondary waste deposits. Flake concentrations within this portion of Horizon 2 appear to represent both types of deposits, though the majority of flakes were probably recovered from secondary contexts. The results of debitage refitting suggest that secondary displacement of debris was largely limited to scuffing, sweeping, and limited dumping.

Third, the results of the debitage refitting also identified two distinct, vertically segregated deposits of cultural debris that relate to at least two different episodes of site occupation. While this was previously suspected from the overall vertical distribution of lithic material and the presence of superimposed features, the relationship between the upper and lower occupations was more clearly defined by the refitting analysis. These results show that, at least in the northwest quarter of the excavation block, these two zones are distinct and relatively unmixed.

Fourth, the apparent infrequency of refits conjoining finished bifacial tools with their resulting manufacturing debris (and vice versa), may indicate that the duration of the occupations of Horizon 2 was relatively brief. If more protracted occupations were represented, we would expect to see a greater frequency of refits joining finished bifacial tools with flaking debris. Instead, a large number of the finished tools recovered from the site appear to have been made elsewhere, and several of the tools produced on the site were probably carried away when the site was abandoned.

With regard to site occupation span, the results of this refitting study confirm some of the general expectations about Early Archaic settlement in the midwestern United States. Specifically, the presence of primary refuse deposits, as well as the limited amount of movement resulting from secondary displacement, suggest that the occupations of the site were of relatively short duration (e.g., less than one full season). This interpretation is bolstered by the relative frequency of "ghosts" and "orphans" in the assemblage and the scarcity of manufacturing flakes that could be linked to the finished tools recovered from the site.

These are but a fraction of the potential applications of lithic refitting to archaeological research. Lithic assemblages consist of numerous individual pieces that relate to the manufacture, use, recycling, and discard of stone tools. Through refitting, these individual pieces are transformed into recognizable and behaviorally meaningful units for further analysis. Refits represent discrete events; as such, they can form an important bridge between static objects and the behavior that created them, as well as the cultural and natural processes that affected them. In short, lithic refitting provides unique kinds of data that are directly relevant to important research questions.

## ACKNOWLEDGMENTS

I would like to express my enormous appreciation to all of the Center for American Archeology staff and the field school students who were involved in the Twin Ditch excavations—in particular, Tony Schwinghamer, Patti Wright, Katie Egan, and Cameron Quimbach. George Odell and Michael Shott provided valuable comments that have contributed to the improvement of the original draft of this manuscript. Julie Morrow contributed to all phases of this project with assistance, encouragement, and editorial support.

## REFERENCES

Binford, L. R. 1980. Willow Smoke and Dog's Tails: Hunter-Gatherer Settlement Systems and Archaeological Site Formation. *American Antiquity* 45:4–20.

Brown, J. A. 1985. Long-Term Trends to Sedentism and the Emergence of Complexity in the American Midwest. In *Prehistoric Hunter Gatherers: The Emergence of Cultural Complexity,* edited by T. D. Price and J. A. Brown, pp. 201–231. Academic Press, New York.

Brown, J. A., and R. K. Vierra. 1983. What Happened in the Middle Archaic? Introduction to an Ecological Approach to Koster Site Archaeology. In *Archaic Hunters and Gatherers in the American Midwest,* edited by J. L. Phillips and J. A. Brown, pp. 165–195. Academic Press, New York.

Callahan, E. 1979. The Basics of Biface Knapping in the Eastern Fluted Point Tradition: A Manual for Flintknappers and Lithic Analysts. *Archeology of Eastern North America* 7:1–180.

Cziesla, E. 1990. On Refitting of Stone Artifacts. In *The Big Puzzle: International Symposium on Refitting Stone Artefacts,* edited by E. Cziesla, S. Eickhoff, N. Arts, and D. Winter, pp. 9–44. Studies in Modern Archaeology, Volume 1, Holos, Bonn, Germany.

Fisher, J. W., and H. C. Strickland. 1991. Dwellings and Fireplaces: Keys to Efe Pygmy Campsite Structure. In *Ethnoarcheological Approaches to Mobile Campsites,* edited by C. S. Gamble and W. A. Boismier, pp. 215–236. International Monographs in Prehistory, Ann Arbor.

Gifford-Gonzalez, D. P., D. P. Damrosch, D. R. Damrosch, J. Pryor, and R. L. Thunen. 1985. The Third Dimension in Site Structure: An Experiment in Trampling and Vertical Dispersal. *American Antiquity* 50:803–818.

Hajic, E. R. 1987. *Geoenvironmental Context for Archeological Sites in the Lower Illinois River Valley.* U.S. Army Corps of Engineers, St. Louis District, Historic Properties Management Report 34.

Hassen, H., and E. R. Hajic. 1984. *Shallowly Buried Archeological Deposits and Geologic Context: Archeological Survey in the Eldred and Spankey Drainage and Levee District, Greene County, Illinois.* Contract Archeology Program, Center for American Archeology, Reports of Investigations 139. Kampsville, Illinois.

Henry, D. O., C. V. Haynes, and B. Bradley. 1976. Quantitative Variations in Flaked Stone Debitage. *Plains Anthropologist* 21:57–61.

Johnson, J. K. 1979. Archaic Biface Manufacture: Production Failures, a Chronicle of the Misbegotten. *Lithic Technology* 8:25–35.

Kelly, R. L. 1988. The Three Sides of a Biface. *American Antiquity* 53:717–731.

Kent, S. 1981. The Dog: An Archeologist's Best Friend or Worst Enemy—the Spatial Distribution of Faunal Remains. *Journal of Field Archeology* 8:367–372.

McKellar, J. A. 1983. Correlates and the Explanation of Distributions. *Atlatl*, University of Arizona, Occasional Papers 4.

Meyers, J. T. 1970. *Chert Resources of the Lower Illinois Valley.* Illinois State Museum, Reports of Investigations 18, Springfield.

Morrow, T. A. 1987. Twin Ditch: Investigations at a Multicomponent Archaic Site in the Lower Illinois River Valley. *Illinois Archeological Survey Newsletter* 2(2):2–4.

Morrow, T. A. 1988. Thebes Knives: Experimental Applications to Archeological Data. *Twentieth Century Lithics* 1:8–23. Moundbuilder Books, Branson, MO.

Morse, D. F. 1971. Recent Indications of Dalton Settlement Pattern in Northeast Arkansas. *Southeastern Archeological Conference Bulletin* 13:5–10.

Morse, D. F. 1975. Reply to Schiffer. In *The Cache River Archeological Project: An Experiment in Contract Archaeology*, assembled by M. B. Schiffer and J. H. House, pp. 113–119. Arkansas Archeological Survey, Research Series 8, Fayetteville.

Morse, D. F., and A. C. Goodyear. 1973. The Significance of the Dalton Adze in Northeast Arkansas. *Plains Anthropologist* 18:316–322.

Murray, P. 1980. Discard Location: The Ethnographic Data. *American Antiquity* 45:490–502.

Newcomer, M. H. 1971. Some Quantitative Experiments in Hand-Axe Manufacture. *World Archeology* 3:85–94.

Odell, G. H. 1980. Toward a More Behavioral Approach to Archaeological Lithic Concentrations. *American Antiquity* 45:404–431.

Patterson, L. W., and J. B. Sollberger. 1978. Replication and Classification of Small Size Lithic Debitage. *Plains Anthropologist* 23:103–112.

Schiffer, M. B. 1972. Archeological Context and Systemic Context. *American Antiquity* 37:156–175.

Schiffer, M. B. 1975a. An Alternative to Morse's Dalton Settlement Pattern Hypothesis. *Plains Anthropologist* 20:253–266.

Schiffer, M. B. 1975b. Some Further Comments on the Dalton Settlement Hypothesis. In *The Cache River Archeological Project: An Experiment in Contract Archaeology*, assembled by M. B. Schiffer and J. H. House, pp. 103–112. Arkansas Archeological Survey, Research Series 8, Fayetteville.

Schiffer, M. B. 1975c. The Effects of Occupation Span on Site Content. In *The Cache River Archaeological Project: An Experiment in Contract Archaeology*, assembled by M. B. Schiffer and J. H. House, pp. 265–269. Arkansas Archaeological Survey, Research Series 8, Fayetteville.

Schiffer, M. B. 1987. *Formation Processes of the Archeological Record.* University of New Mexico Press, Albuquerque.

Schultz, J. M. 1992. The Use-Wear Generated by Processing Bison Hides. *Plains Anthropologist* 37:333–351.

Sheets, P. D., and G. Muto. 1972. Pressure Blades and Total Cutting Edge: An Experiment in Lithic Technology. *Science* 175:632–634.

Shott, M. J. 1989. On Tool Class Use-Lives and the Formation of Archaeological Assemblages. *American Antiquity* 54:9–30.

Stevenson, M. G. 1985. The Formation Processes of Artifact Assemblages at Workshop/Habitation Sites: Models from Peace Point in Northern Alberta. *American Antiquity* 50:63–81.

Stevenson, M. G. 1991. Beyond the Formation of Hearth-Associated Artifact Assemblages. In *The Interpretation of Archeological Spatial Patterning*, edited by E. M. Kroll and T. D. Price, pp. 269–299. Plenum, New York.

Stockton, E. D. 1973. Shaw's Creek Shelter: Human Displacement of Artifacts and Its Significance. *Mankind* 9:112–117.

Villa, P., and J. Courtin. 1983. The Interpretation of Stratified Sites: A View from Underground. *Journal of Archeological Science* 10:267–281.

Winter, D. 1990. "The Refitters Failure"—Some Criteria Concerning the Duration of Settlement through Refitting. In *The Big Puzzle: International Symposium on Refitting Stone Artefacts*, edited by E. Cziesla, S. Eickhoff, N. Arts, and D. Winter, pp. 477–492. Studies in Modern Archaeology, Volume 1, Holos, Bonn, Germany.

Wood, W. R., and D. L. Johnson. 1978. A Survey of Disturbance Processes in Archaeological Site Formation. In *Advances in Archaeological Method and Theory*, Vol. 1, edited by M. B. Schiffer, pp. 315–381. Academic Press, New York.

# Part VI

# Conclusion

## Chapter 13

# Some Comments on a Continuing Debate

GEORGE H. ODELL, BRIAN D. HAYDEN, JAY K. JOHNSON, MARVIN KAY, TOBY A. MORROW, STEPHEN E. NASH, MICHAEL S. NASSANEY, JOHN W. RICK, MICHAEL F. RONDEAU, STEVEN A. ROSEN, MICHAEL J. SHOTT, AND PAUL T. THACKER

The chapters in this volume were discussed thoroughly during the four days of the conference that spawned them. As with most discussions, the tide ebbed and flowed, sometimes centering on the details of a particular paper but often contributing to a better understanding of a more general issue. In this final chapter we have compiled the principal points made, in as readable and logically organized a format as we could muster. Arguments do not necessarily appear here in their original order, and the points made in any particular section may not even have occurred in the same discussion. Every attempt has been made, however, to retain the original intent and context in which the remarks were delivered.

Our objectives are to be useful and engaging, rather than pedantically rigid; our goal is stimulation, not absolute truth. The comments that follow are not meant to be critiques of the papers—in fact, most of the truly paper-specific remarks have been omitted. Rather, these thoughts pertain to general subjects, and are meant to place this set of papers and issues in perspective. We hope they retain some of the flavor of the original interchange.

*Stone Tools: Theoretical Insights into Human Prehistory*, edited by George H. Odell. Plenum Press, New York, 1996.

## CLASSIFICATION

### Design Theory

The prospect of being faced with a vast quantity of stone tools and having to make some behavioral and conceptual sense of them can be daunting. Tackling an entire collection at once is counter-productive, and for this reason the most sensible way of proceeding is by breaking the collection down into smaller, more manageable categories. This is the rationale behind all classification, and is exemplified by the design theory principles articulated in this volume (Hayden et al.), in which the categories used were adaptive strategies.

Design theory has the advantages of combining the functional possibilities inherent in tool form with relevant ethnographic information to provide a sounder basis for behavioral interpretations than is normally the case with straight tool typology. Some potential for circularity exists in the argument, as there may be a tendency to construct categories from ethnography and then "fit" archaeological tools into those categories whether or not the fit is justified. This is less of a problem with specialized tools that are appropriate for only a limited array of tasks, such as drills or microliths.

Another potential problem lies in the implicit assumption that design implies intent—i.e., that something that looks like an axe, for example, was always intended to chop things. The modern screwdriver can also be used to open paint cans (in fact, many paint cans *are meant* to be opened with screwdrivers), and this might become its principal function for, say, a house painter. Such potential uncertainties with respect to prehistoric stone tools suggest that it is important for the analyst to provide independent confirmation for the categories employed. Use-wear analysis and stratigraphic association can provide such confirmation, but are not the only appropriate methods.

### Newly Introduced Organizational Structures

Concepts such as "reliability," "maintainability" (Bleed 1986), "flexibility" and "versatility" (Nelson 1991) have recently been introduced, and these are starting to be employed to organize entire collections of prehistoric material. This trend has several disturbing aspects. The distinction between "maintainable" and "reliable" tools, while conceptually defensible and intriguing at first blush, has turned out to be too abstract to operationalize. Every tool possesses elements of both qualities, rendering most distinctions ambiguous; and quantifying, or even consistently distinguishing, these concepts has proven exceedingly difficult. It is not clear whether or not all of Bleed's criteria need to be satisfied in any specific case. And individual criteria such as "overdesigned" are not defined clearly enough to be useful. In fact, they are not observations but inferences, and are therefore acutely subjective. Such difficulties are exemplified by the drill, which by design was easily maintainable (sharpenable), but does not appear to have been "over-designed." In addition, when contemplating any given piece of stone, it is usually difficult to tell whether that piece (if it was a tool at all) was manipulated in the hand or secured in a haft, let alone what that haft looked like. To place a rubric of "maintainability" or "reliability" on most individual tools or collections would therefore appear to be speculative.

The terms "flexibility" and "versatility" possess many of the ambiguities discussed above. Their main problem, though, is that they appear to encapsulate concepts that are already expressed by terms like "multifunctionality" and "recyclability." Since the latter terms are already in use among archaeologists and are better understood, it makes sense to continue to employ them in favor of the new terminology, unless the latter can be clearly defined and differentiated for each usage.

## Projectile Point Systems

Pointed, bifacially worked stone objects found on archaeological sites typically occasion considerable terminological disjuncture, as they are alternatively called "points" by some and "bifaces" by others. Those who call them points argue that this is the established term for such objects in Europe and much of North America; that it distinguishes pointed specimens from generic, non-pointed ones; and that to call every piece of bifacially worked flint a "biface" invites confusion. The minority opposition argues that the use of a common term for all bifacially flaked artifacts is technologically accurate; that a gray area of ambiguous pointedness exists between the two classes; and that the alleged confusion is not all that acute. Conference participants generally favored the use of the term "bifacial tools" to refer to any or all bifacially worked artifacts in general, and "points" remained a popular designation.

Projectile point classifications were discussed only with reference to the system that Rick (this volume) proposed, and much of that discussion was particularistic. Rick's classification is innovative, flexible to specific levels of analysis, and responsive to issues of continuity and assemblage richness. It also appears to manifest a higher level of chronological sensitivity (whatever that represents) than most typologies or seriations currently in use.

Rick's system did, however, engender a modicum of discomfort among the participants, partly on account of its complexity. It is possible, for example, that simple measurements and/or nominal states such as the presence of basal grinding might provide distinctions as fine-grained as those produced by Rick's analysis. And most of Rick's types have been determined on the basis of outline form, but without control over resharpening (see Hoffman (1985). These are potential (not established) drawbacks and should not detract from the usefulness of the results, which document interpretable breaks in the projectile point sequence.

## ANALYTICAL TECHNIQUES

### Piece Refitting

Morrow's study (this volume) provides a concise rationale for the refitting of flakes onto one another, onto cores, and onto formal tools. It is clear that these techniques can assist in answering many of the questions with which our discussions were concerned, including issues involving reduction sequences, classification, and curation. Essentially, piece refitting involves long hours of relative quiet, punctuated by bursts of activity. Because it is labor-intensive, insights that could save hours of work should be welcomed by all who might engage in such activity.

Morrow specified some operational shortcuts that would be useful to pass along. First, it is crucial to have an explicit goal with which to guide the analysis; at the Twin Ditch site, Morrow's was the technical reconstruction of cores and biface manufacturing debris. Second, it is useful to proceed according to a series of logical steps. One can initiate an analysis by sorting the material into raw material categories. Then one should sort broken flakes into distal and proximal ends. This procedure may enable one to reunite several of them, resulting in being able to work with a greater percentage of complete flakes in subsequent procedures. Then the real work of refitting flakes onto cores, tools and other flakes begins. There is no shortcut to simply spreading the (labelled) material out onto tables and intuitively conjoining associated pieces. It is important not to bias the analysis by focusing on transient variables such as color which, in Morrow's case, was affected by heat alteration and proximity to a sterile sand layer. One can more productively key into characteristics such as texture and fossil inclusions. Finally, rubber cement is a more effective adhesive than Elmer's glue, because the former can be removed more easily if mistakes have been made or if a more detailed analysis of individual flakes is desired. For more information on this topic, see articles in Cziesla et al. (1990) and Hofman and Enloe (1992).

The Twin Ditch refitting analysis established several facts about the site that would have been difficult, if not impossible, without this procedure. For example, the occupants were regularly and demonstrably selecting flakes from a bifacial reduction sequence for use as tools. In at least one instance, smaller flakes were concentrated around a prominent hearth, while larger flakes, which were able to be refitted to some of the smaller pieces, were isolated, removed, and dumped in a different location. In addition, refitting established the existence of two distinct occupations in the upper component, thus corroborating other evidence from the site. And the presence of many tools with no refitted debris, and vice versa, suggested a short occupation of the site (shorter, at any rate, than the use-life of the tools). These results demonstrate that, although refitting techniques are labor-intensive, they can provide valuable information for reconstructing relevant behavioral parameters.

## Use-Wear Analysis

Interpreting the genesis of use-wear, a subject that was broached only briefly at this conference, involves specialized studies. Marvin Kay's research (this volume) supports mechanical abrasive causes for polish formation. All polish formed in this manner would necessitate the presence of particles as hard as, or harder than, the material being polished. This scenario emphasizes the role of autopolishing, in which particles of the tool itself break off and rub against its surface. In this regard it may be significant that those areas where Kay's experimental tools showed developed polish formation were usually associated with an abundantly fractured edge. This mechanical abrasion model is supported by lapidary studies and by Knutsson's (1988) scanning electron microscopic study at very high magnifications of experimental quartz tools. Competing theories (surficial silica gel, etc.) were not discussed at the conference.

Kay's type of use-wear research is very labor-intensive. It can provide answers to detailed and specifically directed questions but, under normal circumstances, should not be attempted until less labor-intensive studies have been conducted. Given this situation, substantial effort should be expended in drawing an effective sample.

In the future it is possible that the most accurate techniques for ascertaining worked material will involve quantitative approaches by comparing complex patterns derived from digital image processing techniques. These techniques are currently in a primitive state of development and will not be usable for several years. Even if they are developed, appropriate screening methods would have to be applied to archaeological samples before this stage is reached in an analysis.

# CURATION

The concept of "curation" is currently rampant in archaeology, and is embedded in the Cultural Resource Management literature. The term originated in connection with Binford's forager-collector conceptualization (see Nash and Odell articles, this volume). It is not generally employed for complex societies, for two reasons: (1) curation as an analytic concept is simply not useful for societies in which trade, long-distance transport, and specialization are recognized parts of the system; and (2) complex societies are by nature sedentary, a quality that negates certain definitions of the term (e.g., transport from one site to another among the same group of people).

There is general agreement that scholars have come to employ Binford's schema too rigidly and in ways he never intended; accordingly, the concept has taken on a life of its own. In reality, hunter-gatherers can rarely be characterized as exclusively foragers or collectors, and it is now apparent that between the two poles exists a continuum on which groups occasionally changed their strategies according to local conditions (Chatters 1987).

Although curation has been associated with hunter-gatherer mobility organization, it exists independently of the forager-collector scenario; in fact, Odell argued that there may be little or no correlation between curation and either residential or logistical mobility strategies. Although it is unfair to criticize the term because it has been made to carry so much baggage, it is this very baggage that renders the concept so unwieldy today. It is essentially being asked to provide answers to questions whose ramifications are only dimly perceived. Most of the conference participants agreed that it would be best to suspend usage of the term, employing instead a description of the type of behavior that appears most relevant to the specific case being described (e.g., transport, recycling).

Curation has been applied at the level of the tool, the technology, and the assemblage. Some agreement exists that it is safest to apply it, if at all, at the lowest possible level, i.e., the tool or tool class. Such an approach is also easier to operationalize. Of course, this approach is open to the criticism that prehistoric peoples did not necessarily distinguish the same artifact classes as we archaeologists do, and that classes so established were not necessarily employed for internally consistent types of tasks. Shott (1989) has defined curation of individual tools in terms of their "potential utility." The usefulness of this definition depends on assumptions involving the derivation of the original dimensions of the tool and its presumed resharpening sequences.

Thacker (this volume) applied curation to the issue of transport costs, distinguishing between Gravettian reduction near quarries and Magdalenian transport of whole nodules away from quarries. This approach engendered a discussion of possible transport aids, including travois, sleds, horses, etc., indicating the possible utility of the concept when used in this way. Other participants, however, pointed out that this definition diverts attention from the *overall* meaning of curation, which should be a continuous, not a presence/absence variable.

Curation is similar to "use-life," which can be defined as the rate at which people exhaust tools and the impact of that on assemblage formation. Difficulties exist in distinguishing between the two concepts operationally, however. Some would measure use-life in terms of reduction, whereas others would call this curation, use-life being measured preferably in terms of time. There must be a relationship between tool use-life (measured in real time) and the amount of maintenance that a tool has been put through. Perhaps, with a large body of good experimental tool use-life data, thorough use-wear analysis, and carefully constructed models of the expected frequencies of various tool-using tasks, it may be possible to translate the number of tools in an archaeological assemblage into the person/hours of tool use represented and even the person/days of a site's occupation.

The role of raw material availability in all this (Bamforth 1986; Odell, this volume) is still being debated. Many societies practiced economizing behavior, a clear example of which was offered by Johnson (this volume) for obsidian use among the Maya. This behavior was demonstrated through plunging blades (i.e., production failures), tiny flakes and striking platforms, abundant core rejuvenation flakes, and lateral recycling from blade to flake cores. It is apparent that economizing behavior can influence mobility patterns, and vice versa, even when raw material is not scarce (e.g., Israel's Negev Desert: Marks et al. 1991).

## HUNTING WEAPONRY

### Dart/Arrowhead Distinctions

Distinguishing dart from arrow points has plagued analysts for years. Most interested scholars have had to employ the results of Thomas' (1978) study of ethnographic specimens, although Shott (this volume) discusses Thomas's relatively low accuracy for dart points and the ambiguous results of transitional forms. In the North American midcontinent, dart point size usually decreased in later prehistory, a phenomenon documented in Shott's article. In some regions this decline appears to have been more gradual than previously thought, and it may have involved a miniaturization of forms already in existence (Hall 1980). In Arkansas, a gradual decrease in the size of Gary points has been documented (Schambach 1982:173–177). However, the introduction of arrow points in central Arkansas did not involve a miniaturization of forms already in existence, a point supported by the very different reduction trajectories employed in the production of dart and arrow points (Nassaney, this volume). Trends in the size of hunting weaponry have been effectively characterized by neck and basal widths of stemmed and notched points.

Shott (this volume) explained this change in technology through an offshoot of Optimal Foraging Theory, the Diet Breadth Model. In essence, the decrease in diet breadth caused by an ever greater dependence on starchy seed crops, especially maize, also appears to have occurred with animals. That is, hunters increasingly emphasized a limited number of prey species, dropping less preferred species from their list of targeted game. To effect this they had to decrease pursuit time, which they accomplished by increasing the efficiency of their technology. One way to increase range and accuracy of weaponry may have been to reduce the length and width of their projectile shafts, which would have affected the size of tip placed on those shafts. Another may have been the adoption of—or greater reliance on—the bow-and-arrow.

Along with projectile points, snares, traps, and blunt arrows were also used on game, and it would be interesting to investigate the relationships among the various hunting methods. Prehistoric people in the American midcontinent appear to have focused more and more on deer, although turkeys and aquatic species also increased in importance in later prehistory. There appears to have been little population pressure on the American Bottom inhabitants studied by Shott; thus hunters could effectively specialize and still have enough to eat. Why they chose to specialize rather than to vary their diet is an unanswered question within this paradigm. A variable that may provide some enlightenment, but remains insufficiently studied, is warfare, for which evidence in the form of fortified sites and skeletons with embedded arrows has been documented in the Late Woodland of eastern North America (Blitz 1988).

One way of approaching dart/arrowhead distinctions might be through use-wear analysis of projected missile tips. It is possible that the presence of Wallner lines on the surface of impact fractures could provide a direct measure of missile velocity. If this is true, this variable will be difficult to record in the vast majority of archaeological specimens, as Wallner lines are visible only on the most fine-grained materials. It might be more effective to investigate hackles and lateral fissures, which also constitute reactions to internal stress and are observable on a wider variety of material classes. However, if velocity (arrow) and mass (dart) cancel each other out, this method may not work at all. Another potentially instructive line of inquiry might be to investigate temporal trends in the nature of blows, angle of entry, etc., for those skeletons that exhibit evidence of violent trauma.

In North America the frequency of bifacial forms decreased as society tended toward food production, but this was primarily a result of a general deemphasis on formal stone tools with increasing sedentism (Parry and Kelly 1987). This pattern is also documented in central Arkansas, where Nassaney (this volume) demonstrates a shift from core to flake tool reduction strategies from the Middle Woodland through Mississippian periods. In the southern Levant, stone arrowheads first appeared with the rise of agriculture in the early Neolithic. They continued in abundance even with the domestication of animals. Given the total absence of skeletal evidence for violent death, fortified sites, and other weaponry, or of any other evidence for warfare, they seem not to have been associated with war. Convincing evidence for warfare did not appear until the Bronze Age. Thus, beyond the technical similarities between stone and metal arrowheads, their actual roles in society seem to have differed significantly, a result that was probably more a function of chronology than technology.

## Tool Refurbishing

The Flenniken-Thomas debate was mediated in our discussions by Rondeau (this volume), who sought to take the point reworking issue in a different direction. Rondeau maintained that neither side dealt very effectively with the archaeological evidence. Flenniken's argument, that sharpening was one of the options open to prehistoric tool users who appear to have exercised that option frequently, is well taken. On the other hand, Elko corner-notched points do not appear to have experienced significant typological change from reworking. This finding appears to be consistent with behavior in the North American midcontinent, where repair often followed the principal outlines of the original point, though a few southeastern examples of changing a notched to a stemmed point were noted. Thus if the issue is typological change, Flenniken appears to have overshot the mark. To establish his point would have required refurbishing and typing by independent parties in such a manner that partisanship was removed from the issue and some quantitative measure was employed. A good strategy for researching this issue would be to search out locales that were responsive to lithic mophological change and rejuvenation. These might be located distant from raw material sources in areas of high mobility.

The rejuvenation of projectile points and resulting typological ramifications discussed by Flenniken have obvious parallels with the sharpening of Middle Paleolithic scrapers discussed by Dibble (1987). In both cases a straw man—the prevailing typological system—was set up. Dibble was reacting to the traditional Bordesian system for which criticism was justifiable, yet Dibble appears to have overreacted in attributing so much of the observed cultural change to flawed scraper morphology. On the other hand, the variability in Bordes's curves was dependent on a small number of types, and it was those types in which Dibble was primarily interested.

## CRAFT SPECIALIZATION AND TRADE

### Some Examples of Specialized Lithic Production

Although the lithic data base is not optimal for distinguishing prehistoric craft specialization, it has yielded several examples of this practice. The best known in Mesoamerica is Colha, a site that is crucial for understanding Maya lithic specialization. Johnson (this volume) provided an instructive comparison between Colha and sites of the Middle Archaic Benton period of the American Southeast, in which craft specialization probably did not exist. Colha possessed specific lithic workshops; and, although other areas at Colha contain habitation debris, the workshop areas do not. Benton sites contain standardized tools, but no real workshops. Standardization in Benton times was probably a result of shared mental templates or specific functional exigencies.

This example emphasizes that standardization does not necessarily imply specialization. A similar situation can be seen in certain Mexican pottery that exhibits a high degree of metric standardization, but in a context in which pots are produced by the household, for household consumption (Arnold 1991). Analysis of such pottery for eventual sociopolitical interpretation should be multiscalar and contextual and include not only metric data, but the quality of the final product.

It is interesting that one finds few or none of the classic Colha lithic types, such as tranchet axes and oval bifaces (Potter 1991; Shafer 1991), at centers such as Tikal, Cuello or Nohmul, even though Colha knappers must have been producing these tools for trade. In fact, some distinctive sharpening flakes of Colha chert have been found at habitation sites. But more importantly, exhausted tools of this chert have been discovered in areas of raised fields such as Pulltrouser Swamp. These tools, then, were probably traded to municipal areas where they were maintained and resharpened, but they were actually used out in the fields where we find the exhausted products.

Another example of craft specialization occurs in the Near East. On most sites of the Chalcolithic and Early Bronze Age of the Levant, ad hoc tools and flakes manufactured from locally abundant flint are ubiquitous. They were obviously the tool of choice for most tasks for which people still used stone tools. Flint sickle segments were also common, apparently manufactured by specialists. The evidence for this consists of the large number of these sickles found at most sites; the total absence of diagnostic waste, especially cores, at the vast majority of these sites; the discovery of a few workshops, usually at satellite locales; and the recovery at four sites of caches of unworked blades, perhaps representing trade packets (Rosen 1989).

In some parts of the world, stone is still being used for certain tasks, or has been until recently. For example, Mexicans still employ Teotihuacan obsidian for shaving; some Guatemalans use basalt chopping tools for working stone; and a few Turks continue to process grain with flint-embedded threshing sledges (the flint blades having been manufactured by specialists). Stone continues to be employed because efficiency and quality are not important considerations, and it costs less than the alternatives.

## Specialization and Complexity

Although a discussion of craft specialization is often confined to societies with a state system of government, it has also been practiced in less centrally organized political systems (see Cross 1993). Sociopolitical complexity is usually not measured by complexity in the use, technology or distribution of stone tools, which may have followed other exigencies, although elites often manipulate the organization of production to mobilize surplus. It is probable that economic specialization followed from (or was caused by) sociopolitical complexity, not the other way around (Earle 1987:75). This relationship may have been a product of lag time in secondary state evolution, in which economic factors may commonly have lagged behind complexity in other elements of society.

Stone tool production probably engendered some prehistoric specialization, though not an excessive amount (Cross 1993). On the basis of ethnographic examples from Australia, it takes most people a relatively short time to knap a stone tool, but not everybody did it. A few semi-specialists appear to have flaked stone for the entire group. This example may be useful for prehistoric groups in similar contexts, though we should not forget that most ethnographic accounts of stoneworking were recorded from groups that had already been exposed to metal. What appeared to be craft specialists in these cases may actually have been those stubborn few who had not yet abandoned the old ways.

## SOCIOPOLITICAL CONTROL

Craft specialization is often accompanied by economic or sociopolitical control and a substantial amount of effort, at this conference and elsewhere, has been expended in attempting to understand this phenomenon. The lithic data base has been employed for these ruminations only infrequently, but it may have something to contribute.

Where elites developed, they did not control the distribution of every single item of exchange, but concentrated on certain ones that were critical to the reproduction of the biological and social systems (e.g., salt in the Maya highlands: Andrews 1980, 1983) or the maintenance of the ritual apparatus on which political fortunes depended (e.g., personal adornment and dress among the Aztec: Brumfiel 1987). Elite control can be monitored through either specific material items or the differential distribution of lithic technological classes such as blades and flakes, which imply a spatial or technical division of labor (Nassaney, this volume). For example, Jackson and McKillop (1989) isolated apparent stopovers under elite control on the Maya coast that were characterized by large quantities of obsidian blades, whereas other locations contained ordinary flakes. Similar situations occurred at Palenque and Kaminaljuyu.

Chiefly or priestly activities have been validated through ritual, and occasionally lithic data can be employed in comprehending ritual context. Such a study was recently conducted at the Illinois Hopewell ceremonial precinct of Napoleon Hollow, where stone bladelets were determined to have been employed for ceremonial functions (Odell 1994). Ritual considerations may also be important in ascertaining why, in certain circumstances, people would import blade cores rather than the blades themselves. One answer is that access to imported blades would be widespread, whereas importing cores would necessitate employing a specialist to manufacture blades from those cores. It would be easier for an elite to control blade production (and producers) for ritual purposes than to control the use of blades if they were manufactured elsewhere (see Clark 1987).

Elite control may be detected by looking for redistribution areas around larger market centers. Some caution is necessary here, because primitive markets without elite control also exist, and these would exhibit similar distributions around their peripheries. In fact, one can point to numerous examples of craft specialization and market economies without direct elite control over craft specialists. Contemporary village-level lithic specialists in Turkey, unbeholden to any elite, serve as a good model for this type of craft specialization.

The Plum Bayou culture of central Arkansas (Nassaney, this volume) provided a forum for discussing issues of control, resulting in titillating, if not totally satisfying, conclusions. After analysis of several lithic raw materials, the only one that exhibited some evidence of control was quartz crystal. But why would anybody want quartz crystal, or wish to control its distribution? Was it a currency, used in transactions? And if control can be established, was the controlling party an elite class or a more egalitarian source?

Although many details of the Plum Bayou settlement system remain unknown, some interesting relationships obtain. A linear falloff from the source of quartz crystal suggests down-the-line trade, with minor exceptions. These exceptions, which contain a disproportionate amount of quartz, may constitute loci of labor mobilization. The Toltec Mounds site—paramount center of the Plum Bayou culture—is one of these quartz concentrations, but one could not conclude that ultimate control was being exercised from this locale. First, it is only one of the concentrations; and second, the distibution of quartz does not fall off from Toltec.

Yet some interesting anomalies have been unearthed from the Toltec site itself. For example, a bone bed in a small mound near a plaza was recently discovered, possible evidence of feasting and/or mound building activities. And disproportionate amounts of quartz crystal have been recovered from low mounds east of the plaza, suggesting some ceremonial activity involving quartz crystal. These disparate facts indicate ritual activity at Toltec and some control of quartz crystal. The nature and context of that control remain under investigation.

## INCURSION OF METALS

### Context of Early Metal Use

The cultural relationships of early competing technologies have been addressed infrequently in the archaeological literature, and this is especially true of stone. One of the most poignant arguments made by Rosen (this volume) is that, for many tasks (e.g., sickle use), flint tools were no less efficient than metal tools; for some tasks, they were actually *more* efficient. The factors that explain replacement in these cases are not efficiency or practicality, but trade, control, and other elements related to increasing cultural complexity.

Contact Period literature from North America also suggests that initially some of the invading European products were no better than the Native technologies they replaced (this is certainly true of firearms: e.g., Hudson 1976:245). They were traded for to gain access to other European products, to acquire power, to create alliances, and to circumvent traditional ways of obtaining status. One does not want to pursue this argument too far, however, because for most tasks metals were unquestionably superior to stone.

Metals themselves had been introduced to Europe and the Near East long before their use became widespread. Rosen maintains that metal, some of which was produced with sophisticated technologies, was probably introduced for cultic reasons. This word is carefully chosen, as specific caches of objects interpreted as cult-associated have been unearthed, one recently in a cave in the Judean Desert (Bar-Adon 1980; Moorey 1988). Metals existed for millennia in places like the Anatolian Plateau without causing the host society to become sociopolitically complex. In fact, the operative sequence appears to have been the opposite: complexity stimulated metallurgy.

Metals conformed to a more general principle, which might be termed "Rosen's Law" or "The Law of Unintended Consequences," i.e., that once a new technology or engineering concept appears, for whatever reason, it will take on added dimensions until it achieves its potential. Refinements to the technology may have nothing to do with its betterment; indeed, they often take it on a totally different trajectory. In this scenario, metals quickly gained ascendancy over stone, because (1) the initial increase in popularity and volume of metal production immediately lowered production costs, and (2) metal production and trade were quickly integrated into existing socioeconomic systems, which needed to be maintained. In this case, then, the new technology was not necessarily better than its competitor in terms of practical uses when it was first traded into the region. It was initially accepted for reasons unrelated to efficiency, and then was established as a practical technology as it became cheaper and technically improved.

Metallurgy was important in the Old and New Worlds, but it was perceived differently in each. In South and Central America, the results of the production trajectory were usually decorative items. Metals appear frequently in the archaeological record of these regions, but the techniques by which they were made were usually not very efficient, because the objects were used as prestige items. And because metallurgical techniques did not develop in linear fashion, they were not used to define archaeological stages, as was the case in Europe.

In the Near East chronologies have been tied to political events, such as the destruction of the Temple in Jerusalem. This has had some bizarre consequences. For example, in the Early Bronze Age of the Near East, there was no bronze in the conventional sense; tin-alloyed bronze does not enter the cultural repertoire until the Middle Bronze Age. However, in the Chalcolithic some people employed an antimony/copper alloy, which was, in fact, bronze. Iron technology entered the region in the Late Bronze Age, but did not become popular until the middle of the Iron Age. Iron was probably not adopted initially for its superior qualities; rather, it was a suitable material that could be readily acquired as a means to circumvent the costs associated with bronze.

### The Effect of Metals on Stone Technologies

The effects of these changes on existing stone technologies were, as Rosen explains, episodic and task-specific, each task seemingly recounting a different scenario. For example, in the Levant the chipped stone axe disappeared with the end of the Chalcolithic period. However, metal axes had existed earlier than that, so we should have seen a stone axe decline earlier in the Chalcolithic, yet they were abundant at that time. They declined not because of the introduction of metal per se, but because of sociopolitical factors related to complexity, such as urbanism and the ready availability of metals procured through expanded trade networks. By the Middle Bronze Age in the Levant, all stone tool forms except sickles and ad hoc tools had been ejected from the cultural repertoire. The continuation of ad hoc tools may have been an effect of sedentism in the Old World, responding to the same factors as similar Late Prehistoric New World forms discussed in Parry and Kelly (1987). Additionally, the rise of specialized industries was probably accompanied by the parallel and complementary rise of ad hoc industries as the more esoteric forms of lithic reduction became more restricted.

Thus, during the era of the last widespread use of stone tools in the Near East, the Bronze Age, societies possessed two primary kinds of stone technologies: an informal one (ad hoc tools) and a formal one (sickles). These forms were eventually deleted as metals became cheaper, as popular acceptance of metal for all tasks became universal (by the middle of the Iron Age), and as the last lithic technologists died out.

## CONCLUDING REMARKS

This chapter has distilled a considerable amount of argumentation and conversation, but it constitutes only a microcosm of the possible conversations that lithic analysts might undertake at any one time. A fortunate outcome of this conference was the resulting breadth of topics, from the more commonly considered subjects of hunting weaponry and tool classification to sociopolitical control and the onset of metallurgy, topics that are rarely discussed from the perspective of the lithic data base. Few of these discussions elicited total agreement, though this outcome might be expected from any group of academicians. In fact, there was general consensus on a number of issues, and these have been enumerated above.

Pursuing theoretical considerations entails both the presentation of models and innovative approaches, and the evaluation of those models and approaches to achieve some direction. In discussing the disparate topics introduced by the papers, we have evaluated a number of positions of relevance to archaeologists, and have offered some considerations for future research. Few people will agree with all of the proposals offered here, just as we were not in unanimous agreement on many of them ourselves. But if these deliberations provide some dire archaeological theory building, they will have been worth the effort.

## REFERENCES


Andrews, A. 1980. The Salt Trade of the Maya. *Archaeology* 33:24–33.

Andrews, A. 1983. *Maya Salt Production and Trade.* University of Arizona Press, Tucson.

Arnold, P. 1991. *Domestic Ceramic Production and Spatial Organization: A Mexican Case Study in Ethnoarchaeology.* Cambridge University Press, Cambridge.

Bamforth, D. B. 1986. Technological Efficiency and Tool Curation. *American Antiquity* 51:38–50.

Bar-Adon, P. 1980. *The Cave of the Treasure.* Israel Exploration Society, Jerusalem.

Bleed, P. 1986. The Optimal Design of Hunting Weapons: Maintainability or Reliability. *American Antiquity* 51:737–747.

Blitz, J. H. 1988. Adoption of the Bow in Prehistoric North America. *North American Archaeologist* 9:123–145.

Brumfiel, E. M. 1987. Elite and Utilitarian Crafts in the Aztec State. In *Specialization, Exchange, and Complex Societies,* edited by E. Brumfiel and T. Earle, pp. 102–118. Cambridge University Press, Cambridge.

Chatters, J. C. 1987. Hunter-Gatherer Adaptations and Assemblage Structure. *Journal of Anthropological Archaeology* 6:336–375.

Clark, J. E. 1987. Politics, Prismatic Blades, and Mesoamerican Civilization. In *The Organization of Core Technology,* edited by J. K. Johnson and C. A. Morrow, pp. 259–284. Westview Press, Boulder.

Cross, J. R. 1993. Craft Specialization in Nonstratified Societies. *Research in Economic Anthropology* 14:61–84. JAI Press, Greenwich, CT.

Cziesla, E., S. Eickhoff, N. Arts, and D. Winter (editors). 1990. *The Big Puzzle: International Symposium on Refitting Stone Artefacts.* Studies in Modern Archaeology, vol. 1, Holos, Bonn, Germany.

Dibble, H. L. 1987. The Interpretation of Middle Paleolithic Scraper Morphology. *American Antiquity* 52:109–117.

Earle, T. K. 1987. Specialization and the Production of Wealth: Hawaiian Chiefdoms and the Inka Empire. In *Specialization, Exchange, and Complex Societies,* edited by E. Brumfiel and T. Earle, pp. 64–75. Cambridge University Press, Cambridge.

Hall, R. L. 1980. An Interpretation of the Two-Climax Model of Illinois Prehistory. In *Early Native Americans,* edited by D. Browman, pp. 401–462. Mouton, the Hague.

Hoffman, C. M. 1985. Projectile Point Maintenance and Typology: Assessment with Factor Analysis and Canonical Correlation. In *For Concordance in Archaeological Analysis,* edited by C. Carr, pp. 566–612. Westport Publishers, Kansas City.

…fman, J. L., and J. G. Enloe (editors). 1992. *Piecing Together the Past: Applications of Refitting Studies … Archaeology.* BAR International Studies 578, Oxford.

… *The Southeastern Indians.* University of Tennessee Press, Knoxville.

…llop. 1989. Defining Coastal Maya Trading Ports and Transportation Routes. … edited by H. McKillop and P. Healy, pp. 91–110. Trent University, …logy no. 8, Petersborough.

…*ning Electron Microscopy of Experimental Quartz Tools.* … Uppsala.

J. Shockler, and J. Zilhao. 1991. Raw Material Usage in the Paleolithic. The Effects of Local Availability on Selection and Economy. In *Raw Material Economies among Prehistoric Hunter-Gatherers*, edited by A. Montet-White and S. Holen, pp, 127–139. University of Kansas, Publications in Anthropology, no. 19.

Moorey, P. R. S. 1988. The Chalcolithic Hoard from Nahal Mishmar, Israel, in Context. *World Archaeology* 20:171–189.

Nelson, M. C. 1991. The Study of Technological Organization. In *Archaeological Method and Theory*, edited by M. Schiffer, pp. 57–100. University of Arizona Press, Tucson.

Odell, G. H. 1994. The Role of Stone Bladelets in Middle Woodland Society. *American Antiquity* 59:102–120.

Parry, W. J., and R. L. Kelly. 1987. Expedient Core Technology and Sedentism. In *The Organization of Core Technology*, edited by J. Johnson and C. Morrow, pp. 285–304. Westview Press, Boulder.

Potter, D. R. 1991. A Descriptive Taxonomy of Middle Preclassic Chert Tools at Colha, Belize. In *Maya Stone Tools: Selected Papers from the Second Maya Lithic Conference*, edited by T. Hester and H. Shafer, pp. 21–29. Prehistory Press, Madison.

Rosen, S. A. 1989. The Analysis of Early Bronze Age Chipped Stone Industries: A Summary Statement. In *L'urbanisation de la Palestine a l'age du bronze ancien*, edited by P. Miroschedji, pp. 199–222. BAR International Series 527, Oxford.

Schambach, F. F. 1982. An Outline of Fourche Maline Culture in Southwest Arkansas. In *Arkansas Archaeology in Review*, edited by N. L. Trubowitz and M. D. Jeter, pp. 132–197. Arkansas Archaeological Survey, Research Series no. 125, Fayetteville.

Shafer, H. J. 1991. Late Preclassic Formal Stone Tool Production at Colha, Belize. In *Maya Stone Tools: Selected Papers from the Second Maya Lithic Conference*, edited by T. Hester and H. Shafer, pp. 31–44. Prehistory Press, Madison.

Shott, M. J. 1989. On Tool-Class Use Lives and the Formation of Archaeological Assemblages. *American Antiquity* 54:9–30.

Thomas, D. H. 1978. Arrowheads and Atlatl Darts: How the Stones Got the Shaft. *American Antiquity* 43:461–472.

# Index

## *INTERDISCIPLINARY CONTRIBUTIONS TO ARCHAEOLOGY*

***Chronological Listing of Volumes***

---

THE PLEISTOCENE OLD WORLD
Regional Perspectives
Edited by Olga Soffer

HOLOCENE HUMAN ECOLOGY IN NORTHEASTERN NORTH AMERICA
Edited by George P. Nicholas

ECOLOGY AND HUMAN ORGANIZATION ON THE GREAT PLAINS
Douglas B. Bamforth

THE INTERPRETATION OF ARCHAEOLOGICAL SPATIAL PATTERNING
Edited by Ellen M. Kroll and T. Douglas Price

HUNTER–GATHERERS
Archaeological and Evolutionary Theory
Robert L. Bettinger

RESOURCES, POWER, AND INTERREGIONAL INTERACTION
Edited by Edward M. Schortman and Patricia A. Urban

POTTERY FUNCTION
A Use-Alteration Perspective
James M. Skibo

SPACE, TIME, AND ARCHAEOLOGICAL LANDSCAPES
Edited by Jacqueline Rossignol and LuAnn Wandsnider

ETHNOHISTORY AND ARCHAEOLOGY
Approaches to Postcontact Change in the Americas
Edited by J. Daniel Rogers and Samuel M. Wilson

THE AMERICAN SOUTHWEST AND MESOAMERICA
Systems of Prehistoric Exchange
Edited by Jonathon E. Ericson and Timothy G. Baugh

FROM KOSTENKI TO CLOVIS
Upper Paleolithic–Paleo-Indian Adaptations
Edited by Olga Soffer and N. D. Praslov

EARLY HUNTER–GATHERERS OF THE CALIFORNIA COAST
Jon M. Erlandson

HOUSES AND HOUSEHOLDS
A Comparative Study
Richard E. Blanton

THE ARCHAEOLOGY OF GENDER
Separating the Spheres in Urban America
Diana diZerega Wall

ORIGINS OF ANATOMICALLY MODERN HUMANS
Edited by Matthew H. Nitecki and Doris V. Nitecki

PREHISTORIC EXCHANGE SYSTEMS IN NORTH AMERICA
Edited by Timothy G. Baugh and Jonathon E. Ericson

STYLE, SOCIETY, AND PERSON
Archaeological and Ethnological Perpspectives
Edited by Christopher Carr and Jill E. Neitzel

REGIONAL APPROACHES TO MORTUARY ANALYSIS
Edited by Lane Anderson Beck

DIVERSITY AND COMPLEXITY IN PREHISTORIC MARITIME SOCIETIES
A Gulf of Maine Perspective
Bruce J. Bourque

CHESAPEAKE PREHISTORY
Old Traditions, New Directions
Richard J. Dent. Jr.

PREHISTORIC CULTURAL ECOLOGY AND EVOLUTION
Insights from Southern Jordan
Donald O. Henry

STONE TOOLS
Theoretical Insights into Human Prehistory
Edited by George H. Odell

CPSIA information can be obtained
at www.ICGtesting.com
Printed in the USA
LVHW021659020423
743254LV00003B/787

9 780306 451980